T0235163

Lecture Notes of the Institute for Computer Sciences, Social Informatics and Telecommunications Engineering   183

More information about this series at http://www.springer.com/series/8197

Xin-Lin Huang (Ed.)

Machine Learning and Intelligent Communications

First International Conference, MLICOM 2016
Shanghai, China, August 27–28, 2016
Proceedings

 Springer

Editor
Xin-Lin Huang
Tongji University
Shanghai
China

ISSN 1867-8211 ISSN 1867-822X (electronic)
Lecture Notes of the Institute for Computer Sciences, Social Informatics
and Telecommunications Engineering
ISBN 978-3-319-52729-1 ISBN 978-3-319-52730-7 (eBook)
DOI 10.1007/978-3-319-52730-7

Library of Congress Control Number: 2016963658

Printed on acid-free paper

This Springer imprint is published by Springer Nature
The registered company is Springer International Publishing AG
The registered company address is: Gewerbestrasse 11, 6330 Cham, Switzerland

Preface

Along with the fast development of mobile communications technologies, the amount of high-quality wireless services required is increasing exponentially. According to the prediction of Cisco VNI Mobile Forecast 2016, global mobile data traffic will increase nearly eightfold between 2015 and 2020, and mobile network connection speeds will increase more than threefold by 2020. Hence, there are still big gaps between future requirements and current communication technologies, even using 4G/5G. How to integrate the limited wireless resources with some intelligent algorithms or schemes and boost potential benefits is the focus of the conference. As an emerging discipline, machine learning is a subfield of computer science that evolved from the study of pattern recognition and computational learning theory in artificial intelligence, and explores the study and construction of algorithms that can learn from and make predictions about complicated scenarios. In communication systems, the previous/current radio situations and communication paradigms should be well considered to obtain a high quality of service (QoS), such as the available spectrum, limited energy, antenna configurations, and heterogeneous properties. Machine learning algorithms facilitate the analysis and prediction of complicated scenarios, and thus to make an optimal actions in OSI seven layers. We hope the integrating of machine-learning algorithms into communication systems will improve the QoS and make the systems smart, intelligent, and efficient.

December 2016   Xin-Lin Huang

Organization

Steering Committee Chair

Imrich Chlamtac   University of Trento, Create-Net, Italy

Steering Committee

Xin-Lin Huang   Tongji University, China

General Chairs

Daqiang Zhang	Tongji University, China
Qingquan Sun	California State University, San Bernardino, USA
Rui Wang	Tongji University, China
Xin-Lin Huang	Tongji University, China

Technical Program Committee Chairs

Xin Liu	Dalian University of Technology, China
Hui-Ming Wang	Xi'an Jiaotong University, China
Wei Wang	Zhejiang University, China
Feng Li	Zhejiang University of Technology, China
Weidang Lu	Zhejiang University of Technology, China
Bo Li	Harbin Institute of Technology, China
Mu Zhou	Chongqing University of Posts and Telecommunications, China
Zhian Deng	Dalian Maritime University, China
Nan Zhao	Dalian University of Technology, China
Guanglin Zhang	Donghua University, China
Deli Qiao	East China Normal University, China

Web Chairs

Hang Dong	Tongji University, China
Xin Zhang	Tongji University, China

Publicity and Social Media Chair

Min Wang   Tongji University, China

Sponsorship and Exhibits Chair

Songlin Chen   Tongji University, China

Publications Chairs

Xin Liu   Dalian University of Technology, China
Feng Li   Zhejiang University of Technology, China

Panels Chair

Lihua Ai   Tongji University, China

Tutorials Chair

Dian Liu   Tongji University, China

Demos Chair

Wei Yu   Tongji University, China

Posters and PhD Track Chair

Teng Zi   Tongji University, China

Local Chairs

Rui Wang   Tongji University, China
Guanghui Zhu   Tongji University, China
Wenfeng Li   Tongji University, China

Conference Manager

Lenka Laukova   EAI - European Alliance for Innovation

Contents

Machine Learning and Information Processing in Wireless Sensor Networks

Machine Learning for Multimedia

Main Track

Invited Paper

Data Mining in Heterogeneous Networks

An Emergency Event Driven Routing Algorithm for Bi-directional Highway in Vehicular Ad Hoc Networks

Yajie Yang[1,2], Demin Li[1,2], Guanglin Zhang[1,2]([✉]), Chang Guo[1,2], and Saifei Jin[1,2]

[1] College of Information Science and Technology, Donghua University, Shanghai 201620, People's Republic of China
{yangyajie,guochang}@mail.dhu.edu.cn, {deminli,glzhang}@dhu.edu.cn
[2] Engineering Research Center of Digitized Textile and Apparel Technology, Ministry of Education, Donghua University, Shanghai 201620, People's Republic of China

Abstract. Vehicular Ad Hoc Networks (VANETs) play a significant role in preventing traffic accidents on the highway. But it is a challenge to reduce the messages transmission delay under emergency condition. In this paper, an emergency event driven routing algorithm for bi-directional highway is proposed. Each vehicle maintains a real-time special neighbor nodes set which includes the next-hop vehicle and vehicle ID in three different directions. In particular, when there is no vehicle ahead or behind the accident vehicles we use the vehicles from reverse direction to reduce the intermittent link. And the emergency events of vehicles are divided into two types, according to the different influence of events on vehicles ahead and behind. And different emergency events launch different transmission algorithms. This ensures that the emergency messages (EMs) can transmit to the vehicles affected much more by the emergency events. Furthermore, we derive the transmission delay formula based on the proposed algorithm. Finally, the algorithm is verified by the simulation of the transmission delay formula. The results show that the proposed emergency event driven routing algorithm can reduce the transmission delay effectively.

Keywords: Vehicle Ad Hoc Networks · Emergency messages · Event-driven · Routing algorithm

1 Introduction

In recent years, with more vehicles on the road, traffic accidents are increasing. However, if the vehicles which are surrounding the accident vehicle can receive the emergency messages (EMs) in time and take corresponding measures, more serious accidents can be reduced and even be avoided. Vehicular Ad Hoc Networks (VANETs) [1] show great potential in reducing traffic accidents. VANETs

© ICST Institute for Computer Sciences, Social Informatics and Telecommunications Engineering 2017
X.-L. Huang (Ed.): MLICOM 2016, LNICST 183, pp. 3–12, 2017.
DOI: 10.1007/978-3-319-52730-7_1

is an important part of Intelligent Transportation System (ITS) [2] and the Internet of things (IoT) [3]. The applications of VANETs mainly include two aspects, security applications [4,5] and user applications [6]. Security applications cover accident pre-warning, intersection-driving and route planning. Internet communication, resources sharing and commercial advertising belong to user applications. There are three kinds of information transmission modes in the VANETs. Respectively, they are Vehicle-to-Vehicle (V2V) [7], Vehicle-to-RSU (V2R) [8,9], and hybrid communication [10].

This paper mainly discusses how to reduce the EMs transmission delay on the highway. In [11], the authors achieve broadcast redundancy, transmission latency by choosing the furthest broadcast vehicle in a queue on the basis of directive broadcast. Minimizing duplicate retransmissions by combining location-based method and time reservation-based method with the aid of Global Positioning System (GPS) of neighboring nodes is proposed in [12]. [13] designs an event-driven Inter-Vehicle Communication protocol that learns about traffic conditions ahead and recommends optimal velocities in order to prevent the formation of vehicular shock waves. And this approach of reacting in case of traffic fluctuations leads to significant improvements in overall traffic flow. Protocol in [14] sends the periodic safety message and event driven safety message by using priority in the messages and context-based communication. Through these papers, we think that making the algorithm adapt to the changeful road scene is a way to reduce the transmission delay and intermittent link.

In this paper, we propose an emergency event driven EMs transmission algorithm for bi-direction highway. The routing algorithm not only reduce the transmission delay but also reduce the intermittent link. The contributions of this paper are as follows.

(1) Each vehicle needs to maintain a real-time special neighbor nodes set. These sets can help to adapt to the diversity of the highway scene and find the next-hop node quickly to reduce the transmission delay.
(2) When there is no next-hop vehicle in the same direction, we use the vehicles from reverse direction to transmit EMs. Because vehicles from reverse direction will meet the accident vehicle at a time. In this way, the intermittent link can be reduced.
(3) The emergency events of vehicles are divided into two types. This ensures that the EMs can be transmitted to the greater affected vehicles firstly.

The rest of context is organized as follows. In Sect. 2, the system model is introduced. The Sect. 3 is the specific design of routing scheme and the estimation of EMs transmission delay. The analysis of the scheme through the simulation results is introduced in Sect. 4. Finally, We conclude this paper in Sect. 5.

2   System Model

The high speed of vehicles on the highway leads to the frequent change of the vehicle density. This phenomenon leads to that the EMs can't transmit timely and effectively. So finding the next-hop vehicle quickly is an important factor to achieve uninterrupted communication. In this paper, we propose a real-time special neighbor nodes set for each vehicle. These sets are mainly used to help each vehicle to judge whether there exist vehicles that can communicate with in three different directions: ahead, behind and reverse line.

In general, the highway is a bi-directional road. If the vehicles from reverse direction can be reasonably utilized, the intermittent link can be reduced effectively. For example, in low traffic density, there may not exist a vehicle in the communication radius of the accident vehicle for a long time. This would cause the EMs can not transmit, which means that the communication link is interrupted. But the accident vehicle will meet the vehicles from reverse direction at a time, the EMs can transmit to the vehicles from reverse direction. Then the vehicles from reverse direction transmit EMs to the target vehicles. As shown in Fig. 1, the circle represents the vehicle transmission range, and road width is negligible compared to the vehicle transmission radius. That is to say, the adjacent vehicles in two lanes can communicate with each other. The intersecting circle represent the vehicle can communication with each other. When the accident vehicle needs to transmit EMs to vehicles ahead, the EMs will transmit as yellow arrow. Although there is no vehicles ahead the accident vehicle with in the transmission radius, the vehicles beside the accident vehicle can help transmit the EMs. Doing so ensures that the transmission link would not be broken.

The accidents are divided into two categories according to different accidents having different influence on the surrounding vehicles. For instance, traffic jam, sudden deceleration and braking. When these accidents occur, the influence on the behind vehicles are much more serious than on the ahead vehicles. So the EMs should transmit to the behind vehicles firstly. When other accidents such as brake failure, overtaking and accelerating occur, the influence on the ahead vehicles are much more serious than on the behind vehicles. So the EMs should transmit to the ahead vehicles firstly. Therefore, the type of events determines the EMs transmission direction. In this paper, we design different transmission algorithm for the two categories events, so the transmission delay can be reduced effectively.

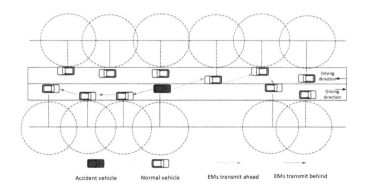

Fig. 1. EMs transmit model on bidirectional highway. (Color figure online)

Meanwhile, we assume that each vehicle has a unique ID. So ID is the unique identifier to distinguish the vehicles. The vehicle transmission radius is the same. The location of vehicles can be known by GPS measurement instrument. This paper uses a two-dimensional highway model. And we stipulate one of the direction for the positive direction, the other direction is the negative direction. Positive direction and negative direction are used to judge whether the vehicles are in the same direction.

3   Proposed Scheme

In this section, the proposed routing scheme is described. First, we describe the problem formulation. Then, the EMs transmission algorithm is proposed. Finally, the transmission delay formula is derived.

3.1   Problem Formulation

Each vehicle has a vehicle packet, the packet data will be updated and saved in real time. But when a vehicle occurs some accident, this vehicle packet will become an EMs. EMs transmit among vehicles to remind the surrounding vehicles. The packet contains five items:

(1) Attribute value. When this value is 0, this indicates that the vehicle is in normal condition. And the packet is vehicle packet. When it is 1, this indicates that the vehicle is under emergency condition. And the packet become EMs.
(2) Parameter 1. Vehicle's ID, which is the only item to identify the vehicle. We use the vehicle's license plate as vehicle's ID.
(3) Parameter 2. Vehicle's driving direction, its main role is to distinguish the vehicles in the same direction or reverse direction.
(4) Parameter 3. Vehicle's location, which can be obtained from GPS data. The distance between vehicles can be roughly calculated by using GPS data.
(5) Parameter 4. EMs Transmission direction, Transmission direction is the direction that the EMs need to transmit to. This item depends on the type of the emergency events. Under normal condition, it is null.

The format of vehicle packet and EMs show in Table 1. The second line is the format of vehicle packet, and the third line is the format of EMs:

Table 1. The format of vehicle packet and EMs.

Attribute value	Parameter 1	Parameter 2	Parameter 3	Parameter 4
0	Vehicle ID	Driving direction	Location	Null
1	Vehicle ID	Driving direction	Location	EMs transmission direction

In addition, each vehicle needs to maintain a real-time special neighbor nodes set. The special neighbor nodes set contains three items. The first item is (ρ_f, ID_f). ρ_f stands whether there exist vehicles ahead itself, if there exist, the item is 1, otherwise, is 0. And ID_f is the next-hop vehicle ID. If there are more than one vehicle ahead of the vehicle, the vehicle select the furthest vehicle within the communication radius as the next-hop vehicle. The second item is (ρ_b, ID_b). ρ_b stands whether there exist vehicles behind itself, if there exist, the item is 1, otherwise, is 0. And ID_b is the next-hop vehicle ID. If there are more than one vehicle behind the vehicle, the vehicle select the furthest vehicle within the communication radius as the next-hop vehicle. The third item is (ρ_s, ID_s). ρ_s stands whether there exist vehicles beside itself, if there exist, the item is 1, otherwise, is 0. And ID_s is the next-hop vehicle ID. If there are more than one vehicle beside the vehicle, the vehicle will select the furthest vehicle within the communication radius as the next-hop. The format of special neighbor nodes set, namely, $\{(\rho_f, ID_f), (\rho_b, ID_b), (\rho_s, ID_s)\}$.

In this paper, we adopts dynamic programming to select the next-hop vehicle in three directions for each vehicle. r is the transmission radius of vehicles. $d(ID_x, ID_i)$ is the distance between ID_x and ID_i, $i = 0, 1, 2...N$. $D(ID_x)$ is the distance between ID_x and the furthest vehicle ID_i in ID_x communication radius. $M(ID_x, ID_x)$ is the derivative of $D(ID_x)$.

Objective function:

$$D(ID_x) = max\{d(ID_x, ID_0), d(ID_x, ID_1), d(ID_x, ID_2), ..., d(ID_x, ID_i)\} \quad (1)$$

Constraint Conditions:

$$0 < D(ID_x) < 2r \quad (2)$$

Decision Variable:

$$M(ID_x, ID_i) = \lim_{\Delta d \to 0} \frac{\Delta d(ID_x, ID_i)}{\Delta d} \quad (3)$$

If there exist $D(ID_x)$ and $M(ID_x, ID_i) < 0$, this means vehicle ID_x and ID_i are in the different direction and ρ_s is 1. If there exist $D(ID_x)$ and $M(ID_x, ID_i) > 0$, this means vehicle ID_x and ID_i are in the same direction and ρ_f or ρ_b is 1.

3.2 EMs Transmission Algorithm

When a vehicle occur a certain accident, it will produce an EMs. And the transmission of EMs should reference the special neighbor nodes set. According to the EMs package the EMs transmission direction can be known. If the EMs should transmit to the vehicles ahead, the ρ_f will be viewed first. If $\rho_f = 1$, the EMs will transmit to the accident vehicle's next-hop vehicle. Otherwise, the ρ_s will be viewed. If $\rho_s = 1$, the EMs will transmit to the vehicles on the reverse lane. If $\rho_s = 0$, EMs should wait for a while. If the EMs should transmit to the behind vehicles, the ρ_b will be viewed first. If $\rho_b = 1$, the EMs will transmit to the accident vehicle's next-hop vehicle. Otherwise, the ρ_s will be viewed. The specific algorithm is shown in Algorithm 1:

Algorithm 1. Event-driven transmission algorithm

if {Attribute value=1} then
    if {EMs' driving direction = vehicle's driving direction} then
        if { EMs needs to transmit ahead } then
            if { ρ_f =0 } then
                if { ρ_s=1 } then
                    { transmit EMs to the vehicles from reverse direction }
                else
                    { wait for a time interval t}
                end if
            else
                { transmit the EMs to the vehicle' next-hop vehicle directly}
            end if
        else
            { EMs need to transmit behind }
            if { ρ_b =0 } then
                { whatever ρ_s is, transmit the EMs to vehicles from reverse direction}
            else
                { transmit the EMs to the vehicle's next-hop vehicle directly}
            end if
        end if
    else

        if { ρ_s=1} then
            { transmit the EMs to vehicles from reverse direction }
        else
            { wait for a time interval t }
        end if
    end if
else
    { Vehicle is in a normal condition}
end if

3.3 The Estimation of EMs Transmission Delay

The time of EMs transmission includes two parts: carry time and forwarding time. Calculation formula is as follows:

$$T = t_1 + t_2, \tag{4}$$

T denotes the total transmission delay, t_1 is the carry time, t_2 is the forwarding time. According to the algorithm, we know there are five kinds of transmission delay.

Case 1: EMs need to transmit ahead. $\rho_f = 0$, $\rho_s = 1$.

Accident vehicle transmit the EMs to the vehicle from the reverse direction. So there only exists forwarding time. t_{hop} denotes an interaction time.

$$T_1 = t_2 = 2t_{hop}, \tag{5}$$

Case 2: EMs need to transmit ahead. $\rho_f = 1$.

Accident vehicle can transmit the EMs to its next-hop vehicle directly. The next-hop vehicle ID can be find in the accident vehicle's special neighbor nodes set. So there only exists forwarding time (Fig. 2).

$$T_2 = t_2 = t_{hop},\tag{6}$$

Case 3: EMs need to transmit behind. $\rho_b = 0$, $\rho_s = 0$.

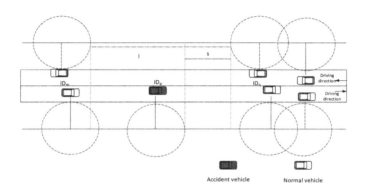

Fig. 2. EMs transmit model when $\rho_b = 0$, $\rho_s = 0$.

If the EMs need to transmit ahead it is a pursuit problem but if EMs need to transmit behind it is a meeting problem. It is assumed that vehicle ID_p need to transmit EMs, vehicle's transmission radius is r. v_{ID_x} is the vehicle ID_x's speed. Vehicle ID_p need drive s to get into the communication range of vehicle ID_q which is on the reverse lane. The distance between two vehicles ID_p and ID_m in the same direction is l; According to algorithm the total time is as following:

$$T_3 = t_1 + t_2 = \frac{s}{v_{ID_p} + v_{ID_q}} + \frac{2r + l}{v_{ID_q} + v_{ID_m}} + 2t_{hop},\tag{7}$$

Case 4: EMs need to transmit behind. $\rho_b = 0$, $\rho_s = 1$.

This case is the same as case 2. Accident vehicle transmits the EMs to the vehicle on the reverse lane. So there only exists forwarding time.

$$T_4 = t_2 = 2t_{hop},\tag{8}$$

Case 5: EMs need to transmit behind. $\rho_b = 1$.

This case is the same as case 3. Accident vehicle will transmit the EMs to the next-hop vehicle directly. And the vehicle ID can be find in the accident vehicle's special neighbor nodes set. So there only exists forwarding time.

$$T_5 = t_2 = t_{hop},\tag{9}$$

Assuming the probability of occurrence of case i is P_i, then the total time is as follows:

$$T = \sum_{i=0}^{4} P_i T_i. \tag{10}$$

4   Performance Evaluation

In this paper, we use MATLAB to carry on the comparative test and confirmatory test for the model and formulas given above. We assume that s is 150, l is 60 and the EMs should tranmit to the n hop vehicles in ahead or behind the accident vehicle.

Assuming the five cases obey uniform distribution, so $P_i = \frac{1}{5}$. The transmission delay of event driven routing scheme is shown in Fig. 3. With the increase of n, T increased too. But when n is constant, as the increase of r, T changed little.

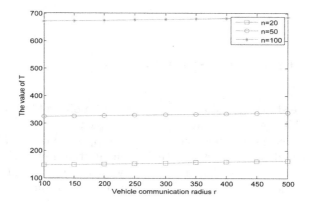

Fig. 3. The time T of event-driven algorithm.

In this paper, we use directive broadcast model to be compared. And assume $n = 50$, the T of directive broadcast model and event driven model are shown in Fig. 4.

Both the directive broadcast model and event driven model transmission delay will increase with the increase of vehicle transmission radius r. The time delay of event-driven model is lower than directive broadcast models. And with the increase of r event-driven's advantage is increasing too. So the proposed event driven routing althgrithm have an advantage in reducing transmission delay and intermittent link.

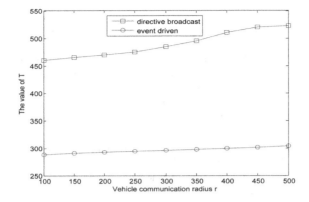

Fig. 4. The contrasting result of time T.

5 Conclusion

In this paper, we put forward an emergency event driven routing algorithm for bi-directional highway. The real-time special neighbor nodes set makes the algorithm more adaptable to the changeful highway. Through using the vehicles from reverse direction the intermittent link reduce effectively. Event classification makes the EMs transmit quickly to the greater affected vehicles. Besides, we derive transmission delay formula. Through the simulation results, we verify our proposed algorithm, which achieves better performance on reducing transmission delay. In terms of future work, we will devote to reduce transmission delay on the network layer and analyse capacity on the basis of the proposed algorithm.

Acknowledgement. This work is supported by the NSF of China under grant 71171045, 61301118; the Innovation Program of Shanghai Municipal Education Commission under Grant No. 14YZ130; the Fundamental Research Funds for the Central Universities and a DHU Distinguished Young Professor Program.

References

1. Li, Y., Wang, W.: Horizon on the move: geocast in intermittently connected vehicular ad hoc networks. In: Proceedings of the IEEE INFOCOM, pp. 2553–2561 (2013)
2. Blum, J., Eskandarian, A., Hoffman, L.: Challenges of inter-vehicle ad hoc networks. IEEE Trans. Intell. Transp. Syst. **5**(4), 347–351 (2004)
3. Sotiriadis, S., Bessis, N., Asimakopoulou, E., Mustafee, N.: Towards simulating the internet of things. In: Workshop Paper for the 28th IEEE International Conference on Advanced Information Networking and Applications, 13–16 May, Victoria, Canada, pp. 444–448 (2014)
4. Abboud, K., Zhuang, W.: Modeling and analysis for emergency messaging delay in vehicular ad hoc networks. In: Proceedings of IEEE GLOBECOM, pp. 1–6 (2009)

5. Almulla, M., Wang, Y., Boukerche, A., Zhang, Z.: A fast location-based hand-off scheme for vehicular networks. In: IEEE ICC Ad-Hoc and Sensor Networking Symposium, pp. 1464–1468 (2013)
6. Lee, K., Lee, S.-H., Cheung, R., Lee, U., Gerla, M.: First experience with cartor-rent in a real vehicular ad hoc network testbed. In: 2007 Mobile Networking for Vehicular Environments, pp. 109–114 (2007)
7. Bazzi, A., Masini, B., Pasolini, G.: V2V and V2R for cellular resources saving in vehicular applications. In: Wireless Communications and Networking Conference (WCNC), pp. 3199–3203. IEEE (2012)
8. Paier, A., Faetani, D., Mecklenbruker, C.: Performance evaluationof IEEE 802.11p physical layer infrastructure-to-vehicle real-world measurements. In: 2010 3rd International Symposium on Applied Sciences in Biomedical and Communication Technologies (ISABEL), pp. 1–5 (2010)
9. Villas, L., Leandro, A.: Network partition-aware geographical data dissemination. In: 2013 IEEE International Conference on Communications (ICC), pp. 1439–1443 (2013)
10. Jer, M., Marlier, P., Senouci, S.M.: Experimental assessment of V2V and I2V communications. In: IEEE International Conference on Mobile Ad Hoc and Sensor Systems (MASS) (2007)
11. XU, S., Zhou, H., Li, C., Zhao, Y.: A multi-hop V2V broadcast protocol for chain collision avoidance on highways. In: IEEE Proceedings of ICCTA, pp. 110–114 (2009)
12. Nuri, D.M., Nuri, H.H.: Strategy for efficient routing in VANET. In: IEEE, pp. 903–908 (2010)
13. Forster, M., Frank, R., Engel, T.: An event-driven inter-vehicle communication protocol to attenuate vehicular shock waves. In: International Conference on Connected Vehicles and Expo (ICCVE), pp. 540–545 (2014)
14. Kumar, A., Nayak, R.P.: An efficient group-based safety message transmission protocol for VANET. In: International conference on Communication and Signal Processing, pp. 270–274 (2013)

Decentralized Learning for Wireless Communication Systems

A Novel Constellation Shaping Method to Reduce PAPR for Rate Compatible Modulation

Min Wang, Qin Zou[✉], and Xiaoqiang Tu

School of Mathematics and Computer Science, Gannan Normal University,
Ganzhou 341000, China
mwangcs@163.com, zou_qin1979@163.com, ren2005zhe@163.com

Abstract. Although rate compatible modulation (RCM) achieves continuous and efficient spectrum efficiency, it has high peak to average power ratio (PAPR) because that its modulation constellation is a dense rectangular. To reduce PAPR of RCM, this paper proposes a novel constellation shaping method, namely transform from rectangular to circle (TR2C), which different to the traditional constellation shaping methods. The TR2C approach compress the points, that with higher amplitude and lower probability, into a circle, such that it can reduce the PAPR of shaping constellation. Furthermore, to deal with the problem of AWGN noise amplification in inverse transform of TR2C, we give out the geometrical method of estimating noise variance. Simulations show that, comparing with original constellation, TR2C scheme can achieve $1\,dB$ performance gain at $BER = 10^{-5}$ by improving the transmission power with $1.8\,dB$. And TR2C achieves the capacity performance close to original constellation when SRN from $5\,dB$ to $30\,dB$.

Keywords: Constellation shaping · Rate compatible modulation · PAPR

1 Introduction

Rate adaptation technology is the effective means to achieve high spectral efficiency under varying channel conditions, in physical layer of modern wireless communications. In recent year, three receiver adaptation schemes [1–4] were proposed to tackle the problem, that sender needs accurate and immediate channel feedback from the receiver in traditional adaptive modulation coding technology. Rate compatible modulation (RCM) [1] is a novel rate adaptation technology based on compressive sensing (CS) [5], and it is a promising application of the emerging CS theory in wireless communications. In RCM scheme, modulated symbols are incrementally generated from information bits through weighted mapping. Therefore, rate adaptation is achieved by varying the number of modulated symbols.

RCM achieves continuous and efficient spectrum efficiency. However, its peak to average power ratio (PAPR) is very high. In modern wireless communications,

© ICST Institute for Computer Sciences, Social Informatics and Telecommunications Engineering 2017
X.-L. Huang (Ed.): MLICOM 2016, LNICST 183, pp. 15–24, 2017.
DOI: 10.1007/978-3-319-52730-7_2

constellation shaping is the major mean to reduce PAPR and approach channel capacity. Constellation shaping technique partition the basic constellation into several subconstellations, so that the lower energy signals are transmitted more frequently than their higher energy counterparts [6]. Constellation shaping technique can be used in BICM and OFDM. A subset of the interleaved bits output by a binary LDPC [7] encoder are passed through a nonlinear shaping encoder, whose output is more likely to be a zero than a one. The shaping bits are used to select from among a plurality of subconstellations, while the unshaped bits are used to select the symbol within the subconstellation. Because the shaping bits are biased, symbols from lower-energy subconstellations are selected more frequently than those from higher-energy sub-constellations [8–11].

To reduce PAPR of RCM, this paper proposes a novel constellation shaping method named transform from rectangular to circle (TR2C). The TR2C approach compress the points, that with higher amplitude and lower probability distribution, into a circle, such that it can reduce the PAPR of RCM constellation. Furthermore, to deal with the problem of AWGN noise amplification in inverse transform of TR2C, we give out the geometrical method of estimating noise variance. Simulations show that, comparing with original constellation, TR2C scheme can achieve 1 dB performance gain at BER $= 10^{-5}$ by improving the transmission power. And TR2C achieves the capacity performance close to original constellation when SRN from 5 dB to 30 dB.

The rest of this paper is organized as follows. Section 2 gives out the system model based on constellation shaping, and briefly reviews the rate adaptive scheme of RCM. Section 3 presents the proposed constellation shaping methods for RCM. The simulation evaluations are included in Sect. 4. Finally, Sect. 5 concludes this paper with some discussions on future work.

2   System Model

In this section, we first describe a system model which combines RCM and novel constellation shaping. Then we we briefly review RCM and its standard decoding algorithm [1].

2.1   System Model

To reduce PAPR of RCM, we design a system model as shown in Fig. 1. Form the diagram, we can see that this communication system combines RCM component indicated in the rectangle of red dashed line and shaping component indicated in the rectangle of blue dashed line. Here, shaping component includes constellation shaping subcomponent and its inverse transform component. It is should be noted that the system only with RCM component also can work. In this paper, we only focus on shaping component.

This system model includes transmitter and receiver. At transmitter, the binary data stream is divided into blocks with length N_b. Each block b is encoded by RCM encoder, and we get a real vector denoted by u. In order to make use of

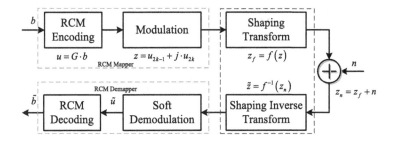

Fig. 1. System model of constellation shaping with RCM (Color figure online)

two dimensional modulation, two consecutive symbols are combined together to form In-phase and Quadrature-phase modulation symbol z, i.e., $z_k = u_{2k-1} + j \cdot u_{2k}, k = 1, 2, \cdots, N_b/2$. Then, Each complex symbol is transformed by shaping transformation that named TR2C. After passing the standard AWGN channel model, receiver gets $z_n = z_f + n$, where n is Gaussian noise vector with $n \sim N\left(0, \sigma^2\right)$. At receiver, through shaping inverse transform, soft demodulation, and RPC decoding operator, we get the estimation of $\boldsymbol{b}$.

2.2 Rate Compatible Modulation

RCM Encoding. A bipartite graph representation of RCM encoding is provided in Fig. 2. Square and circle denote symbol nodes and bit nodes, respectively. Each edge is assigned with a weight $w \in W$, where $W = \{w_1, w_2, \cdots, w_L\}$ is the weight set with length L. The bipartite graph can be represented by $G = (U, V, E)$, where $U = b_j, j = 1, \cdots, N$ is the source bits, $V = u_i, i = 1, \cdots, M$ is the set of measurements representing modulated symbols, and E defines the connection between the two sets. The RCM encoding maps binary bits into modulation symbols. Each modulation symbol is calculated by $u_i = \sum_{l \in N(i)} w_{il} \cdot b_{i_l}$, where w_{il} is the weight corresponding to i_l^{th} bit b_{i_l}, and $N(i)$ denotes the set of neighbors of the ith symbol node. The RCM encoding process can be described by

$$\boldsymbol{u} = \boldsymbol{\Phi} \cdot \boldsymbol{b}, \tag{1}$$

where $\boldsymbol{\Phi}$ is a random projection matrix with dimension $M \times N$, and $\boldsymbol{u}$ is the RCM encoded symbols vector with length M.

RCM Decoding. RCM decoding uses the belief propagation (BP) algorithm like LDPC decoding. Since RCM is employing weighted sum check, probability convolution operation is used in horizontal iteration instead of $log(tanh)$ operation. The decoding algorithm of RCM is depicted as factor graph shown in Fig. 2. An edge with a weight w_{ij} denotes the connection between symbol node i and bit node j, and arrow line denotes the probability message flow. $r_{ij}^{(t)}$ defines

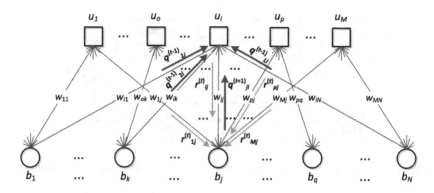

Fig. 2. Bipartite graph of ARC code.

the probability message from the ith symbol node to the jth bit node in the tth iteration. $q_{ji}^{(t+1)}$ defines the probability message from the jth bit node to the ith symbol node in the $(t+1)^{th}$ iteration. The RCM decoding algorithm includes initialization, horizontal decoding, vertical decoding and decision steps. The details of the RCM decoding algorithm are shown in [1].

3   Shaping Method

In this section, we firstly propose the constellation shaping method TR2C. Secondly, we present the geometrical method of estimating noise variance. Finally, we give out the modification of RCM decoding.

3.1   TR2C

The idea of RCM constellation shaping is that points in rectangle constellation is compressed into a circle via geometry transform, which is used generally in digital image processing. So, we call this approach as transform from rectangular to circle. TR2C reduces the amplitude of points with higher amplitude and lower probability, such that increase average amplitude of all constellation points. Therefore, TR2C reduces the PAPR of the whole constellation. It is noted that TR2C is different with traditional constellation shaping technology, which partitions a constellation into a few sub-constellations. The details of TR2C is following.

Assume that the original constellation point is $z = x + i \cdot y$, the point shaped by TR2C is $z_f = x_f + i \cdot y_f$, the point with awgn noise is $z_n = x_n + i \cdot y_n$, the point of TR2C inverse transform is $\hat{z} = \hat{x} + i \cdot \hat{y}$. Obviously, the 2-D coordinates of above four points are (x, y), (x_f, y_f), (x_n, y_n) and $(\hat{x}, \hat{y})$, respectively.

- TR2C Transform if $x == 0 \&\& y == 0$, then $x_f = 0, y_f = 0$. if $|x| \geq |y|$, then $r = \sqrt{1 + \left(\frac{y}{x}\right)^2}$, otherwise $r = \sqrt{1 + \left(\frac{x}{y}\right)^2}$. Finally, we calculate $x_f = \frac{x}{r}$ and $y_f = \frac{y}{r}$.

– TR2C Inverse Transform if $x_n == 0 \&\& y_n == 0$, then $\hat{x} = 0, \hat{y} = 0$. if $|x_n| \geq |y_n|$, then $\tilde{r} = \sqrt{1 + (\frac{y_n}{x_n})^2}$; otherwise $\tilde{r} = \sqrt{1 + (\frac{x_n}{y_n})^2}$. Finally, we calculate $\hat{x} = x_n \cdot \tilde{r}$ and $\hat{y} = y_n \cdot \tilde{r}$.

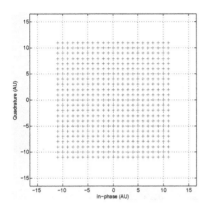

Fig. 3. Original constellation

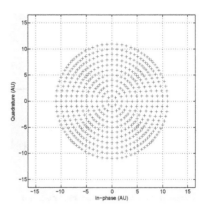

Fig. 4. Shaping constellation

Figures 3 and 4 show the original constellation and shaped constellation with a weighted set $W = \{\pm 1, \pm 2, \pm 4, \pm 4\}$, respectively. We can observe three features from the two figures. The first one is that shaped constellation is circle. The second one is that the points in first and third quadrant are symmetrical about the line $y = x$, and the points in second and fourth quadrant are symmetrical about the line $y = -x$. The last one is that the distribution of points closer to symmetrical line is dense, while the distribution of points farther from symmetrical line is sparse.

3.2   Noise Variance Estimation

Figure 5 describes the change process of constellation points in different stages of shaping transform. If shaping transform is operated on Z, we can get formula (2).

$$Z_f = f(Z) = \frac{Z}{r}. \tag{2}$$

After Z_f through AWGN channel, we can get its noise version

$$Z_n = Z_f + n, \tag{3}$$

where $n \sim CN(0, 2\sigma_n^2)$. Then, shaping inverse transform is operated on Z_n, we get

$$\tilde{Z} = f^{-1}(Z_f) = \frac{\tilde{r}}{r} Z + \tilde{r} n. \tag{4}$$

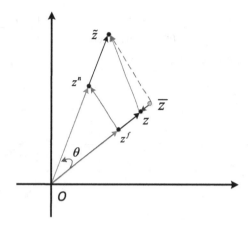

Fig. 5. The change process of constellation points in different stages of shaping transform

After simple deformation for formula (4), we finally have

$$\tilde{Z} = Z + \left(\frac{\tilde{r}}{r} - 1\right) Z + \tilde{r} n, \tag{5}$$

where the second item of right side in formula (5) is cause by transform and AWGN noise, The third item of right side in formula (5) is the scale of AWGN noise.

In Fig. 5, $z_f \rightarrow z_n$ denotes AWGN noise, $z \rightarrow \tilde{z}$ denotes the real noise because of AWGN noise and inverse transform of TR2C. We can observe two features from the Fig. 5. Firstly, The transform and inverse transform of TR2C are not change the phase of constellation point, and only change its amplitude. Therefore, z_f and z are on the line Oz, z_n and $\tilde{z}$ are on the other line $O\tilde{z}$. Secondly, AWGN noise is the factor that result in the change of phase and amplitude of constellation points. Especially, the inverse transform of TR2C amplify the noise, and it introduces the error of transform in communications system based on RCM.

How to estimate accurately noise variance is a key step in TR2C. Because that it is hard to give out the closed-form expression of probability distribution function (PDF) of noise, we propose geometrical method to estimate approximately the noise variance of scaling up. In practice, we only make use of received symbols vector z_n and AWGN noise variance σ_n^2 to estimate. The geometrical method of estimating noise variance includes two steps. Firstly, according to $\tilde{z}$, we construct triangle $\Delta_{\tilde{z}O\tilde{z}}$ similar with triangle $\Delta_{z_nOz_f}$, as shown in Fig. 5. Then, using the proportional relationship among edges in similar triangles, we can approximately calculate the variance of real noise, i.e., $\tilde{z} \rightarrow \tilde{z}$.

After analysis, we find that our estimation method abandons the noise expressed by $\tilde{z} \rightarrow z$, which corresponding to the second item of right side of formula (5). Furthermore, the variance of AWGN affects the accuracy of estimating for real noise variance. The AWGN noise variance greater, the estimation of real noise variance is more inaccurate. On the contrary, the AWGN noise variance smaller, the estimation is more accurate.

3.3 Modification of RCM Decoding

There is two places that should be modified in standard RCM decoding algorithm. The one is that real noise variance is should be estimated via geometrical estimation method. The another one is that each soft demodulation symbol uses itself noise variance to compute Eq. (6) in horizontal iteration.

$$P(Y_{ij}) = \left\{ \bigotimes_{m \in R_i \setminus j, l_m \neq l} P(w_{l_m} x_m) \right\} \bigotimes P(n_i), \tag{6}$$

where $\bigotimes$ is the convolution of PDFs. The distribution of weighted variables should be $P(w_{l_m} x_m = 0) = q_{mi}^{(t-1)}(x_m = 0)$, $P(w_{l_m} x_m = w_{l_m}) = q_{mi}^{(t-1)}(x_m = 1)$, and $P(n_i) \sim \mathcal{N}(0, \sigma^2)$ for zero-mean Gaussian channel.

4 Simulation Results

In this section, we present the performance of proposed TR2C method in terms of PAPR, BER and capacity metrics.

In PAPR simulation, we use Eqs. (7) and (8) to evaluate the PAPR performance of TR2C. PAPR is calculated by

$$PAPR \ (dB) = 10 \log_{10} \left(\frac{\max\limits_{0 \leq i \leq N} \{|x(i)|^2\}}{\frac{1}{N} \sum\limits_{i=1}^{N} |x(i)|^2} \right), \tag{7}$$

where x_i is modulation symbol, N is the length of modulation symbol vector. Furthermore, complementary cumulative distribution function (CCDF) is calculated by

$$CCDF(PAPR) = \Pr(PAPR \succ PAPR_0)$$
$$= 1 - (1 - e^{-PAPR_0})^N, \tag{8}$$

where $PAPR_0$ is the threshold value of PAPR, and N is the number of count. The simulation runs 10^6 frames random data. Figure 6 shows PAPR performance comparison between original and shaped constellation of W. In this figure, "W" indicates the original constellation, and "$TR2C - W$" indicates the shaped constellation. We can observe that "$TR2C - W$" achieves significant gain compare with "W". "$TR2C - W$" reaches maximum PAPR gain 1.8 dB at CCDF $= 10^{-4}$.

In BER simulation, we test TR2C at transmission rates as 1 *bps*. The channel SNR is from 5 dB to 15 dB. The size of bit block is 400. The BER is calculated after transmitting 10^5 bit blocks. Figure 7 shows the BER performance comparison between original constellation and shaping constellation. From the results

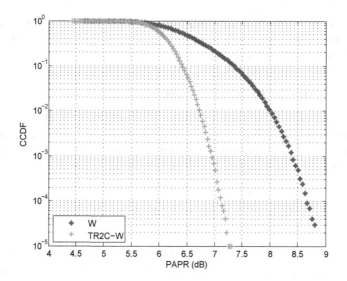

Fig. 6. PAPR performance comparison between original constellation and shaped constellation

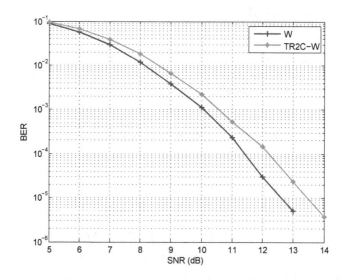

Fig. 7. BER performance comparison between original constellation and shaped constellation

in Fig. 7, we can observe that original constellation has better BER performance then shaped constellation. Comparing with "W", "$TR2C - W$" suffers from a performance loss about $0.8\,\text{dB}$ at $\text{BER} = 10^{-5}$. The BER performance loss is increase slightly with the improvement of SNR. But the loss can be accepted

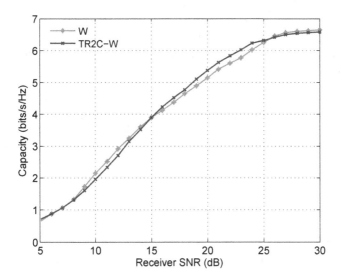

Fig. 8. Capacity performance comparison between original constellation and shaped constellation

relative to PAPR gain. And, the overall performance of our proposed system can be improved by improving the transmission power.

Finally, we evaluate the capacity performance of TR2C. Figure 8 shows the results under AWGN channel models. In this figures, the x-axis is receiver SNR. We run simulation on every integer SNR from 5 to 30 dB. At each SNR, 1000 data frames are transmitted, each with size 400 bits. The y-axis in figures are goodput. It is computed by transmission rate times 1-PER for 802.11a modulation and coding schemes. From Fig. 8, we can observe that shaped constellation suffers from a little capacity performance loss compare with original constellation when SNR $\leq$ 15 dB. Specially, the maximum loss gain is 0.2014 $bits/s/Hz$ at SNR = 12 dB. This is conformity with the BER performance comparison shown in Fig. 7. In other hand, shaped constellation achieves a little capacity performance gain when SNR in [15, 26] dB, and the maximum capacity gain is 0.236 $bits/s/Hz$ at SNR = 23 dB. On the whole, TR2C scheme achieves the capacity performance that close to original constellation when SRN from 5 dB to 30 dB.

5   Conclusion

This paper presents an novel constellation shaping method TR2C to reduce PAPR for RCM, and gives out the geometrical method of estimating noise variance. TR2C can achieve better overall system performance via improving transmission power, and achieves the capacity performance that close to original RCM in SNR range [5,30] dB. One thing to be noted is that our proposed shaping scheme makes the Euclidean distance of points in constellation smaller. In the future, we will work on the optimization design of our proposed constellation shaping method. Especially, we will focus on the design that can enlarge the Euclidean distance of constellation points, improve capacity performance and reduce PAPR for RCM.

Acknowledgment. The authors would like to thank the anonymous reviewers. This work was supported in part by the Nature Science Foundation of China (No. 61305052), the NSF of Jiangxi province of China (No. 20142BAB217007), the Science and Technology Plan Funding of Jiangxi province of China (No. 20151102040042), and Science and Technology Projects of education bureau of Jiangxi province of China (No. GJJ150984, No. GJJ151001).

References

1. Cui, H., Luo, C., Tan, K., Wu, F., Chen, C.W.: Seamless rate adaptation for wireless networking. In: Proceedings of the 14th ACM International Conference on Modeling, Analysis and Simulation of Wireless and Mobile Systems, MSWiM 2011, pp. 437–446. ACM, New York (2011)
2. Perry, J., Balakrishnan, H., Shah, D.: Rateless spinal codes. In: Proceedings of the 10th ACM Workshop on Hot Topics in Networks, pp. 6:1–6:6 (2011)
3. Perry, J., Iannucci, P.A., Fleming, K.E., et al.: Spinal codes. In: Proceedings of the ACM SIGCOMM 2012 Conference on Applications, Technologies, Architectures, and Protocols for Computer Communication, pp. 49–60 (2012)
4. Gudipati, A., Katti, S.: Automatic rate adaptation. In: Proceedings of the Ninth ACM SIGCOMM Workshop on Hot Topics in Networks, Hotnets 2010, pp. 14:1–14:6. ACM, New York (2010)
5. Donoho, D.: Compressed sensing. IEEE Trans. Inf. Theory **52**(4), 1289–1306 (2006)
6. Calderbank, A., Ozarow, L.: Monequiprobable signaling on the gaussian channel. IEEE Trans. Inf. Theory **36**(4), 726–740 (1990)
7. MacKay, D., Neal, R.: Near shannon limit performance of low density parity check codes. Electron. Lett. **33**(6), 457–458 (1997)
8. Yankov, M., Forchhammer, S., Larsen, K., Christensen, L.: Rate-adaptive constellation shaping for nearcapacity achieving turbo coded BICM. In: 2014 IEEE International Conference on Communications (ICC), pp. 2112–2117, June 2014
9. Khoo, B.K., Le Goff, S., Sharif, B., Tsimenidis, C.: Bit-interleaved coded modulation with iterative decoding using constellation shaping. IEEE Trans. Commun. **54**(9), 1517–1520 (2006)
10. Goff, S.Y.L., Khoo, B.K., Tsimenidis, C.C., Sharif, B.S.: An improved bit-interleaved turbo-coded modulation scheme using constellation shaping and iterative decoding. In: 2006 4th International Symposium on Turbo Codes Related Topics; 6th International ITG-Conference on Source and Channel Coding (TURBOCODING), pp. 1–6, April 2006
11. Valenti, M., Xiang, X.: Constellation shaping for bit-interleaved LDPC coded apsk. IEEE Trans. Commun. **60**(10), 2960–2970 (2012)

Application of Four-Channel Broadband Transmitter in Coal Mine

Xiaobing Han[✉], Qi Li, and Chenglin Fu

College of Communication and Information Engineering,
Xi'an University of Science and Technology,
Xi'an 710054, China
870095119@qq.com

Abstract. For the situation of water bursting and gas security in coal mine, a wireless transmitter system was designed to survey the water, gas and other geological structures in the coal seam. The Four-channel broadband transmitter is to generate multiplexed high-frequency pulse signal from the FPGA, through power amplifiers respectively, and the high frequency signals are fed via antennas into the ground, then completing the launch function. The results showed that application of Four-channel broadband transmitter technology is achievable, and the antenna act as the most important part of transmitter is also tested by HFSS simulation and network analyzer, the testing result is better. Studies suggest that the design of transmitter can be applied to the advanced detection of coal seam, which has played a leading role for the application of wireless detection technology in the promotion.

Keywords: Transmitter · FPGA · Four-channel · Antenna

1 Introduction

With the rapid development of wireless communication technique, the wireless detection technology has been more and more widespread used in the mining filed. Especially during the process of mining, the advanced detection of the conditions of coal seam can reduce the water inrush accident and gas leak [1–3]. The wireless detection technology use a transmitter to send signals during the antennas. Then the structure, type and distribution of underground media is deduced according to the difference between the various medias' electrical parameters and the receiver signals.

In this paper, we design a transmitter, which can be used in the wireless detection technology for coal mining. The Shape structure, types and distribution of the underground medium can be detected by using the technique of multi-frequency &Four-channel. It has important practical significance for learning the surrounding of coal seam ahead of time and reducing coal mining accident [11]. Considering the problem of the signal stability and multiple frequency, a FPGA development board is used to generate signals. The presented system of transmitter can not only meet the requirement of advanced detecting coal seams, but also be applied in other fields.

© ICST Institute for Computer Sciences, Social Informatics and Telecommunications Engineering 2017
X.-L. Huang (Ed.): MLICOM 2016, LNICST 183, pp. 25–32, 2017.
DOI: 10.1007/978-3-319-52730-7_3

2   The System Structure

The transmitter is made up of three main parts, the signal generator, the power amplifier and the antenna. Firstly a FPGA Cyclone IV is used to generate Four-channel high frequency pulsed signals. Then the signals are power amplified by a RF Power Transistor. A quarter-wave patch antenna is applied to transmit the signals output by the power amplifiers. Figure 1 shows the structure of a kind of wideband transmitter with four-channels.

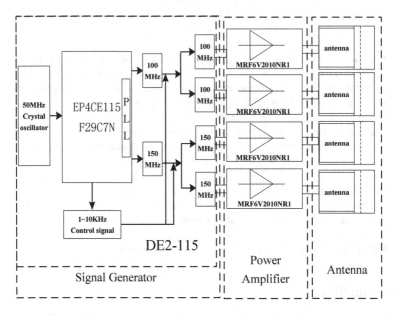

Fig. 1. Structure diagram of Four-channel broadband transmitter

2.1   The Signal Generator

The transmitter will firstly generate wideband pulse signal, and the range of signal frequency is from 50 MHz to 200 MHz. Taking advantage of avalanche transistors, the nanosecond pulse generator has been frequently-used. Though it can generate signals with large amplitude, the circuit design and adjustment process are complex. To get more stable and smart pulse signals, a FPGA development board, the present popular tool, is used as the signal generator. Figure 2 shows the main structure the generator. The multiple-frequencies signal and output control signal can be produced simultaneously. And the delayed signal, coded by Verilog Language, is used to control the signal generated.

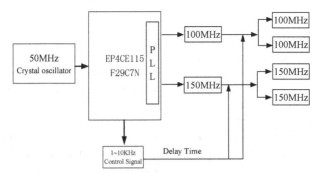

Fig. 2. Signal generator

2.2 The Power Amplifier

Due to the power is too small, the signal outgoing signal of the FPGA development board can not be sent out directly. A power amplifier is needed to improve the signal power. According to the power and frequency of the output of the signal generator, the MRF6V2010NR1 Field Effect Transistor developed by Freescale combined with a peripheral circuit is designed in Fig. 3 as the power amplifier (Table 1).

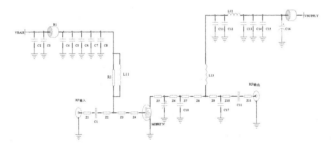

Fig. 3. The application circuit of MRF6V2010NR1

Table 1. The parameters of RF power amplifier MRF6V2010NR1

Parameters	Value
Working frequency range	10 ~ 450 MHz
Input power	7–15 dBm
Output power	30–40 dBm
Power gain	18 ~ 22 dB
Noise figure	4 dB
Three-order intercept point	40.5 dBm

2.3   Antenna

The function of the antenna is to convert electromagnetic energy into electromagnetic field, and the electromagnetic field in the space is converted to electromagnetic energy to receive electromagnetic wave, which is the transmission and sending electromagnetic energy of the conductive element [9]. As the most important component of the transmitter, the performance of the antenna can play a decisive role in the correct realization of the whole system. The design requirements of the transmission frequency in 50 MHz ∼ 200 MHz, microstrip patch antenna is not only able to meet the frequency requirements, but also to compare with the general antenna, with a light weight, small size and easy to achieve, and so on. Therefore, the design of microstrip patch antenna, through its continuous optimization to achieve system functions.

The following is the central frequency of 150 MHz patch antenna design process:

(a) Theoretical calculation of the various parameters of the patch.

The design uses 1/4 wavelength microstrip patch antenna, Fig. 4 is the principle of 1/4 wavelength patch antenna:

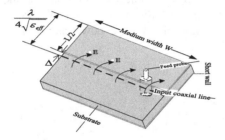

Fig. 4. Theory of quarter-wavelength patch antenna

A virtual short circuit is formed on the surface of the middle of the two radiation edge when the patch antenna is working at the lowest die. So the patch can be used in half of the short circuit to produce the antenna [5]. Then the E surface direction map is stretched to a single gap. The resonance length is about 1/4 of the wavelength in the medium of the substrate. Using the patch width W for the effective dielectric constant $\varepsilon_{有效}$ of the microstrip line and the equivalent radiation width ΔL are used to calculate the resonance length L of the 1/4 wavelength patch.

$$\frac{L}{2} = \frac{\lambda}{4\sqrt{\varepsilon_e}} - \Delta L \tag{1}$$

The patch width, effective dielectric constant and the equivalent radiation width are calculated as follows:

$$W = \frac{c}{2f_o}\left(\frac{\varepsilon_r + 1}{2}\right)^{-\frac{1}{2}} \tag{2}$$

$$\varepsilon_{有效} = \frac{\varepsilon_r + 1}{2} + \frac{\varepsilon_r - 1}{2}(1 + 12\frac{h}{W})^{-\frac{1}{2}} \tag{3}$$

$$\Delta L = 0.412\frac{(\varepsilon_e + 0.3)(W/h + 0.264)}{(\varepsilon_e - 0.258)(W/h + 0.8)}h \tag{4}$$

At the same time, a series of probes or etched with the patch between the hole to achieve short circuit. The transmission line model of the antenna increases the inductance component. The equivalent extra length Δl can be obtained from a parallel plate model, which is equivalent to a uniform spacing of the probes. The distance of the probe center is S. Radius is r, and the wavelength of the medium is $\lambda_d = \lambda_0/\sqrt{\varepsilon_r}$, which is calculated by the reduction of the length of the patch:

$$\Delta l = \frac{S}{2\pi}\left[\ln\frac{S}{2\pi r} - \left(\frac{2\pi r}{S}\right)^2 + 0.601\left(\frac{S}{\lambda_d}\right)^2\right] \tag{5}$$

As for the feed position, the formula (6) gives an approximate feed position from the side of the short circuit.

$$x = \frac{L}{\pi}\sin^{-1}\sqrt{\frac{R_i}{R_e}} \tag{6}$$

Because of the $R_e = \frac{1}{2G}$, so the radiation conductance of the single edge is:

$$G = \frac{\pi W}{\eta \lambda_0}\left[1 - \frac{(KH)^2}{24}\right] \tag{7}$$

With the movement of the feed position, the resonant frequency will be slightly deviated (Table 2).

Table 2. 150 MHz center frequency of the quarter-wavelength antenna parameters

Parameters	Value
Effective dielectric constant ε_e	4.071
Length L/2	178.955 mm
Width W	140 mm
Thickness h	7 mm
Equivalent radiation width ΔL	3.221 mm
Single edge radiation conductance G	0.562 mS
Feed coordinate x	37.33 mm

(b) Model simulation.

The patch antenna is simulated by ADS software of Agilent company.

(c) Material selection

For the patch antenna, dielectric material is FR4, selecting patch for the copper, choosing a thick copper for wire probe.

(d) Making antenna

The antenna production process is complex, using AB glue to paste and press, looking for the feed point and joining the coaxial line, making the antenna finally.

(e) Theoretical and practical comparison

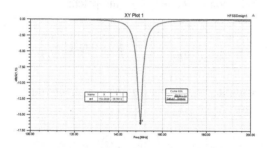

Fig. 5. Comparison chart of simulation and physical results

HFSS simulation software is used to simulate the model, and the simulation results are compared with the simulation results of the network analyzer, as shown in Fig. 5.

Figure 5 shows that there is a litter error between the antenna and the theoretical analysis of the actual production, but the error range is small, so it is necessary to test and optimize the actual production of the antenna, so that its performance is the best.

Figure 6 is the simulation results of the antenna in the 50 MHz $\sim$ 200 MHz frequency gain by HFSS simulation software. The analysis shows that the total gain effect is remarkable, and it is consistent with the broadband characteristic.

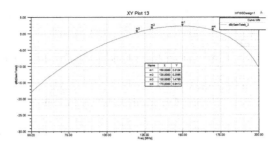

Fig. 6. Gain Total of 100 MHz ~ 200 MHz

3 Conclusions

This article describes the overall design of the Four-channel broadband transmitter, In particular, this paper has completed the design of the antenna portion, which plays a crucial role in the performance of the entire transmitter. The results show that Four-channel broadband transmitter can achieve the basic requirements for the coal seam detection, and reach the designed requirements. The author believes that the wireless detection technology has a great advantage in the coal seam detection. Four-channel, multi-frequency point selection, it can increase the capacity of the coal seam detection, In addition, the design and implementation of the band broadband antenna can improve detection the coal seam depth. It will not only be able to deter-mine the thickness of the seam, but also to achieve the coal seam geological structure detection, This reduces the occurrence of water inrush, methane gas and other accidents in the coal mining process. The study result suggests that the designed transmitter can be used in the coal seam leading exploration, It promotes the application of wireless advanced detection technology.

References

1. Daniels, D.J.: Ground penetrating radar. In: Introduction to ground penetrating radar, pp. 183–194
2. Zhang, P., Li, Y., Zhao, Y., et al.: Application and analysis on structure exploration of coal seam by mine ground penetrating radar. In: International Conference on Ground Penetrating Radar, pp. 469–472 (2012)
3. Chen, B., Hu, Z., Li, W.: Using ground penetrating radar to determine water content of rehabilitated coalmine soils treated by different methods. In: Tenth International Conference on Ground Penetrating Radar, pp. 513–516. IEEE (2004)
4. Taiquan, L.: Design, Implementation and Optimization of Antenna System Used in Impulse Ground Penetrating Radar. Wuhan University, Wuhan (2004)
5. Milligan, T.A.: Modern Antenna Design, 2nd edn (2005)
6. Wenguang, L.: Research and Design of RF Power Amplifier. Huazhong University of Science and Technology (2006)

7. Elkorany, A.S., Elhalafawy, S.M., Shahid, S., et al.: UWB integrated microstrip patch antenna with unsymmetrical opposite slots. In: Antennas and Propagation in Wireless Communications. IEEE (2015)
8. Stutzman, W.L., Thiele, G.A.: Antenna theory and design. Microwave antenna theory and design, p. 267
9. Chufang, X., KeJin, R.: Electromagnetic field and wave. Higher Education (2006)
10. Carr, J.J.: Microwave and Wireless Communications Technology. Butterworth-Heinemann, Newton (1996)
11. Xiaobing, H., Shoujie, L.: Application of ground penetrating radar receiver system in coal mine. Coal Eng. (10), 26–29 (2015)

Intelligent Cooperative/Distributed Coding

A CRC-Aided LDPC Erasure Decoding Algorithm for SEUs Correcting in Small Satellites

Hao Zheng[1(✉)], Zinan Song[1], Shuyi Zhang[1], Shuo Chai[1],
and Liwei Shao[2]

[1] School of Information and Electronics, Beijing Institute of Technology,
Beijing 100081, China
3120130330@bit.edu.cn
[2] Research Institute of BIT in Zhongshan, Zhongshan 528400, China

Abstract. Bit-flip caused by SEUs is one of the main reasons resulting in small satellites malfunction. This paper proposes a LDPC erasure decoding method aided by CRC to detect and correct errors in stored data. The key idea is that the encoded message is divided into multiple fragments protected by individual CRC so that fragments with error could be detected and corrected. Moreover, CRC is also used as an early stop criterion of decoding. Simulation and implementation results show that the proposed method has better performance compared with MS decoding both in error correcting and hardware requirements.

Keywords: LDPC erasure codes · Decoding · CRC · SEU

1 Introduction

Small satellites that could be launched in short time-scales and tight budgets have provided fast and cheap accesses to space [1]. These low-cost satellites make it possible for commercial organizations and emerging space nations to conduct independent low earth orbits (LEO) missions. Therefore, there is a growing need of small satellites.

In small satellites applications, stored digital data in random-access memory (RAM) or flash is easily suffered from single-event upsets (SEUs) caused by radiation which is impossible to be shielded [2]. More than half of system malfunctions in satellites are the result of bit-flips in stored data caused by SEUs [3].

To secure the data, many methods have been proposed, which could be divided in to three categories: (1) Multiple modular redundancy methods, such as those proposed by P.K. Samudrala in [4] and S. Hao in [5]. Methods in this category take advantages of making multiple copies of data so that voting logic can be used to detect and correct errors. However, these schemes can only offer protection against the effect of bit-flips when less than a half of data copies are affected and require larger RAM or flash. (2) Error-Correcting code based methods, such as RS codes used by Y. Zhang in [6] and G.C. Cardarilli in [7], LDPC codes used by B. Vasic in [8]. These methods can provide excellent correcting performance against bit-flips, but their decoders are usually complicated to implement. (3) Mixed methods, such as Hsiao codes and triple

© ICST Institute for Computer Sciences, Social Informatics and Telecommunications Engineering 2017
X.-L. Huang (Ed.): MLICOM 2016, LNICST 183, pp. 35–43, 2017.
DOI: 10.1007/978-3-319-52730-7_4

modular redundancy method proposed by Y. Zhao in [9], which in fact are compromise between implementation complexity and error correcting performance. Yet aforementioned methods either require multiple storage space or high complexity.

In this paper, we propose a new method using CRC codes and low-density parity-check (LDPC) erasure codes to detect and correct data errors caused by SEUs. This method applies CRC codes to detect errors in each data fragments that are divided from original data segments encoded by LDPC erasure codes and terminate the decoding process early, while the correcting step of the method is more accurate and requires less on hardware.

The rest of this paper is organized as follow. Section 2 will briefly introduce the LDPC erasure codes adopted and describe the system. Section 3 will gives the CRC-aided LDPC erasure decoding algorithm and discuss the CRC selection problem. Simulation and implementation results are presented in Sect. 4, and Sect. 5 concludes this paper.

2 The LDPC Erasure Codes and System Description

2.1 LDPC Erasure Codes and Encoding

LDPC erasure codes are a kind of concatenated code with low-density parity-check matrix, and its outer code and inner code are random LDPC code and irregular-repeat-accumulate (IRA) LDPC code, respectively.

The parity-check matrix of LDPC erasure codes with code length n_i and information bits length k_o is described below

$$\begin{cases} H = \begin{bmatrix} H_o & 0 \\ H_{i,u} & H_{i,p} \end{bmatrix} \\ H_o = \begin{bmatrix} H_{o,u} | H_{o,p} \end{bmatrix} \end{cases} \tag{1}$$

where H_o is a $(n_o - k_o) \times n_o$ parity-check matrix of outer code with code length n_o, $H_{i,u}$ and $H_{o,u}$ are binary random matrices with size $(n_o - k_o) \times k_o$ and $(n_i - k_i) \times k_i$, $H_{o,p}$ and $H_{i,p}$ are $(n_o - k_o) \times (n_o - k_o)$ and $(n_i - k_i) \times (n_i - k_i)$ matrices in *dual-diagonal* pattern

$$H_{*,p} = \begin{bmatrix} 1 & 1 & 0 & \cdots & 0 & 0 \\ 0 & 1 & 1 & \cdots & 0 & 0 \\ 0 & 0 & 1 & \cdots & 0 & 0 \\ \vdots & \vdots & \vdots & \ddots & \vdots & \vdots \\ 0 & 0 & 0 & \cdots & 1 & 1 \\ 0 & 0 & 0 & \cdots & 0 & 1 \end{bmatrix} \tag{2}$$

where $*$ is o or p. Moreover, the parity-check matrix of inner code is $H_i = \begin{bmatrix} H_{i,u} | H_{i,p} \end{bmatrix}$, whose size is $(n_i - k_i) \times n_i$, $k_i \geq n_o$. As $H_{o,p}$ and $H_{i,p}$ are both in *dual-diagonal* pattern, both H_o and H_i could easily transformed into systematic form, from which generator matrices of both outer code and inner code, denoted as G_o and G_i, can be obtained. With G_o and G_i, codewords c could be gained through following steps

Step 1.   Gain the outer codeword $c_o = m \cdot G_o$.
Step 2.   Gain the inner codeword, namely the codeword of LDPC erasure codes, c. If $k_i = n_o$, then $c = c_o \cdot G_i$. Otherwise, fill zeroes at the end of c_o to make it length equal to n_o, and let $c = c_o \cdot G_i$.

2.2   System Description

Figure 1 depicts the procedure that data input into and output from RAM/Flash. The input data m is encoded by LDPC erasure encoder and become a codeword c so that errors can be corrected at output stage, then c can be decomposed into different data fragments p, each of which can be protected by individual CRC, and written into RAM/flash. When stored data need to be read, data fragments p' are read out and checks by CRCs, and data fragments with errors will be erased. Then c' can be composed by various data fragments, and all erased data will be recovered by LDPC erasure decoder and get the output data m'.

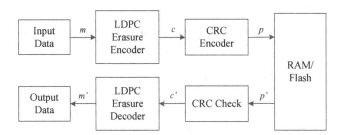

Fig. 1. In the write-in part (top): the input data is encoded by LDPC Erasure Encoder, then decomposed into different fragments protected by individual CRC. In the read-out part (button): the read-out packages that do not pass CRC check will be erased and recovered by LDPC erasure decoder.

Figure 2 illustrates the decomposition of LDPC erasure codewords and multiple data fragments that protected by individual CRC. The original data segment is decomposed into l fragments protected by individual CRC respectively, while the parity segment is decomposed into f fragments. Therefore, length of all fragments is $n_i + (l+f)k_{CRC}$ bits, in which length of data segment and parity segment is $k_o + lk_{CRC}$ bits and $n_i - k_o + fk_{CRC}$ bits respectively.

In this paper, CRCs in p will be used in three ways below. The first is to detect errors in fragments so that the decision that whether the decoding procedure is necessary could be made; the second is to decide which fragments should be erased or used as correct information in the decoding procedure; the last is to early terminate the decoding process by checking whether all data fragments could satisfy CRCs. If a

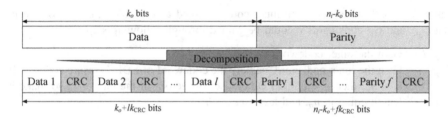

Fig. 2. Decompose of LDPC erasure codewords (top) into multiple fragments (button) protected by individual CRC.

fragment is decided as correct one by its CRC, it is assumed to be correct despite the undetected error of its CRC. Then, when the decoder is initialized, bits inside this fragment will be treated as correct bits. Due to the small probability that bit-flips caused by SEUs happened more than once in such a small fragment of data, the probability of undetected error is sufficiently small.

3 LDPC Erasure Decoding Based on CRCs

3.1 CRC-Aided LDPC Erasure Decoding

The proposed CRC-Aided LDPC erasure decoding algorithm is based on the peeling decoding (PD) in [10]. To describe PD, an LDPC erasure code should be represented as a Tanner graph [11, 12], which is constituted by a set of variable nodes connected to a set of check nodes. The PD is initialized by removing all the variable nodes whose corresponding bits are not erased and complementing the parity of check nodes connected to removed nodes whose value are 1. Then at each iteration, the PD looks for degree-one check nodes and gives their parity value to connected variable nodes, and removes those variable nodes while reverses the parity of check nodes connected to them if their value is 1. The PD will stop decoding if a codeword is obtained (i.e. decoding successes) or there is no more check nodes with one degree (i.e. decoding fails).

To describe the proposed algorithm, some notations should be made. The parity-check matrix of LDPC erasure codes is denoted as H, and described as

$$H = \begin{bmatrix} h_{1,1} & h_{1,2} & \cdots & h_{1,n_i-1} & h_{1,n_i} \\ h_{2,1} & h_{2,2} & \cdots & h_{2,n_i-1} & h_{2,n_i} \\ \vdots & \vdots & \ddots & \vdots & \vdots \\ h_{m-1,1} & h_{m-1,2} & \cdots & h_{m-1,n_i-1} & h_{m-1,n_i} \\ h_{m,1} & h_{m,2} & \cdots & h_{m,n_i-1} & h_{m,n_i} \end{bmatrix} \tag{3}$$

where $h_{r,q} \in \{0, 1\}$ is the elements in H, in which $r = 1, 2, \ldots, m$ and $q = 1, 2, \ldots, n_i$, and $m = n_i - k_o$ is the number of rows in H. Let the decoded codeword and the number of erased bits in each parity-check equation to be $\hat{c} = \{\hat{c}_q\}$ for $q = 1, 2, \ldots, n_i$ and $C_{eb} = \{C_{eb}^r\}$ for $r = 1, 2, \ldots, m$, respectively.

With these notations, the CRC-Aided LDPC erasure decoding algorithm is described in the following steps, where the major contribution of this paper is Step 4.

Step 1.   Initialization. Initialize the decoded codeword $\hat{c} = \{\hat{c}_q\}$ for $q = 1, 2, \ldots, n_i$ using CRC check results

$$\begin{cases} \hat{c}_q = 0, & \text{if } p'_q \text{ is erased} \\ \hat{c}_q = 1 - 2p'_q, & \text{if } p'_q \text{ is correct} \end{cases}, \ p'_q \in \{0, 1\}, q = 1, 2, \ldots, n_i, \quad (4)$$

and set the erased bits counter C_e to the number of bits that are erased, namely the number of zeroes in $\hat{c}$

$$C_e = n_i - \sum_{q=1}^{n_i} |\hat{c}_q|, \quad (5)$$

count the number of erased bits in each parity-check equation

$$C_{eb}^r = \sum_{\substack{ch_{r,q} = 1 \\ \hat{c}_q = 0}} h_{r,q}, \ \text{if} \prod_{h_{r,q}=1} \hat{c}_q \neq 1, r = 1, 2, \ldots, m, q = 1, 2, \ldots, n_i. \quad (6)$$

Step 2.   Correct erased bits. Search in C_{eb} and find r that let $C_{eb}^r = 1$. Update the only erased bits in this parity-check equation

$$\hat{c}_q = \prod_{\substack{h_{r,q'} = 1 \\ q' \neq q}} \hat{c}_{q'}, C_{eb}^r = 1, r = 1, 2, \ldots, m, q = 1, 2, \ldots, n_i. \quad (7)$$

Step 3.   Update C_e and C_{eb} with (5) and (6), respectively.

Step 4.   Stop decoding. If any condition below is satisfied, the decoding process will terminate and declares a success decoding.
(1) $C_e = 0$;
(2) All fragments in data segment can pass CRCs check.

If any condition below is satisfied, the decoding process will terminate and declares a fail decoding.

(1) $C_e \neq 0$, while $C_{eb}^r \neq 1$, for any $r = 1, 2, \ldots, m$;
(2) The number of iterations reaches the max iteration number.

Otherwise, the decoding process will continue and add the number of iterations by 1.

It is noted that the decoded codeword $\hat{c}$ is also an updatable indicator that indicates bits that been erased, when a bits is erased due to CRCs its corresponding $\hat{c}_q$ is 0. Moreover, compared with conventional PD, decoders using the proposed algorithm can employ parallel structure to reduce decoding delay.

3.2 Selection of CRC

The proposed CRC-Aided LDPC erasure decoding algorithm is based on the CRCs, and the performance of it relies on the error detection capability of CRC. Thus, selecting efficient CRC is important. Normally, the undetected error probability P_{ud} is used to evaluate the performance of CRC [13], which is given by

$$P_{ud} = \sum_{j=1}^{N} A_j p^j (1 - p)^{N-j} \tag{8}$$

where N is the length of data that CRC protects, p is the bit error probability, the set $\{A_j\}$ is the weight distribution of generator polynomial $g(x)$, while R is the order of $g(x)$. To find the optimal CRC code, there is the tremendous code space need to be searched, which is not an easy work. However, [13] has reported many efficient and proper CRC codes. From [2], the bit-flips caused by SEUs is at most 4 times per day every 1 Mb, in other words, the bit error probability is less than 10^{-10} such that P_{ud} is even small. Hence, the CRC code used could be selected from [13].

4 Simulation and Implementation Results

Simulations of the CRC-Aided LDPC erasure decoding algorithm regard the code error rate (CER) and average number of iterations (ANI) to verify error performance and decoding speed of it. In all experiments, the erasure channel model describe in [14] is employed, and the maximum number of iteration and codewords are 100 and 10^6. The employed LDPC erasure codes and proper CRC codes are described in the CCSDS standard [15] and [13], respectively.

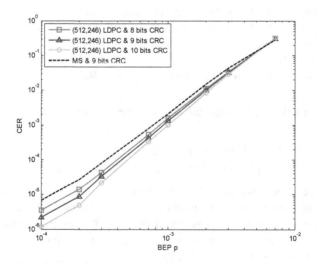

Fig. 3. Codeword error rate (CER) versus bit error probability (BEP) p.

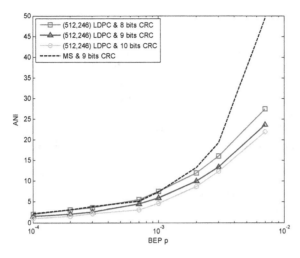

Fig. 4. Average number of iterations (ANI) versus bit error probability (BEP) p.

To confirm that the proposed method has a better performance for correcting storage errors, different configurations have been applied under the same code rate and compared with Min-Sum (MS) Algorithm. Figures 3 and 4 indicate the CER performance and ANI of proposed algorithm, which show that the proposed algorithm has outperformed in both CER and ANI compare to MS. However, CRC codes with different lengths effect the performance of proposed algorithm, longer CRC codes owns better performance, as longer CRC codes owns lower undetected error probability. But longer CRC codes require extra storage space and hardware. Hence, to meet the requirements, multiple simulations, like CER and ANI performance, should be conducted when proposed method is employed.

Table 1. The implementation of proposed algorithm on FPGA

Logic utilization	Proposed	MS
Slices	148	576
LUTs	180	436
Block RAMs	2	4
Max clock	357.3 MHz	280.6 MHz

In order to evaluate the complexity of proposed method, both proposed methods and MS algorithm using (512,246) LDPC aided by 9 bits CRC has been implemented on a Xilinx Virtex5 FPGA. Table 1 shows the implementation results of both methods, from where it is can be observed that the proposed algorithm require less on hardware because of its correcting step could achieved using one-bit addition operations, instead of several multi-bits operations. Thus, the proposed method outperforms MS both in error correcting and hardware requirements.

5 Conclusions

Using CRC codes to detect errors in stored data, a new method is proposed using LDPC erasure codes correct data errors caused by SEUs. In this method, message encoded by LDPC erasure codes are decomposed into several fragments, each of which are protected by individual CRC, so that fragments fail to pass CRC check can be detected and then corrected by LDPC erasure decoding process. Due to the accuracy and implementing fitness of its correcting step, the proposed algorithm is more efficient and requires less on hardware compare with MS algorithm with CRCs.

Acknowledgments.   This research is supported by Opening Fund Project of Space Target Measurement Key Laboratory of PLA General Armament Department and Guangdong Provincial Science and Technology Project (No. 2015B010101002).

References

1. Wicks, A., da Silva Curiel, J., Ward, M.: Fouquet: advancing small satellite earth observation: operational spacecraft, planned missions and future concepts. In: 14th Annual AIIA/USU Conference Small Satellites, pp. 21–24. Logan (2000)
2. Ma, R., Zhang, Y., Bai, Z.: Practice V Satellite and its flight achievements. Aerosp. Chin. **11**, 5–10 (1999)
3. Wang, C.: The influene with reliability of motional satellite by the single-event phenomena. Semicond. Inf. **35**, 1–8 (1998)
4. Samudrala, P.K., Ramos, J., Katkoori, S.: Selective triple modular redundancy (STMR) based single-event upset (SEU) tolerant synthesis for FPGAs. IEEE Trans. Nucl. Sci. **51**(5), 2957–2969 (2004)
5. Hao, S., Yang, Z., Chai, Z.: Study on memory's fault tolerant design based on multiple module redundancy reconfiguration. Comput. Meas. Control **17**, 190–194 (2009)
6. Zhang, Y., Yang, G., Li, H., Chang, L.: Parallel reed-solomon error correction for spaceborne mass memory system. Aerosp. Control **3**, 86–89 (2009)
7. Cardarilli, G.C., Leandri, A., Marinucci, P., Ottavi, M., Pontarelli, S., Re, M., Salsano, A.: Design of a fault tolerant solid state mass memory. IEEE Trans. Reliab. **52**(4), 476–491 (2003)
8. Vasic, B., Ivanis, P., Brkic, S.: Low complexity memory architectures based on LDPC codes: benefits and disadvantages. In: 12th International Conference on Telecommunication in Modern Satellite, Cable and Broadcasting Services (TELSIKS), pp. 11–18. IEEE Press, Serbia (2015)
9. Zhao, Y., Hua, G.: Method of fault tolerant design for memory. Aerosp. Control Appl. **35**(3), 61–64 (2009)
10. Richardson, T., Urbanke, R.: Peeling decoder and order of limits. In: Richardson, T., Urbanke, R. (eds.) Modern Coding Theory, pp. 115–122. Cambridge University Press, Cambridge (2008)
11. Tanner, R.M.: A recursive approach to low complexity codes. IEEE Trans. Inf. Theory **27**(5), 533–547 (1981)
12. Loeliger, H.A.: An introduction to factor graphs. IEEE Signal Process. Mag. **21**(1), 28–41 (2004)

13. Chun, D., Wolf, J.K.: Special hardware for computing the probability of undetected error for certain binary CRC codes and test results. IEEE Trans. Commun. **42**(10), 2769–2772 (1994)
14. Xia, H., Cruz, J.R.: On the performance of soft Reed-Solomon decoding for magnetic recording channels with erasures. IEEE Trans. Magn. **39**(5), 2576–2578 (2003)
15. CCSDS Orange Book: Erasure Correcting Codes for Use in Near-Earth and Deep-Space Communications. http://public.ccsds.org/publications/archive/131x5o1.pdf

An MELP Vocoder Based on UVS and MVF

Tangle Lu and Xiaoqun Zhao[✉]

Tongji University Shanghai, Shanghai, China
{lutangle,zhao_xiaoqun}@tongji.edu.cn

Abstract. Mixed excitation linear prediction (MELP) vocoder is generally used in low bit-rate vocoder, whose target now focuses on overall coding scheme, decrease of coding rate and improvement of robustness. Unvoiced/voiced/silence detective algorithm (UVS) possesses certain robustness and anti-noise property, while voiced excitation model based on maximum voicing frequency algorithm (MVF) is closer to the original speech characteristics. In this paper, the original excitation model of MELP vocoder is replaced and UVS is joined so that an improved 2.4 kbps coding rate vocoder is accomplished. Compared with MELP of federal standards, the improved vocoder owns better synthetic speech quality and robustness.

Keywords: Signal processing · MELP · UVS · MVF · Speech evaluation

1 Introduction

Vocoder makes it possible to transmit speech with limited bandwidth; and receiving end has high intelligibility and naturalness. Military communications need to reduce power consumption; multimedia communication systems need to reduce storage costs; satellite communications are quite short of channel resources in poor communication conditions; underwater communications possess serious signal attenuation. As a result, speech signal should be low bit-rate coded [1, 2].

Research on low bit-rate vocoder focuses primarily on the overall scheme, decrease of coding rate and improvement of robustness. Generally the research object is representative MELP vocoder, which nevertheless has poor synthetic speech naturalness, hum and low robustness in noise environment. Excitation model's performance has a significant impact on synthetic speech quality in encoder, of which multi-band mixed excitation model is superior with small time and space complexity. However sub-band sound intensity error-detection (multi-band causes) can have serious consequences, and literature [3] indicated multi-band model does not match the actual excitation. A better excitation model is required.

Based on the original MELP vocoder, in this paper, new UVS is joined, and excitation model based on MVF closer to original speech characteristic is used. Compared in performance with the one of Federal standard, improved MELP vocoder improves synthetic speech quality and robustness.

© ICST Institute for Computer Sciences, Social Informatics and Telecommunications Engineering 2017
X.-L. Huang (Ed.): MLICOM 2016, LNICST 183, pp. 44–52, 2017.
DOI: 10.1007/978-3-319-52730-7_5

2 Principle of MELP Vocoder

2.1 Vocoder Model

By linear prediction [4] speech waveform is analyzed to create channel excitation and parameters of transfer function so that speech waveform coding turns to parameter coding and the amount of data of speech transmission greatly reduces. The improved MELP vocoder retains original part of parameter extraction and quantification and improves the performance of its excitation model, as shown in Fig. 1. Encoder extracts these parameters, including UVS, pitch period, line spectral pairs (LSF), gain, Fourier series values, MVF, which are vectorial or scalar quantized. Decoder interpolates parameter, generates excitation signal, enhances adaptive spectrum, linear predicts, and generates synthetic speech. With gain suppression, noise suppression, pulse discrete filtering and so on, decoder improves the quality of synthetic speech.

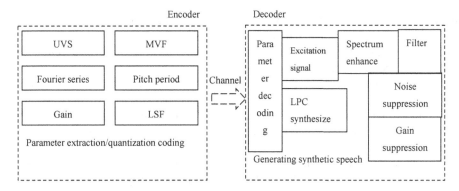

Fig. 1. Improved MELP vocoder

UVS and MVF are extracted according to method in literature [5, 6]. After extracting pitch period, integer pitch period is calculated by normalized correlation function, which fraction pitch period amends to search eventual integer pitch period by recursion. LSF parameter is produced by 15 Hz bandwidth spread on linear forecast coefficients. 10 LSF parameters' interval is at least 50 Hz. If interval is smaller, computation offset needs to be increased between former and latter component. Gain parameter used mean square value of windowing signal. Window's length is pitch period; centers respectively are located in two position of current frame. Calculate gain value two times to prevent frame gain located in the transition part; Fourier series retain the top 10 largest harmonics value, obtained by FFT calculation of residual signal.

Parameters extracted in encoder need further compression at a relatively small number of bits. Pitch period, gain, and Fourier series use scalar quantization; while LSF uses four-stage vector quantization. UVS judgment results and MVF value in improved algorithm use scalar quantization, and UVS uses only 2 bits to fully express three types of frame. According to the distribution of harmonic number, MVF value can be uniform quantized by interval. For high probability numerical interval [0, 10], quantitative interval is 1; interval (10, 30] 4. If harmonic number is greater than 30, MVF value can be considered as full band distribution, so 16 numerical distributions are quantized by 4 bits. Combining UVS judgment results, MVF quantitative bit numbers are further reduced: MVF value of silent frame is 0 and only needs 1 bit to quantize; unvoiced frame ranges from [1, 4] and only 2 bits; voiced frame [8, 24] and 3 bit. MVF with UVS judgment results quantitation can reduce numbers of bits.

Decoder synthesizes speeches. Firstly, parameter interpolation converts frame parameter to pitch period for synchronization interpolation with pitch period. Secondly, excitation signal filtered by adaptive spectrum enhancement is constructed, which makes synthetic speech better match formant waveform. LPC coefficients restored by LSF parameters construct synthetic filter. In addition, synthetic speech's coherence is improved by gain correction, pulse shaping filter.

2.2 Performance of Vocoder

Improved vocoder adds UVS and MVF to enhance anti-noise property of the vocoder. UVS judgement filters some stationary noise interference from time domain. At the same time, quality of synthetic speech in noise environment is improved according to unvoiced frame length that can help preliminarily decide background noise or unvoiced frame of objective speech. From frequency domain MVF technologies filter aliasing noise in speech frames. Harmonics as excitation signal are extracted to avoid impact on synthetic speech from noise component of other frequencies. In addition, for high frequency noise harmonic component, MVF technologies filter by counting harmonic component.

Coding rate of the vocoder remains 2.4 kbps; however, improved schedule owns more redundant bits that can effectively enhance data accuracy in decoder. Tables 1 and 2 respectively are bits allocation of quantized parameters of the original MELP and the improved. Total number of bits of the original MELP vocoder is 54 bits. The improved is only 53 bits in unvoiced frame and about 11 bits in silent frame. Redundant bits that are essential are for error-correcting coding of critical parameters to effectively enhance the robustness of the vocoder.

Table 1. Bits allocation of the original MELP vocoder

	Voiced	Unvoiced
UV	1	1
Pitch period	7	7
LSF	25	25
Gain	8	8
Fourier series	8	–
Sound intensity of sub-band	4	–
Non-cyclical sign	1	–
Forward error correcting coding	–	13

Table 2. Bits allocation of improving the MELP vocoder

	Voiced	Unvoiced	Silent
UVS	2	2	2
Pitch period	7	7	–
LSF	25	25	–
Gain	8	8	8
Fourier series	8	8	–
MVF	3	2	1
Non-cyclical sign	1	1	–
Forward error correcting coding		1	

Computational complexity of the improved vocoder is slightly lower than the original, mainly from the following aspects:

(1) Eliminates the process of five sub-bands, that is to say, sound intensity pf five sub-bands does not need to be calculated.

(2) Reduces the computation of pitch period of silent frames, LSF, Fourier series and other parameters.

(3) Only half frame after some low-complexity operation such as add, minus and square is used to extract UVS judgement eigenvalue, including mean and variance.

(4) For the calculation of peaks, the original MELP used dual hard decision of each sub-band; MVF uses soft decision, only calculating harmonic peaks.

(5) Combining with the UVS judgement, MVF consider judgement result as a prior probability, which greatly reduces computational complexity.

Delay of speech information is reduced because of low computational complexity of the vocoder. In addition, in the process of UVS judgement and MVF excitation model, intermediate variables need less storage space. For UVS, single frame can decide the frame type, where CAMDF mean, CAMDF variance and 4 thresholds are stored. For MVF, space complexity of sub-band voiced/unvoiced decision is close to the original. Intermediate variables are mainly accumulated energy value of candidate harmonics, whose number is about half length of the pitch period.

3   Experiment and Analysis

3.1   Simulation

Different coding schemes are implemented in Matlab, including coding schemes of federal standards MELP, TBE as well as improved MELP. By comparing the quality of synthetic speech, optimization of the improved vocoder is proved.

For all speech in corpus, synthetic speech is obtained by three coding schemes above. As shown in Fig. 2, speech of improved MELP can better restore the original speech with its higher similarity with original speech than TBE. At the same time, as shown in dashed line of Fig. 2, part of defect of 2.4 kbps encoder of the Federal standard is effectively improved, with improved the MELP accurately enhancing default high frequency components of voiced frames of the original MELP vocoder.

Speeches in corpus are pure. Through MELP, TBE, and improved MELP vocoder, subjective MOS [7] scores are respectively 3.94, 4.12 and 4.29. If noise library speech is added to pure speech with certain SNR, MOS scores turn to 2.95, 3.21 and 3.32. PESQ [8, 9] scores are also in line with scores rule above, respectively 2.27, 2.30 and 2.31. For noise speech, scores are 1.90, 1.96, and 1.97.

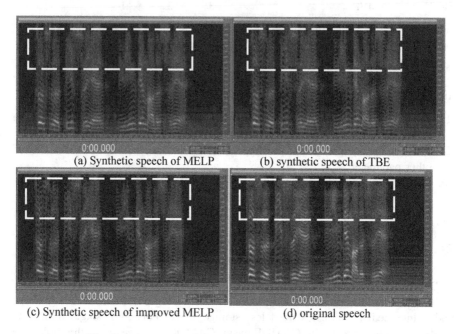

(a) Synthetic speech of MELP   (b) synthetic speech of TBE

(c) Synthetic speech of improved MELP   (d) original speech

Fig. 2. spectrogram of synthetic speech and original speech

3.2 Data Analysis

Robustness of encoder is analyzed from perspectives of gender of the speaker, speech length and noise characteristics.

(1) Gender robustness

Figures 3 and 4 represent PESQ score distribution of part of female and male speaker's speech respectively. Ordinate of each sample corresponds to PESQ scores of three coding schemes.

According to the scores of female speakers, improved MELP coding schemes generally owns higher scores than original MELP coding scheme, with sometimes difference about 0.4, while scores of TBE coding scheme are between MELP and improved MELP, slightly lower than improved MELP. Difference can be less than 0.001 for part of sample points. This is because the part of speech owns voiced features obviously, and few unvoiced frame, leading to similar advantage between UVS judgment and original voiced/unvoiced judgment.

Compared with female speakers, male speakers' synthetic speech of improved MELP is similar to the original MELP, mainly because base frequency of male speaker is lower than female speaker. Most high harmonic of male speaker concentrates about 2 kHz. Voiced frame for high frequency component cannot be accurately divided, so, adaptive MVF and spectrum division of fixed sub-band is close to performance of pure male speech.

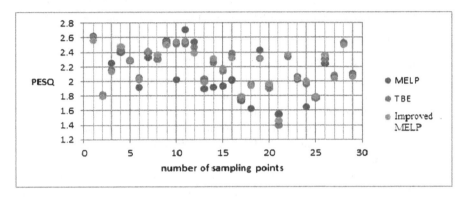

Fig. 3. Distribution of PESQ scores (female speaker)

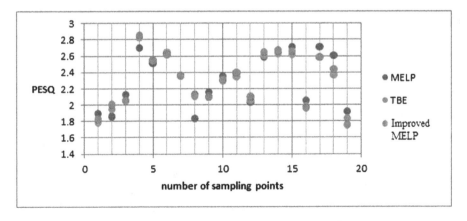

Fig. 4. Distribution of PESQ scores (male speaker)

(2) Time length robustness

Compared are PESQ scores of synthetic speech in corpus of different time length through MELP, TBE and improved MELP vocoder. For example, PESQ of long speech (120 s) of schemes above are respectively 2.87, and 2.88 and 2.89. PESQ of short speech (5 s), which are a little lower than long speech, are respectively 2.20, and 2.23, and 2.24.

For time differences, MELP, TBE and improved MELP vocoder embody the same rule: PESQ score difference is 0.6. For improved MELP vocoder, a long speech is useful for MVF threshold training, which makes the spectrum boundary of voiced/unvoiced frame closer to the original speech. And there is a lot of codebook training of vocal tract parameters in the original MELP vocoder. Long speech is conducive to optimal codebook search, of which improved MELP takes advantage.

In actual communication, length of speech is generally about 5 min [10]. Therefore, improved MELP coding scheme can be applied in the area with good performance.

(3) Noise robustness

Figure 5 and 6 shows PESQ scores in daily noise and military noise with SNR 15 dB to −10 dB. Because of military secrets military noise involved, only F16 cockpit noise data is listed, similar to other military noise in noise robustness.

Data is also divided into male and female, and then scores of the synthetic speech in different noise type are compared. It can be seen that female speech's score is higher than male. This is because the pitch frequency of female speech is higher. In the frequency domain, harmonic intervals of female speech are greater than those of male speech, so noise effect in low frequency for random scatter is relatively small.

MELP vocoder suffers the greatest impact by the noise type of babble and volvo, while improved MELP coding scheme enhance communication quality in these noise environments. For noise type of white and pink environment, improved MELP also enhance score of original MELP, which verifies improved MELP has better noise robustness. For noise type of factory1 F16 where males often appear, improved MELP has performance advantages.

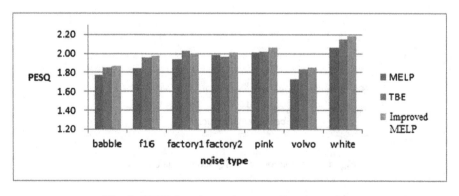

Fig. 5. PESQ in noise environment (female speaker)

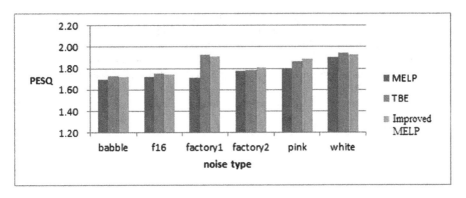

Fig. 6. PESQ in noise environment (male speaker)

4 Conclusion

This paper introduces an improved coding scheme from encoder and decoder, involving coding principle and application of the key technology, including the UVS judgement and MVF excitation model.

Differences in performance between improved MELP coding scheme and the original MELP vocoder are compared from many aspects: anti-noise property, coding rate, computational complexity, and delay and space complexity. In theory the performance advantage of the improved MELP vocoder is verified.

Schemes of MELP, TBE, and improved MELP are implemented in Matlab. Corpus uses PKU-SRSC, noiseX-92, and recorded speech, including diversities of gender, age, rate, intensity, duration, noise environmental and so on. Three coding schemes respectively encode and decode speeches, give synthetic speech PESQ scores, classify/evaluate scores, and verify the robustness of improved MELP coding scheme.

References

1. C114 communications network in China. Development and application of low bit-rate speech coding (2015). http://market.c114.net/154/a190256.html
2. Underwater acoustic communication. http://wiki.dzsc.com/info/7374.html
3. Degottex, G., Stylianou, Y.: Analysis and synthesis of speech using an adaptive full-band harmonic model. IEEE Trans. Audio Speech Lang. Process. **21**(10), 2085–2095 (2013)
4. Zongfu, L.: Multimedia Technology. Tsinghua University Press, Beijing (2009)
5. Jingyun, X., Xiaoqun, Z., Rongyun, L., Jiao, W.: Vocoder excitation model based on voicing cut-off frequency of speech. J. Beijing Univ. Posts Telecommun. **03**, 28–33 (2015)
6. Rongyun, L.I., Xiaoqun, Z., Jingyun, X.U.: Adaptive anti-noise unvoiced/voiced/silence detection algorithm. J. Yanshan Univ. **02**, 133–138 (2015)
7. Rothauser, E.H., et al.: IEEE recommended practice for speech quality measurements. IEEE Trans. Audio Electroacoust. **17**, 227–246 (1969)

8. Conway, A.E.: Output-based method of applying PESQ to measure the perceptual quality of framed speech signals. Wirel. Commun. Netw. Conf. **4**, 2521–2526 (2004)
9. Hines, A., Skoglund, J., Kokaram, A., et al.: Robustness of speech quality metrics to background noise and network degradations: comparing ViSQOL, PESQ and POLQA. In: IEEE International Conference on Acoustics, Speech and Signal Processing (ICASSP), pp. 3697–3701 (2013)
10. Wei, D.: Establishment and application of call duration model. Telecommun. Technol. **10**, 58–60 (2001)

Optimization of Voiced Excitation Model by MVF Algorithm

Bing Xue and Xiaoqun Zhao[✉]

College of Electronic and Information Engineering,
Tongji University, Shanghai, China
{xuebing8529,zhao_xiaoqun}@tongji.edu.cn

Abstract. In mixed excitation linear prediction (MELP) vocoder incentive model, the decision error of traditional maximum voicing frequency (MVF) algorithm is relatively large. In this paper, it proposes a method to optimize MVF algorithm. Firstly, in many MVF algorithms, considering about algorithm accuracy, applicability and real-time, It selects the most appropriate algorithm (cumulative harmonic scoring algorithm) for very low rate speech coder. And then, the method optimize the definition and noise immunity of the algorithm, which use adaptive MVF value to divided spectrum into two sub-bands. Finally, using pitch as performance parameters of incentive model, it simulates voiced incentive model, and compares the pitch error rate of two band excitation (TBE) model and MELP model. A conclusion is generated that TBE model is closer to the original speech features, and its performance is better than that of the original MELP incentive model.

Keywords: MVF · Residual signal · Pitch · Voicing excitation model · TBE

1 Introduction

In the (very) low rate speech coding, as an important part of speech in the model, incentive model is one of the important breakthrough. The speech model of audio segment consists of periodic and aperiodic component. The spectrum ratio of those two parts has two different calculation ways. One is multiband; the other is dual-band. In the dual-band calculation, low frequency band is regard as periodic components, with the method of impulse excitation; high frequency band is regard as aperiodic components, with the method of noise excitation. The demarcation point of the dual-band is called maximum voicing frequency (MVF). This way of division is closer to actual phonetic pronunciation characteristics. Two band excitation (TBE) model can effectively improve the quality of synthesized speech [1], at the same time, its performance depends on MVF extraction accuracy.

MVF often used in harmonic and noise model (HNM) [2, 3], deterministic and stochastic model (DSM) [4], excitation signal model and so on. In recent years, it has sprung up a large number of MVF calculation methods, including peak valley value (P2V) [5], the amplitude spectrum-phase spectrum [2], multi parameter correlation method, iterative judgment method, cumulative harmonic scoring (CHS) [6, 7] and so on. Considering the accuracy, applicability and real time performance of MVF, it is the

© ICST Institute for Computer Sciences, Social Informatics and Telecommunications Engineering 2017
X.-L. Huang (Ed.): MLICOM 2016, LNICST 183, pp. 53–61, 2017.
DOI: 10.1007/978-3-319-52730-7_6

most appropriate algorithm that CHS is applied to mixed excitation linear prediction (MELP) vocoder incentive model.

Based on the peak of the energy spectrum, CHS method respectively calculate the scoring of periodic part and aperiodic part and the weighted value from each harmonic to this two parts. The final total harmonic score is used to filter the MVF values.

2   The Optimization of MVF Algorithm

2.1   Optimize the Definition of MVF Algorithm

Amplitude spectrum-phase spectrum method use hanning window, but CHS method uses rectangular window. Although the rectangular window may bring side lobe leakage, the algorithm that we use mainly focused on characteristics of main lobe. At the same time, compared to hanning window, rectangular window's main lobe energy is more concentrated, so it uses rectangular window in optimized MVF algorithm.

For silent voice segment, MVF is set to 0 Hz. For the speech segment, the window length is twice the pitch period length. Assuming each frame of the pitch period is f_0, the estimate of each frame' MVF is mf_0 (m is the largest number of harmonic). The original speech signal is $s(k), k = 1, 2, \ldots 2N + 10$ (N is pitch period length).

Ten samples before the current sample is used to predict the current sample, which is express as below:

$$s_{pre}(k) = -\sum_{i=1}^{10} a_i s(k - i), \, k = 1, 2, \ldots, 2N \tag{1}$$

In order to facilitate representation, the expression of predictive value selects negative sign. After linear prediction (LPC) treatment, residual signal (res(k), $k = 1, 2, \ldots 2N$) is defined as below:

$$\text{res}(k) = s(k) - s_{pre}(k) = \sum_{i=0}^{10} a_i s(k - i) \tag{2}$$

As for periodic signal, its spectrum is discrete, therefore, after the discrete Fourier transform processing, the direct component of residual signal (res(i)) is removed.

$$R(k) = \sum_{i=1}^{2N} \text{res}(i) e^{-\frac{\pi j(k-1)(i-1)}{N}}, \, k = 1, 2, \ldots, 2N \tag{3}$$

Energy spectrum density is defined as: $P(k) = |R(k)|^2, k = 1, 2 \ldots 2N$. It increases the range of spectrum amplitude and makes peak features more dramatically. It is assumed that A is an even number, then, $P(k), k = 1, 3, \ldots 2N - 1$ is the pitch harmonic power components of current frame residual signal (res(i)); $P(k), k = 2, 4 \ldots, 2N$ is noise power components. Because $P(k)$ is a point-symmetrical with respect to N, it only need to deal with $P(k), k = 1, 2 \ldots, N$, at the same time, the largest number of harmonic m should be determined in the harmonic power components in $k = 1, 3 \ldots, N - 1$. The normalized power function is defined as follow:

$$P_n(k) = \begin{cases} 1, k = 1 \\ \dfrac{[P(k) + P(k+2)]/2}{[P(k) + P(k+2)]/2 + P(k+1)}, k = 2, 4 \ldots, N \\ \dfrac{P(k)}{[P(k-1) + P(k+1)]/2 + P(k)}, k = 3, 5 \ldots, N - 1 \end{cases} \quad (4)$$

The comprehensive accumulated energy E consists of the accumulation of harmonic energy E_h and noise energy E_a.

$$E_h(x) = \sum_{k=3,5\ldots2x+1} \max[0, P(k)(P_n(k) - P_n(k-1))], x = 1, 2 \ldots N/2 \quad (5)$$

$$E_a(x) = 2 \sum_{k=2x,2x+2\ldots,N} P(k)P_n(k), x = 1, 2 \ldots, N/2 \quad (6)$$

$$E(x) = E_h(x) + bE_a(x), x = 1, 2 \ldots N/2 \quad (7)$$

Among them, A is adjustment parameter. The scope of it generally takes to: $0.2 \sim 0.6$ [8]. MVF is the xth harmonic frequencies when accumulated comprehensive energy E obtain the maximum value. $m = x$.

2.2 Optimize the Noise Immunity of MVF Algorithm

Original speech, $s(k), k = 1, 2, \ldots N$, in which N is the length of pitch period, consists of effective voice information $u(k)$ and noise information $n(k)$. In the short-time signal processing, $u(k)$ is periodic signal, $n(k)$ is aperiodic signal, therefore, $s(i)$ is processed with DFT, as shown below.

$$S(k) = \sum_{i=1}^{N} u(i)e^{-j\frac{2\pi ki}{N}} + \sum_{i=1}^{N} n(i)e^{-j\frac{2\pi ki}{N}}, k = 1, 2, \ldots N \quad (8)$$

Harmonic components are mutually orthogonal, so the energy spectrum is as follow:

$$P(k) = |S(k)|^2 = \sum_{i_1=N_x}^{N'} u(i_1)^2 + \sum_{i_2=N_y}^{N''} n(i_2)^2 + \sum_{i_3=N_z}^{N'''} [u(i_3) + n(i_3)]^2 \quad (9)$$

As is shown in the above type, N times harmonic is divided into three types of interval: i_1, i_2 and i_3. i_1 represents a harmonic interval which only distributes effective information. i_2 represents a harmonic interval which only distributes noise information. i_3 represents an interval which distributes both effective and noise information. In the harmonic interval of i_2 and i_3, the amplitude of $S(k)$ can accumulate noise information, and the amplitude of energy spectrum will change. It will also affect the outcome when the noises accumulate to a certain extent. As shown in Table 1, we assume that it happens in the noise environment. N_0 is the harmonic which correspond to the standard value of MVF. E_0 is comprehensive accumulated energy. In interval i_1, N_1 is some harmonic and E_1 is comprehensive accumulated energy. Similarly, N_2 and E_2 is in i_2. N_3 and E_3 is in i_3.

Table 1. Compare the differences of adding noise verdict

	Comprehensive accumulated energy	The largest harmonic number	Result of judgment	Comment
Noise environment 1	$E_1 = E_0$	N_1	Accuracy	The signal is not affected by noise
Noise environment 2	$E_2 > E_0$	N_2	Serious miscalculation	Adding some other harmonic components, noise can impact excitation signal.
Noise environment 3	$E_3 > E_0$	$N_3 > N_0$ $(N_3 < N_0)$	High (Low)	

The original speech signal is processed with LPC (linear predictive coding) and inverse filtering, and then, predicting residual signal can be got. In the processing of add noise speech, LPC can analyze and extract format, sound loudness, pitch period and some other key messages. Inverse filter can filter high frequency noise; therefore, residual signal is used for frequency domain processing instead of the original speech signal. Residual signal effectively avoid some of the noise. As is shown in Fig. 1, adding automobile noise, SNR = −4 dB, residual signal fully retain the periodicity and peak character of the original speech without noise. However, with the loss of SNR, time-domain signals of the original speech with noise will be seriously deformative and noise interference will also be serious.

The anti-noise performance of modified MVF algorithm is superior to primitive CHS algorithm. Compare the MVF values of voiced frame before and after adding noise. For example, in automobile noise, as shown in Fig. 2, it is spectrum difference before and after adding noise. Even under low SNR circumstance, peak characteristics of harmonic component are still obvious. MVF of the original speech frame is 1440 Hz. MVF of the adding noise speech frame which is calculated by original CHS algorithm is 240 Hz. However, it is 1440 Hz which is calculated by modified MVF algorithm. It can be concluded that using residual signal, modified MVF upgrade the anti-noise performance of the algorithm.

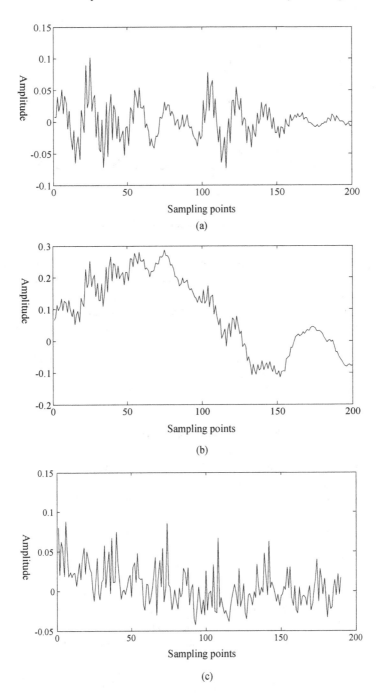

Fig. 1. Time domain waveform of voiced frame ((a) original speech signal with no noise (b) original speech signal with noise (c) residual signal with no noise)

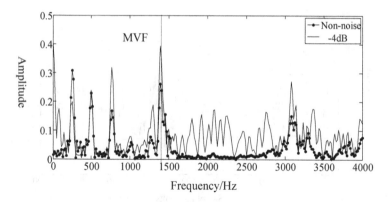

Fig. 2. Single frame spectrum with/without noise

To improve the anti-noise performance of modified MVF algorithm, the frequency spectrum distribution of noise should be considered. In principle, different frequency spectrum distribution of noise has some differences between low frequency and high frequency. In the processing of frequency domain, when different kinds of noise are superposed in target voice, it can be achieved by the improved MVF algorithm to suppress noise.

The improved MVF algorithm has good noise resistance to the general background noise, but human background noise has potential interference. When human background noise exists, UVS Judgment can be used to filter out some interference.

3 The Performance Simulation of Residual Excitation Model

The effect of motivation model, produced in vocoder, can provide the original voice frequency and sound loudness features. It can also be termed pitch. From the point of synthetic speech's subjective judgment, pitch is also one of the intuitive factors to decide the voice quality. It can directly affect the hearing feeling. For example, bass and soprano have obvious difference on the pitch. Except for accurate pitch period, it can effectively improve the naturalness of synthesized speech that coding pitch information of the closer original voice and compound incentive model in the decoder. In addition, the change of pitch directly affects the meaning expression, which is Chinese tone. Different tones correspond to different pitch changes. In the incentive model, the accuracy of extracting pitch corresponds to the naturalness and intelligibility of the synthetic speech, so pitch error rate can be used to evaluate the performance of incentive model. The accuracy of pitch is mainly manifested on the tone and volume. Figures 3 and 4 respectively are spectrograms which are different in tone and volume.

Exciting signal of TBE and MELP can be respectively calculated from original speech signal. The pitch of exciting signal and original speech signal can be extracted and compared in spectrograms. As is shown in Fig. 5, TBE exciting signal and MELP

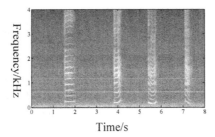

Fig. 3. Spectrogram of different tones

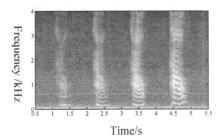

Fig. 4. Spectrogram of different volume

exciting signal both retain the information of pitch change and four tones do not have some confusing phenomena. However, for the voiced frame of speech segment, MELP exciting signal just retains the periodic information under 2 kHz. The exciting signals above 2 kHz are replaced by random signals. As is shown in the spectrogram, there are no clear white horizontal stripes. In the spectrogram of original speech signal, speech segments in $0.7 \sim 1$ s, $3.2 \sim 3.5$ s and $2.2 \sim 2.6$ s both have periodic information in $0 \sim 4$ kHz. In other words, the pitches of those speech segments are at 4 kHz. TBE model can not only accurately extract the pitch of voiced frame, but also the pitch of devoiced frame. As an example, the devoiced frame in the vicinity of 0.5 s should contain the third-harmonic component. TEB spectrogram has three clear horizontal white stripes, but in the MELP spectrogram, the harmonic in sub band of 500 Hz is intercepted and stimulated by pulse signal. It only has two horizontal white stripes, including fundamental frequency and second harmonic. Therefore, there are some errors in MELP exciting signals.

Comparing with the spectrograms of some different speech segments and the frequency distribution of their exciting signals, the pitch error rate of TBE exciting signal is far less than that of the MELP exciting signal in both of devoiced and voiced frames. The fixed subband division method cannot accurately extract the pitch information of speech signal, but the method of adaptive MVF can overcome this problem. TBE incentive model is better than original MELP incentive model.

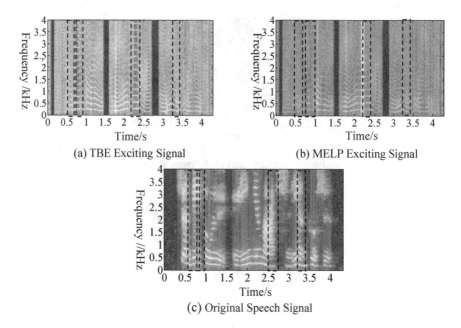

(a) TBE Exciting Signal   (b) MELP Exciting Signal

(c) Original Speech Signal

Fig. 5. Spectrogram of exciting signal and original speech signal

4 Conclusion

Contraposing the shortcomings of MELP vocoder incentive model, in this paper, it adopt the way of dynamic segmentation double band and implement TBE incentive model. Firstly, CHS is chosen in many MVF algorithms, which is one of the most suitable algorithms for very low bit-rate speech coder. Then, using residual signal to analyze the performance of algorithms, it improve the definition and antinoise performance of MVF algorithm, and put forward using pitch as a parameter of ruling the performance of incentive model. Finally, through the simulation on the performance of the incentive model, a conclusion can be drawn that the pitch error rate of TBE exciting signal is far less than that of the MELP exciting signal. Therefore, TBE model is closer to original speech features and it can effectively improve the performance of MELP vocoder incentive model.

References

1. Sang-Jin, K., Jong-Jin, K., Minsoo, H.: HMM-based Korean speech synthesis system for hand-held devices. IEEE Trans. Consum. Electron. **52**(4), 1384–1390 (2006)
2. Erro, D., Sainz, I., Navas, E., et al.: Harmonics plus noise model based vocoder for statistical parametric speech synthesis. IEEE J. Sel. Top. Signal Process. **8**(2), 184–194 (2014)
3. Drugman, T., Dutoit, T.: The deterministic plus stochastic model of the residual signal and its applications. IEEE Trans. Audio Speech Lang. Process. **20**(3), 968–981 (2012)

4. Stylianou, Y.: Applying the harmonic plus noise model in concatenative speech synthesis. IEEE Trans. Speech Audio Process. **9**, 21–29 (2001)
5. Drugman, T., Stylianou, Y.: Maximum voiced frequency estimation: exploiting amplitude and phase spectra. IEEE Signal Process. Lett. **21**(10), 1230–1234 (2014)
6. Hermus, K., Van Hamme, H., Irhimeh, S.: Estimation of the voicing cut-off frequency contour based on a cumulative harmonicity score. IEEE Signal Process. Lett. **14**(11), 820–823 (2007)
7. Hermus, K., Girin, L., Van Hamme, H. et al.: Estimation of the voicing cut-off frequency contour of natural speech based on harmonic and aperiodic energies. In: IEEE International Conference on Acoustics, Speech and Signal Processing, pp. 4473–4476 (2008)
8. Edge, R.L.: Measuring speech naturalness of children who do and do not stutter: the effect of training and speaker group on speech naturalness ratings and agreement scores when measured by inexperienced listeners [DB/OL]. https://getd.libs.uga.edu/pdfs/edge_robin_l_201208_phd.pdf

Intelligent Cooperative Networks

Capacity of Content-Centric Hybrid Wireless Networks

Jian Liu[1,2], Guanglin Zhang[1,2(✉)], Demin Li[1,2], and Jiajie Ren[1,2]

[1] College of Information Science and Technology, Donghua University,
Shanghai 201620, People's Republic of China
{jianliu,renjiajie}@mail.dhu.edu.cn, {glzhang,deminli}@dhu.edu.cn
[2] Engineering Research Center of Digitized Textile and Apparel Technology, Ministry
of Education, Shanghai 201620, People's Republic of China

Abstract. In this paper, we investigate the capacity scaling laws of content-centric hybrid wireless networks, where users aim to retrieve content stored in the network rather than maintain source-destination communication. n nodes that have limited-capacity content store are assumed to be independently and uniformly distributed in the network area. The content store equipped in each node is used to cache contents according to the proposed caching schemes. m base stations are regularly placed and act as relays during the content retrieving process. We consider heterogenous caching access scheme where the cached probability of contents is different and the requested contents follow a Zipf content popularity distribution. We present the closed form capacity formulae for the heterogenous caching access scheme in order sense.

Keywords: Capacity · Wireless hybrid network · Content-centric · Cache

1 Introduction

The number of wireless users is exponentially booming day by day. The scaling behavior of wireless networks fascinates wide interests in both academic and industry communities. In [1], Gupta and Kumar's ground breaking work pioneers the scaling behavior study of wireless networks. By assuming that n nodes are distributed independently and uniformly on a unit area, they show that the per-node throughput capacity in random wireless networks scales as $\Theta(\frac{W}{\sqrt{n \log n}})$, which decreases to zero as the number of nodes goes to infinity. While in this work, nodes can simultaneously serve as sources, destinations, and as relays for other source-destination pairs.

To increase the capacity of wireless networks, researchers study hybrid wireless network which combines base stations and ad hoc mode. Liu et al. [3] first study the scaling behavior of throughput capacity for hybrid wireless networks under two different routing policies. Considering a hybrid wireless network composed of n normal ad hoc nodes and m base stations, they show that under the

X.-L. Huang (Ed.): MLICOM 2016, LNICST 183, pp. 65–75, 2017.
DOI: 10.1007/978-3-319-52730-7_7

K-nearest cell routing strategy, the per-node capacity scales as $\Theta(\mathrm{W}\sqrt{\frac{1}{n\log n/m^2}})$ when m grows asymptotically slower than $\sqrt{n}$; otherwise, the per-node capacity scales as $\Theta(\frac{Wm}{n})$. Kozat et al. [5] investigate the throughput capacity of hybrid wireless ad hoc networks with the support of infrastructures under different assumptions. By assuming that both the ad hoc nodes and base stations are randomly placed, they show that the per-node throughput capacity is $\Theta(W/\log n)$. Recently, network content is accessed increasingly in different ways in wireless environments. While in wireless applications, caching content objects closest to requesters can significantly decrease the delay of content acquiring, which could improve the throughput capacity potentially. In the academic society, by adopting caching technique, new content-centric networking architectures such as Named Data Networking (NDN) [11] and Content-Centric Networking (CCN) [12] have been developed for efficient content distribution. In [2], Liu et al. study the scaling laws of the throughput capacity of cache enabled content distribution wireless ad hoc networks with Nearest Caching Node scheme, and Transparent Enroute Caching scheme. In [6], they study the per-node throughput capacity of an information-centric network when the data cached in each node has a limited lifetime. In [7], Mahdian et al. study the throughput-delay tradeoffs in content-centric Wireless Networks. The paper [10] presents the problem of great growth demand for video content based on femto-like base stations which can cache the popular content.

In this paper, we characterize the throughput capacity of content-centric hybrid wireless networks. We assume that each node is equipped with cache which can store content objects. Cached content objects closest to requesters can significantly decrease the distance between the sources and destinations, which means that the requesting nodes and its desired contents have minimum number of hops using caching contents in the purely ad hoc mode transmissions. In our network model, we assume that the requested content objects can be accessed successfully in ad hoc mode transmissions in the small distance. Otherwise, if the desired contents have not be cached, or the distance between the sources and destinations is larger than the minimum distance, the request should be carried out by the infrastructure mode. We consider heterogenous caching access scheme where the probability that the jth content cached in a node is q_j and the content popularity distribution is p_j, which means the probability that a typical user request a content is p_j.

The main contributions of our paper are as follows.

(1) We investigate the throughput capacity of content-centric hybrid wireless networks under heterogenous caching access scheme and propose the closed form capacity scaling results in order sense, respectively.
(2) We analyze the impact of system parameters such as the number of nodes, the number of base stations, the caching probability of content, as well as the number of content objects on the throughput capacity, which can provide theoretical guidance for content-centric hybrid wireless network design.

The rest of the paper is organized as follows. Section 2 presents the system model. In Sect. 3, we derive the throughput capacity of content-centric hybrid wireless network under heterogeneous caching access scheme. Finally, we conclude the paper in Sect. 4.

2 System Model

2.1 Network Model

We consider a two-tier content-centric hybrid wireless network where n nodes (users) are distributed uniformly and randomly in the low tire, overlaid with infrastructure with m base stations in the high tire on the surface of a torus of unit area. The assumption of torus enables us to avoid edge effects. And for the nodes located on an unit square, the results derived in the paper are applicable as well. We assume each node employs the same transmission range and power. The base stations which are regularly placed in the network divide the area into a hexagonal tessellation. Each hexagon is named a cell which has a base station in its center. And the infrastructure network follows the classic 7-cell reuse model in [8]. These base stations are added as relay nodes instead of data source or data receiver. All the base stations are connected by a wired network. Furthermore, we assume that these base stations have unlimited bandwidth and no power constraints in the wired network.

There is a total number of M distinct content objects in the network. We assume each content object has the same unit size. Each node is assumed to equip with a local cache, which can store copies of content objects. The cache size of each node is C units of content. For the problem of caching contents not to be trivial, it should be $C < M$, so that each node must select which content objects to be cached. In addition, to have large enough memory to store at least one copy of each content object in the network, it has to be $nC \geq M$.

We adopt the Protocol Interference model introduced in [1]. A transmission from node X_i is successfully received by node X_j if the following two conditions are satisfied: (i) The distance between node X_i and node X_j is within the transmission range r, i.e., $|X_i - X_j| \leq r$; (ii) For every other node X_k that is simultaneously transmitting over the same channel, it should be followed, $|X_k - X_j| \geq (1 + \Delta)|X_i - X_j|$ for some $\Delta > 0$.

2.2 Routing Strategy and Content Access Scheme

There are two types of transmissions in the system model: ad hoc mode and infrastructure mode. In the ad hoc mode, we use the content-centric access scheme, where a content is requested successfully by the nearest node that has the copies of desired content in its local cache without using any infrastructure. While in the infrastructure mode, the request is first forwarded from the requesting node to the base station, and then to the caching node. For the sake of simplicity but without loss of generality, the two models are as shown Fig. 1.

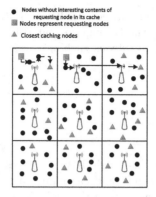

Fig. 1. Illustration of hybrid wireless network architecture.

In this paper, we consider heterogenous caching content access scheme. For the scheme, we assume a random content caching strategy where the jth content object is cached with probability q_j, $0 \leq q_j \leq 1$, $j = 1, \cdots, M$. Then a user requests the jth content object according to the content popularity distribution law p_j, which follows the Zipf laws. In other words, the probability that a user requests a file is p_j. Since we have n nodes, the average number of copies of jth content object is nq_j.

In this paper, we assume channel bandwidth is W bits/second, which is divided into ad hoc mode and infrastructure mode. i.e., the bandwidth is split into W_1 for ad hoc mode, W_2 for the downlink and W_3 for the uplink transmission for the infrastructure mode, respectively. We assume that the uplink bandwidth equals the downlink bandwidth, i.e., $W_2 = W_3$. Hence, $W = W_1 + 2W_2$.

Throughput: The per-node throughput is defined as expected number of bits/second that can be transmitted by each node to its chosen destination. The sum of the per-3node throughput over all the nodes in a network is defined as the aggregate throughput of the network.

Feasible Throughput: We say that the aggregate throughput, denoted by $T(n)$, is feasible if there is a spatial and temporal scheduling scheme that yields an aggregate network throughput of $T(n)$ bits/sec.

Aggregate Throughput Capacity of A Network: We say that the aggregate network throughput capacity is of order $O(f(n))$ bits/sec if there exists deterministic constant $0 < c_1 < +\infty$ such that

$$\lim_{n \to \infty} \mathbf{Prob}(T(n) = c_1 \mathrm{f(n)} \text{ is feasible}) < 1.$$

And is of order $\Omega(f(n))$ bits/sec if there is deterministic constants $0 < c_2 < +\infty$ such that

$$\lim_{n \to \infty} \mathbf{Prob}(T(n) = c_2 \mathrm{f(n)} \text{ is feasible}) = 1.$$

3   Heterogenous Content Access Capacity

In this section, we consider heterogenous content access scheme that the random content caching policy where the jth content is cached in a node with the probability of q_j. Then, the cooperation policy is the content objects popularity distribution, which means that the probability that a user requests the jth content is p_j. The content popularity distribution follows the Zipf law. Hence, $p_j = j^{(-\alpha)}/H_\alpha(M)$, where α is the Zipf's law exponent. And $H_\alpha(M) = \sum_{i=1}^{M} i^{-\alpha}$, it is given by [13].

$$H_\alpha(M) = \begin{cases} \Theta(1), & \alpha > 1; \\ \Theta(logM), & \alpha = 1; \\ \Theta(M^{1-\alpha}), & \alpha < 1. \end{cases}$$

3.1   Ad Hoc Mode Throughput Capacity

For the low tier network component, n nodes are uniformly and independently distributed on a planar torus and a total number of M content objects are requested and cached with different probability. The local cache of each node can store C content units. We first get the achievable lower bound by approaches in [1], shown as follows. $c_i's$ denote deterministic constants independent of n.

Voronoi Tessellation [4]: Given a set of n points in a plane, Voronoi tessellation divides the network area into a set of polygonal cells. The border-line of each region is the vertical bisector of the lines joining the points.

Lemma 1 [1]: For every $\varepsilon > 0$, a Voronoi tessellation has the property that every Voronoi cell contains a disk of radius ε and is contained in a disk of radius 2ε.

Then for n nodes, we can construct a Voronoi tessellation V_n, which satisfies the following property:

(V1) Every Voronoi cell contains a disk of area $100\frac{\log n}{n}$.
(V2) Every Voronoi cell is contained in a disk of radius $2\rho(n)$. Let $\rho(n) :=$ the radius of a disk of area $100\frac{\log n}{n}$.

Adjacent Voronoi Cells: Two cells are called adjacent neighbor if they share a common point (every cell is a closed set).

We assume that the transmission range of each node is $r(n)$, and we have

$$r(n) = 8\rho(n).$$

The transmission range permits direct communication within a Voronoi cell and between adjacent Voronoi cells.

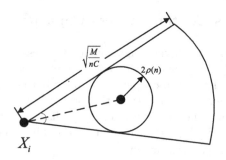

Fig. 2. Illustration for calculate the probability of L_i intersects Voronoi cell V.

Interfering Neighbors: Two Voronoi cells are called interfering neighbors if there is a point in one cell which is within a distance $(2 + \Delta)r(n)$ of some point in the other cell.

Lemma 2 [1]**:** When omnidirectional antennas are used by all nodes in the network, every cell in V_n has no more than c_1 interfering neighbors, and c_1 depends only on Δ and grows no faster than linearly in $(1 + \Delta)^2$.

Proof: We omit the proof due to space limitation.

Lemma 3 [1]**:** In the Protocol Model, there is a schedule for forwarding packets such that in every $(1 + c_1)$ slots, each cell in the tessellation V_n gets one slot for packet transmission, and all transmissions are successfully received within a distance $r(n)$ from their transmitters.

Lemma 4 [9]**:** Based on the assumptions that there is a total number of nq_j cache copies for each content object. Thus, for any node requesting each content object, the average Euclidean distance from the requesting node to the closest copy of desired content object is $\Theta(\frac{1}{\sqrt{nq_j}})$.

We choose the routes of packets to approximate the straight line connecting the requesting node and its closest interesting content objects. Let L_i denotes the straight segment connecting the requesting node and its closest caching node. Under the heterogenous access content pattern, we bound the probability that a line L_i intersects a given cell V in V_n.

Lemma 5: For segment L_i and Voronoi cell V, under the heterogenous content access pattern,

   $P(L_i$ intersects V and L_i uses W_1 transmitting packets successfully$)$ $\le$ $c_3(\frac{1}{q_j})^{\frac{3}{2}} \frac{\sqrt{\log n}}{n^2}$ (Fig. 2).

Proof: We omit the proof due to space limitation.

Since there is a total number of n lines $\{L(i, j_i)\}_{i=1}^n$, connecting the X_i and Y_i, the expected number of lines passing through a Voronoi cell that uses frequency band W_1 is:

$$E(\text{Number of lines in } \{L(i, j_i)\}_{i=1}^n \text{ intersects } V \text{ and}$$

use W_1 transmitting packets successfully)

$$\leq \frac{c_2 \sum_{j=1}^M p_j}{(q_j)^{\frac{3}{2}}} \frac{\sqrt{\log n}}{n}.$$

By using the uniform convergence in large numbers law, we have the following two results.

Lemma 6: There is a $\delta(n)' \to 0$ such that
P($\sup_{V \epsilon V_n}$ (Number of lines L_i intersecting V and L_i uses W_1 transmitting packets successfully)

$$\leq \frac{c_3 \sum_{j=1}^M p_j \sqrt{\log n}}{(q_j)^{\frac{3}{2}}} \frac{\sqrt{\log n}}{n}) \geq 1 - \delta'(n).$$

Note that a cell is proportional to the number of lines passing through it, which can handle the traffic. Since the frequency band W_1 carries traffic of rate $T_a^0(n,m)$ bits/second of each line L_i, we obtain the following:

Lemma 7: There is a $\delta(n)' \to 0$ such that
P($\sup_{V \epsilon V_n}$ (Traffic needing to be carried by cell V) $\leq c_3 T_a^0(n, m)(\frac{c_3 \sum_{j=1}^M p_j}{(q_j)^{\frac{3}{2}}} \frac{\sqrt{\log n}}{n})$
$\geq 1 - \delta'(n).$

Lemma 7 implies that the rate every cell needs to transmit is less than $c_3 T_a^0(n, m)(\frac{c_3 \sum_{j=1}^M p_j}{(q_j)^{\frac{3}{2}}} \frac{\sqrt{\log n}}{n})$ with high probability. This rate can be managed by every cell if it is less than the rate available, i.e., if

$$c_3 T_a^0(n, m)(\frac{c_3 \sum_{j=1}^M p_j}{(q_j)^{\frac{3}{2}}} \frac{\sqrt{\log n}}{n}) \leq \frac{W_1}{c_2}. \tag{1}$$

Hence, we derive a lower bound on the per-node throughput capacity contributed by ad hoc mode transmissions, which is shown in the following lemma by changing Eq. (1).

Lemma 8: For ad hoc mode transmissions, under heterogenous content access scheme, the lower bound of per-node throughput capacity with content-centric is as follows: When $q_j = o(\frac{\log^{\frac{1}{3}} n}{n^{\frac{2}{3}}} (\sum_{j=1}^M p_j)^{\frac{2}{3}})$, there is a deterministic constant $c > 0$ not depending on n, Δ, or W_1, such that $T_a^0(n, m) = \frac{cn(q_j)^{\frac{3}{2}} W_1}{(1+\Delta)^2 \sqrt{\log n} \sum_{j=1}^M p_j}$ bits/second is feasible with high probability, i.e., $T_a^0(n, m) = \frac{n(q_j)^{\frac{3}{2}} W_1}{\sqrt{\log n} \sum_{j=1}^M p_j}$.

When $q_j = \Omega(\frac{\log^{\frac{1}{3}} n}{n^{\frac{2}{3}}} (\sum_{j=1}^{M} p_j)^{\frac{2}{3}})$, there is a deterministic constant $c > 0$ not depending on n, Δ, or W_1, such that $T_a^0(n,m) = W_1$ bits/second is feasible with high probability.

Next, we will derive upper bound on the per-node throughput capacity with content-centric under heterogenous content access scheme.

Lemma 9 [1]: For protocol model, the number of simultaneous transmissions on any particular channel with content-centric for the entire network is no more than $N_{max} = \frac{4}{c_8 \pi \Delta^2 r^2(n)}$. Hence, $T_a^0(n,m) \leq \frac{c_9 W_1}{\Delta^2 (\frac{1}{q_j})^{\frac{3}{2}} \frac{r(n)}{\sqrt{n}} \sum_{j=1}^{M} p_j}$.

It has shown in [9], $r(n) > \sqrt{\frac{\log n}{\pi n}}$ is necessary to guarantee connectivity with high probability, then we obtain $T_a^0(n,m) \leq (q_j)^{\frac{3}{2}} \frac{n}{\sqrt{\log n} \sum_{j=1}^{M} p_j}$.

Proof: We omit the proof due to space limitation.

In addition, $T_a^0(n,m) \leq W_1$, we have the following lemma.

Lemma 10: For ad hoc mode transmissions, under heterogenous content access pattern, the upper-bound of per-node throughput capacity with content-centric has two cases: When $q_j = o(\frac{\log^{\frac{1}{3}} n}{n^{\frac{2}{3}}} (\sum_{j=1}^{M} p_j)^{\frac{2}{3}})$, an upper bound on per-node throughput capacity is $T_a^0(n,m) = \frac{n(q_j)^{\frac{3}{2}} W_1}{\sqrt{\log n} \sum_{j=1}^{M} p_j}$ bit/second, where $c' < +\infty$, not depending on n, Δ, or W_1. When $q_j = \Omega(\frac{\log^{\frac{1}{3}} n}{n^{\frac{2}{3}}} (\sum_{j=1}^{M} p_j)^{\frac{2}{3}})$, an upper bound on per-node throughput capacity is that $T_a^0(n,m) = W_1$.

Thus, the total traffic in ad hoc mode is $\frac{\pi}{q_j} T_a^0(n,m)$. Combining Lemma 8 and Lemma 10, we have

Theorem 1: Under the heterogenous access pattern, the throughput capacity of the network with content-centric contributed by ad hoc mode transmissions is

$$T_a(m,n) = \begin{cases} \Theta(\frac{n\sqrt{q_j} W_1}{\sqrt{\log n} \sum_{j=1}^{M} p_j}), \\ \quad q_j = o(\log^{\frac{1}{3}} n (\frac{\sum_{j=1}^{M} p_j}{n})^{\frac{2}{3}}); \\ \Theta(\frac{W_1}{q_j}), \\ \quad q_j = \Omega(\log^{\frac{1}{3}} n (\frac{\sum_{j=1}^{M} p_j}{n})^{\frac{2}{3}}). \end{cases}$$

3.2 Infrastructure Mode Throughput Capacity

Since the base stations divide the area into a hexagon tessellation, there exists a 7-cell frequency reuse pattern in the infrastructure mode. We know that the bandwidth of uplink for infrastructure mode transmission is W_2 bits per second. Thus, the throughput capacity per cell is lower bounded by $\frac{1}{7} W_2$ and upper bounded by W_2. We derive the following lemma.

Lemma 11: Under the heterogenous access pattern, the throughput capacity of the network with content-centric contributed by infrastructure mode transmissions is

$$T_b(n, m) = \Theta(mW_2).$$

By Theorem 1 and Lemma 11, we have the following results.

Theorem 2: Under the heterogenous access pattern, the throughput capacity of the network with content-centric is

$$T(n, m) = \begin{cases} \Theta(\frac{n\sqrt{q_j}W_1}{\sqrt{\log n}\sum_{j=1}^{M} p_j}) + \Theta(mW_2), \\ q_j = o(\log^{\frac{1}{3}} n(\frac{\sum_{j=1}^{M} p_j}{n})^{\frac{2}{3}}); \\ \Theta(\frac{W_1}{q_i}) + \Theta(mW_2), \\ q_j = \Omega(\log^{\frac{1}{3}} n(\frac{\sum_{j=1}^{M} p_j}{n})^{\frac{2}{3}}). \end{cases}$$

When $q_j = o(\log^{\frac{1}{3}} n(\frac{\sum_{j=1}^{M} p_j}{n})^{\frac{2}{3}})$. According to Theorem 2, We have $T(n, m) = \Theta(\frac{n\sqrt{q_j}W_1}{\sqrt{\log n}\sum_{j=1}^{M} p_j}) + \Theta(mW_2)$. If $m = \Omega(\frac{n\sqrt{q_j}}{\sqrt{\log n}\sum_{j=1}^{M} p_j})$, we can get higher throughput capacity when $W_1 = 0$, i.e., $W_2 = W/2$, and $T_{max}(n, m) = \Theta(mW)$, and therefore, if $m = \Omega(n)$, $T_{max}^0(n, m) = \Theta(W)$, if $m = o(n)$, $T_{max}^0(n, m) = \Theta(\frac{mW}{n})$. If $m = o(\frac{n\sqrt{q_j}}{\sqrt{\log n}\sum_{j=1}^{M} p_j})$, we can get higher throughput capacity when $W_2 = 0$, i.e., $W_1 = W$, and $T_{max}(n, m) = \Theta(\frac{n\sqrt{q_j}W}{\sqrt{\log n}\sum_{j=1}^{M} p_j})$, and therefore, $T_{max}^0(n, m) = \Theta(\frac{\sqrt{q_j}W}{\sqrt{\log n}\sum_{j=1}^{M} p_j})$ When $n \to \infty$, then $\log n \to \infty$, and hence $T_{max}^0(n, m) \to 0$, which implies that the per-node throughput capacity diminishes as n grows. However, we can improve the per-node throughput capacity by increasing the probability of content cached q_j, as shown in Fig. 3.

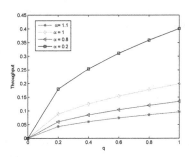

Fig. 3. We show the per-node throughput capacity for various of α vs. caching probability of each content. We assume $M = 500, n = 1000$.

4 Conclusion

In this paper, we studied the through capacity in content-centric hybrid wireless networks under the homogeneous content access scheme and heterogeneous caching scheme, respectively. We shown that under the homogeneous access scheme, the throughput capacity for hybrid wireless networks is a function of the cache size C, the number of contents M, and the number of base stations m. And under the heterogeneous caching scheme, the throughput capacity for hybrid wireless networks greatly depend on the caching probability q_j, the content popularity p_j, and the number of base stations m. We also found that the per-node throughput capacity can scale if the system parameters satisfy certain conditions. As for future work, we plan to investigate the multicast capacity of content-centric hybrid wireless networks.

Acknowledgment. This work is supported by the NSF of China under Grant No. 61301118; the Innovation Program of Shanghai Municipal Education Commission under Grant No. 14YZ130; the International S&T Cooperation Program of Shanghai Science and Technology Commission under Grant No. 15220710600; and the Fundamental Research Funds for the Central Universities under Grant No. 16D210403.

References

1. Gupta, P., Kumar, P.R.: The capacity of wireless networks. IEEE Trans. Inf. Theor. **46**(2), 388–404 (2000)
2. Liu, B.: Capacity of cache enabled content distribution wireless ad hoc networks. In: IEEE Mobile Ad Hoc and Sensor Systems (MASS) (2014)
3. Liu, B., Liu, Z., Towsley, D.: On the capacity of hybrid wireless networks. In: Proceedings of IEEE INFOCOM (2003)
4. Okabe, A., Boots, B., Sugihara, K., Chiu, S.: Spatial Tessellations: Concepts and Applications of Voronoi Diagrams. Wiley Series in Probability and Statistics. Wiley, Hoboken (2000)
5. Kozat, U.C., Tassiulas, L.: Throughput capacity of random ad hoc networks with infrastructure support. In: Proceedings of the 9th Annual International Conference on Mobile Computing and Networking, pp. 55–65. ACM (2003)
6. Azimdoost, B., Westphal, C., Sadjadpour, H.R.: On the throughput capacity if information-centric networks. In: Proceedings of the 25th International Teletraffic Congress (ITC) (2013)
7. Mahdian, M., Yeh, E.: Throughput-delay tradeoffs in content-centric wireless networks (2015). arXiv preprint arXiv:1504.03754
8. Rappaport, T.: Wireless Communications: Principles and Practice, 2nd edn. Prentice-Hall PTR, Upper Saddle River (2002)
9. Gupta, P., Kumar, P.: Critical power for asymptotic connectivity in wireless networks. In: McEneaney, W.M., Yin, G.G., Zhang, Q. (eds.) Stochastic Analysis, Control, Optimization and Applications, A Volume in Honor of W.H. Fleming, pp. 547–566. Springer, Heidelberg (1998)
10. Golrezaei, N., et al.: Femtocaching: wireless video content delivery through distributed caching helpers. In: INFOCOM, 2012 Proceedings IEEE. IEEE (2012)

11. Zhang, L., Estrin, D., Burke, J., Jacobson, V., Thornton, J., Smetters, D.K., Tsudik, G., Zhang, B., Claffy, K.C., Krioukov, D., Massey, D., Papadopoulos, C., Abdelzaher, T., Wang, L., Crowley, P., Yeh, E.: Named data networking (NDN) project (2010)
12. Jacobson, V., Smetters, D.K., Thornton, J.D., Plass, M.F., Briggs, N.H., Braynard, R.L.: Networking named content. In: Proceedings of the 5th International Conference on Emerging Networking Experiments and Technologies CoNEXT 2009, (New York, NY, USA), pp. 1–12. ACM (2009)
13. Gitzenis, S., Paschos, G., Tassiulas, L.: Asymptotic laws for joint content replication and delivery in wireless networks. IEEE Trans. Inf. Theor. **59**(5), 2760–2776 (2013)

D2D-Based Cooperative Uplink Transmission for Vehicular Users

Yun Pan$^{(\boxtimes)}$, Chao Wang, Fuqiang Liu, and Ping Wang

Department of Electronics and Information Engineering,
Tongji University, Shanghai 201804, People's Republic of China
{1433157,chaowang,liufuqiang,pwang}@tongji.edu.cn

Abstract. In this paper, we study a wireless communication network in which multiple mobile users located within a public transportation vehicle intend to send data to the serving base station. Due to the vehicle penetration loss (VPL), conventional direct uplink transmission may demand an unnecessarily high power level to reach a satisfactory communication performance. To address this issue we consider establishing cooperation among the vehicular users through the multiplexing-mode device-to-device (D2D) communication technique before their scheduled uplink transmissions. By these means the users can adopt simple space-time coding to jointly deliver their data, when the transmitter-side channel state information is not available. Via analytical and simulation results, we show that the proposed scheme can enhance the achievable energy efficiency over the conventional uplink transmission approach.

Keywords: VPL · D2D · Energy efficiency · Space-time coding

1 Introduction

Wireless communication technologies have progressed rapidly in recent years. Nowadays mobile Internet connections can be found almost everywhere in modern human societies. This results in significant enhancements in mobile data traffic and also leads to great commercial profits for mobile Internet service providers. The quality of human life and effectiveness of work have been dramatically improved, when people can freely browse Internet, chat with friends, watch video, and download documents no matter whether they are staying at home/office or travelling in cars/trains. It is widely believed that the demand for higher data rate and better service quality will never end. The industry has predicted a 1000-fold increase in mobile data traffic within the next decade [1]. In addition, according to the vision of the EU METIS project, 50 billion devices will be connected to each other in the year 2020 [2]. Hence, efficiently using limited resources to establish reliable and cost-effective wireless communications to satisfy the ever-increasing service demands becomes more and more important.

Due to the facts that mobile Internet applications have penetrated into various aspects of human life and many people need to spend a huge amount of time

© ICST Institute for Computer Sciences, Social Informatics and Telecommunications Engineering 2017
X.-L. Huang (Ed.): MLICOM 2016, LNICST 183, pp. 76–85, 2017.
DOI: 10.1007/978-3-319-52730-7_8

in road traffic every day, it is commonly envisioned that, a typical challenging issue in the future 5G era is to deliver data services to end users located in moving vehicles, for example in public transportation buses. Apart from the mobility of vehicles, the difficulty in providing satisfactory data transmissions to these users comes mainly from an extra *vehicle penetration loss* (VPL) induced by the vehicle metal hull, compared with normal outdoor end users. Measurements have shown that the VPL can be as high as several tens dB [2]. Consequently a relative large transmission power level is usually desired, which would be hard for mobile devices with limited power budget. Therefore, it is not easy to guarantee the data communication quality for uplink transmissions.

To tackle this problem, reference [2] proposes to utilize the direct device-to-device (D2D) communication concept (which will be an inherent part in 5G systems [3]) to establish collaborations between vehicular user equipments (VUEs) within the same vehicle. In other words, the VUEs first share their messages through D2D technology before sending them to the serving base station (BS). By this means, the VUEs can form a virtual antenna array and thus can jointly beamform their messages to the BS. It is shown that this strategy can significantly reduce energy consumption for the VUEs without sacrificing performance.

However, to realize the above benefits, the VUEs are demanded to have a certain level of channel state information (CSI) regarding the uplink channels. Although some practical solutions with limited signalling overhead can be adopted to attain the transmitter-side CSI, they may not be always applicable in all realistic systems, especially when the vehicle velocity is relatively high. In addition, to demonstrate the potential of VUE cooperation, it is assumed that the D2D message sharing is also conducted with transmitter-side CSI and hence can be managed without potential decoding errors. Clearly, since the signal propagation environment inside a vehicle can be very complicated, decoding errors at the VUEs may occur with non-negligible probability and thus would affect the final communication performance.

In this paper, these practical issues are taken into consideration. Specifically, we study the situation that multiple VUEs intend to send their messages to the BS. To compensate the negative effect of the VPL, we allow two VUEs inside the same vehicle to form a cooperation pair, by sharing their messages through D2D communications, before the uplink transmissions start. It is known that operating D2D communications in the multiplex mode can further improve system spectral efficiency. Hence we allow potentially multiple VUE pairs (in different vehicles) to reuse the uplink channel of a normal cellular user. No transmitter-side CSI is available. If the message sharing of a cooperation pair is successful, the two VUEs adopt the Alamouti space-time coding to jointly send their messages. Our analytical and simulation results demonstrate that such a D2D-based cooperative uplink transmission scheme can significantly improve the energy efficiency compared with the conventional direct source-destination transmissions.

2   System Model and Transmission Process

We consider a wireless communication scenario in which several public transportation vehicles are running on a road section within the coverage of a serving
BS, as illustrated in Fig. 1. Inside each vehicle, multiple VUEs intend to send
their data to the BS. In conventional uplink transmissions, the VUEs individually communicate with the BS, using two orthogonal channels (e.g. TDMA time
slots). However, it is known that when a signal travels through the vehicle hull,
a notable reduction in signal strength will be induced. Due to such a VPL problem, the message delivery process would demand a large power level to avoid
encountering poor performance. This would result in a low efficiency of resource
utilization.

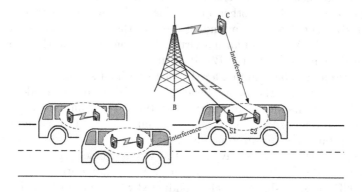

Fig. 1. System model.

To address this issue, in this paper we follow the idea proposed in reference [2] and consider establishing cooperation among the VUEs. Specifically, we
require every two VUEs inside each vehicle to form a cooperation pair. Before
being activated in their scheduled uplink transmission time, the VUEs in each
cooperation pair share their messages via D2D communications. (The activities
of the D2D message exchange process, such as channel and power assignment,
can be monitored/controlled either directly by the serving macro BS or through
the assistance of a micro BS/moving relay attached on the vehicle.) In other
words, an extra channel is allocated to each cooperation pair. This channel can
be divided into two unit time or frequency slots. In each slot, one VUE sends its
data to the other. Albeit that such a D2D transmission channel can be reserved
dedicatedly for these two VUEs, observing that the VPL would dramatically
reduce signal power leakage outside the vehicle, to further exploit the D2D multiplexing gain and improve overall spectral efficiency, we consider the situation
that the message exchange of one cooperation pair in each vehicle can reuse the
uplink channel of a regular cellular user C, as displayed in Fig. 1.

Without loss of generality, we focus on one cooperation pair and denote the
messages of the two VUEs by s_1 and s_2 respectively. Use P_d to denote the

transmit power of the VUEs to exchange data and use R to denote the data rate of s_1 and s_2. For analytical simplicity, we consider a block Rayleigh fading environment. (The proposed transmission scheme is applicable in other fading environments.) Channel fading coefficients are modeled by random variables generated from complex Gaussian distribution $CN(0, \lambda)$, where λ reflects the effect of large-scale path loss (including the VPL if the signal goes through the vehicle hull). Due to the mobility of vehicles, the channel conditions may change rapidly. To avoid a large amount of signalling overhead, the fading coefficients are hence estimated and known at only the corresponding receivers.

In addition, we use x_c and P_c to represent the transmit signal and power of the cellular user C. x_{d_i} denotes the signal from other cooperation pairs that also reuse the uplink channel of C (assume there are a total of N such pairs and hence $i \in \{1, \cdots, N\}$). h'_{jc} (resp. h''_{jd_i}) is used to denote the channel coefficient between C (resp. the ith interfering cooperation pair) and the jth VUE in the considered cooperation pair. The number of superscript $'$ denotes how many times VPL would affect the signal propagation. Now for the D2D message exchange process, denoting the received signals of the two considered VUEs by y_1 and y_2 respectively, we can have

$$y_j = \sqrt{P_d}h_{jk}s_k + \sqrt{P_c}h'_{jc}x_c + \sum_{i=1}^{N} \sqrt{P_d}h''_{jd_i}x_{d_i} + n_j, \tag{1}$$

where $j, k \in \{1, 2\}$ $(j \neq k)$, h_{jk} denotes the channel fading coefficient between the transmitter k and receiver j, and n_j denotes the unit-power additive white Gaussian noise (AWGN). Clearly, the second and third terms on the right hand side (RHS) of (1) are the interference signals from C and other cooperation pairs.

Upon receiving the exchanged data, each VUE tries to carry out decoding by treating all interference signals as noise. If the decoding processes at both VUEs are successful, during their scheduled uplink transmission time slots, the two VUEs utilize the Alamouti space-time coding to send their messages to the BS using two TDMA time slots. Denote the received signals at the BS by r_1 and r_2 respectively. We can have:

$$r_1 = \sqrt{P_a}g'_1 s_1 + \sqrt{P_a}g'_2 s_2 + \tilde{n}_1, \tag{2}$$

$$r_2 = -\sqrt{P_a}g'_1 s_2^* + \sqrt{P_a}g'_2 s_1^* + \tilde{n}_2, \tag{3}$$

where P_a is the transmission power of the VUEs, g'_j denotes the channel fading coefficient between the jth VUE and the BS, and $\tilde{n}_1$ and $\tilde{n}_2$ are the unit-power AWGN.

On the other hand, if any VUE cannot correctly decode the other VUE's message, the conventional direct uplink transmission is adopted. Each VUE uses power P_u, which can be different from P_a in (2) and (3), to transmit its data to the BS. The received signals are hence simply

$$r_k = \sqrt{P_u}g'_k s_k + \tilde{n}_k, \text{ for } k \in \{1, 2\}. \tag{4}$$

3   Energy Efficiency Analysis

To demonstrate that the proposed D2D-based uplink transmission scheme can potentially improve the wireless resource utilization efficiency, in this section we derive the expression of the system's achievable *energy efficiency*, ξ_{EE}. ξ_{EE} is defined as the average successful transmission data rate between each VUE and the BS, in bits per unit of bandwidth by using one joule energy. In what follows, we assume that the channel coding applied in the physical layer is sufficiently strong so that the outage probability dominates error probability at each receiver. Use $\mathcal{P}_{out}$ to denote the outage probability of each VUE, i.e., the probability that the BS cannot correctly decode the messages from that VUE. Then the achievable energy efficiency is calculated as:

$$\xi_{EE} = \frac{(1 - \mathcal{P}_{out})R}{P_{total}}, \tag{5}$$

where P_{total} is the total power consumption of each VUE.

The value of $\mathcal{P}_{out}$ is dependent on whether the D2D message exchange process is successful. Denote the probability that a VUE cannot correctly decode the other VUE's message by $\mathcal{P}_c$. According to Eq. (1), $\mathcal{P}_c$ can be calculated by

$$\mathcal{P}_c = P_r \left\{ \log \left(1 + \frac{P_d|h_{jk}|^2}{P_c|h'_{jc}|^2 + \sum_{i=1}^{N} P_d|h''_{jd_i}|^2 + 1} \right) < R \right\}. \tag{6}$$

In general, it is involved to attain a closed-form expression of $\mathcal{P}_c$ due to the unpredictable nature of the interference from other D2D pairs, i.e. the term $\sum_{i=1}^{N} P_d|h''_{jd_i}|^2$. However, as we mentioned earlier, the VPL can dramatically reduce signal power leakage. The power level P_d is also normally kept small, since the cooperative VUEs are located within the same vehicle. As a result, with high probability the interference signals from VUEs in other vehicles, which experience two times of VPL, would be much weaker compared with the interference from C plus noise. Hence in what follows, we will omit them to simplify the derivation of $\mathcal{P}_c$. In the next section, we will show via simulations that such a consideration does not induce much loss of accuracy because of these reasons.

Now $\mathcal{P}_c$ can be approximated as

$$\mathcal{P}_c \approx P_r \left\{ \log \left(1 + \frac{P_d|h_{jk}|^2}{P_c|h'_{jc}|^2 + 1} \right) < R \right\} = P_r \left\{ \frac{P_d|h_{jk}|^2}{P_c|h'_{jc}|^2 + 1} < 2^R - 1 \right\}. \tag{7}$$

Let $X = P_d|h_{jk}|^2$, $Y = P_c|h'_{jc}|^2 + 1$, and $Z = \frac{X}{Y}$. The probability density function (pdf) of X is $f_X(x) = \frac{1}{\lambda_d P_d} \exp(-\frac{x}{\lambda_d P_d})$, where λ_d denotes the variance of D2D channel coefficient h_{jk}. The pdf of Y is $f_Y(y) = \frac{1}{\lambda_c P_c} \exp(-\frac{y-1}{\lambda_c P_c})$, where λ_c is the variance of the channel from C to VUEs, i.e., h'_{jc}. Using the probability transformation rule [8] one can have the pdf of Z as $f_Z(z) = \frac{\lambda_c P_c + \lambda_c P_c \lambda_d P_d + \lambda_d P_d}{(\lambda_c P_c z + \lambda_d P_d)^2} \exp(-\frac{1}{\lambda_d P_d}z)$. Now the outage probability $\mathcal{P}_c$ is calculated as

$\mathcal{P}_c = P_r\{Z < 2^R - 1\} = F_Z(2^R - 1)$, where $F_Z(z)$ is the cumulative distribution function of Z. Hence $\mathcal{P}_c$ and can be derived as [9]

$$\mathcal{P}_c = 1 - \frac{\lambda_d P_d}{\lambda_c P_c(2^R - 1) + \lambda_d P_d} \exp(-\frac{2^R - 1}{\lambda_d P_d}). \tag{8}$$

Therefore, the system outage probability $\mathcal{P}_{out}$ can be expressed as

$$\mathcal{P}_{out} = (1 - \mathcal{P}_c)^2 \mathcal{P}_{co,out} + \left(1 - (1 - \mathcal{P}_c)^2\right) \mathcal{P}_{di,out}, \tag{9}$$

where $\mathcal{P}_{co,out}$ is the outage probability at the BS when the two VUEs cooperatively transmit their messages, and $\mathcal{P}_{di,out}$ represents the outage probability at the BS when the VUEs transmit their messages individually to the BS.

As we mentioned earlier, if the D2D message exchange process is successfully carried out, the two VUEs send their messages to the BS using the Alamouti space-time coding. The received signals at the BS can be expressed as (2) and (3). In order to decode s_1 and s_2 from the received signals r_1 and r_2, the BS can perform the following transform of the signals:

$$\tilde{s}_1 = \frac{g_1'^*}{\sqrt{|g_1'|^2 + |g_2'|^2}} r_1 + \frac{g_2'}{\sqrt{|g_1'|^2 + |g_2'|^2}} r_2^* = \sqrt{P_a}\sqrt{|g_1'|^2 + |g_2'|^2} s_1 + \hat{n}_1, \tag{10}$$

$$\tilde{s}_2 = \frac{g_2'^*}{\sqrt{|g_1'|^2 + |g_2'|^2}} r_1 - \frac{g_1'}{\sqrt{|g_1'|^2 + |g_2'|^2}} r_2^* = \sqrt{P_a}\sqrt{|g_1'|^2 + |g_2'|^2} s_2 + \hat{n}_2, \tag{11}$$

where $\hat{n}_1$ and $\hat{n}_2$ are complex Gaussian distributed random variables with unit power. It can be seen from the above equations that

$$\mathcal{P}_{co,out} = P_r\left\{\log\left(1 + P_a|g_1'|^2 + P_a|g_2'|^2\right) < R\right\}$$
$$= 1 - \exp(-\frac{2^R - 1}{P_a}) - \frac{2^R - 1}{P_a}\exp(-\frac{2^R - 1}{P_a}). \tag{12}$$

In addition, if any VUE cannot correctly decode the other VUE's message, then they individually transmit their messages. From (4) it is easy to obtain:

$$\mathcal{P}_{di,out} = P_r\left\{\log\left(1 + P_u|g_1'|^2\right) < R\right\} = 1 - \exp(-\frac{2^R - 1}{P_u}). \tag{13}$$

As a result, the system outage probability can be calculated by substituting (7), (12), (13) into (9). Finally, the total power consumption is expressed as

$$P_{total} = P_d + (1 - \mathcal{P}_c)^2 P_a + \left(1 - (1 - \mathcal{P}_c)^2\right) P_u \tag{14}$$

Now, the system's achievable energy efficiency ξ_{EE} can be found by substituting the above equations to (5).

4   Numerical Evaluations

In this section we use simulation results to compare the proposed scheme with the conventional direct uplink transmission. In what follows, we assume the cell radius to be 800 m, and the distance between the road section and the BS is 200 m. The target cooperation VUE pair is assumed to locate in the center of the road section. For presentation simplicity, we choose C to be close to the BS. Hence the distances from the target VUE pair to the BS and C are both 200 m. The road section is one directional. The distance between adjacent vehicles follows an exponential distribution, whose pdf is governed by $f(x) = \theta e^{-\theta x}$ for $x > 0$ (the value of θ can represent the traffic density). In addition, we consider an extreme situation in which every vehicle on the road contains one pair of cooperation VUEs that reuse the uplink channel of C, in order to maximize the overall usage efficiency of system resource and also study the impact of inter-vehicle interference.

The main simulation parameters follow [10]: The bandwidth is 10 kHz and the noise density is 174 dBm/Hz. The transmit power of C is set to be $P_c = 25$ dBm. The path loss between the target VUEs is set to follow the indoor model $PL_d = 38.46 + 20\log_{10} d_d$ dB, where the distance d_d is set to be 5 m. The path loss between C (and also other co-channel cooperation pairs) and the target VUEs follows the outdoor model: $PL_o = 15.3 + 37.6\log_{10} d_o$ dB. Further, the VPL is set to be 20 dB.

In each of the following figures, we plot three performance curves. The first is termed "direct transmission" and represents the energy efficiency of conventional direct uplink transmission. The other two represent the performance of the proposed transmission strategy. In other words, before the scheduled uplink transmission, the two target VUEs share their messages through D2D communication, being potentially interfered by C and other VUEs reusing the same channel. In Sect. 3, we omitted the interference from other VUEs and obtained the approach to derive the closed-form of the achievable energy efficiency. The performance curve plotted following this derivation is termed "cooperation with analytical results." We also carried out simulations that take the interference from other VUEs into consideration, in order to see how such interference actually affects system performance. The associated curve is termed "cooperation with simulation." The second is termed "cooperation with analytical results." This curve is plotted according to the analytical derivations provided in Sect. 3.

Figure 2 plots the achievable energy efficiency versus transmission data rate, when we fix $P_d = 10$ dBm and $P_a = 25$ dBm. First consider the case with light traffic, $\theta = 0.01$ (i.e., the average distance between two adjacent vehicles is 100 m), shown in Fig. 2(a). It can be seen that the simulation results nicely match the analytical results. In other words, the analysis provided in Sect. 3 can be adopted to correctly predict the performance of the considered system. Further, within a large range of transmission rate, the proposed scheme leads to a significant energy efficiency improvement over the conventional uplink transmission. Note that this performance gain is achieved without extra bandwidth consumption, since the D2D transmissions reuse the uplink channel of C.

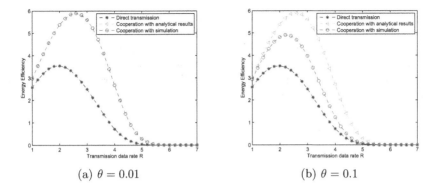

(a) $\theta = 0.01$ (b) $\theta = 0.1$

Fig. 2. ξ_{EE} versus R

Moreover, when the transmission data rate is very large, the fixed power level is hard to support successfully transmission. Hence almost no message can be received. Both schemes attain very small energy efficiency. Figure 2(b) plots the case when the traffic density is high, $\theta = 0.1$ (i.e., the average distance between adjacent vehicles is 10 m). We can see that in this case the interference between VUE pairs will affect the system performance. But the proposed scheme still performs better than conventional direct transmission.

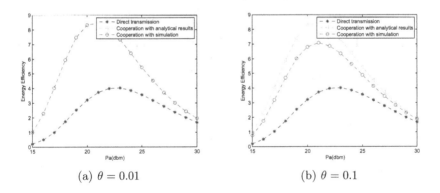

(a) $\theta = 0.01$ (b) $\theta = 0.1$

Fig. 3. ξ_{EE} versus P_a

Figure 3 plots the similar trend as that in Fig. 2. Specifically, when the traffic is low, ignoring the interference between VUE pairs does not lead to much analytical inaccuracy. For a wide range of P_a, the proposed scheme is more efficient than direct transmission. However, if P_a is too small, a large portion of the source messages would be lost, even with user cooperation. If P_a is too large, the outage probability at the BS approaches zero, even with direct transmission. Hence in both cases, the proposed scheme attains similar performance as direct transmission. But it is worth noting that the parameter settings in these extreme cases are not sufficiently good. When P_a is small, a smaller transmission rate should be chosen, to increase the probability of correct decoding, and when P_a is large, the transmission should also be large to maintain a good level of throughput. If we follow such choices, the proposed scheme would exhibits more significant performance advantages again. Finally, when the traffic is very heavy, the analytical result can be a little bit bias compared with the true performance.

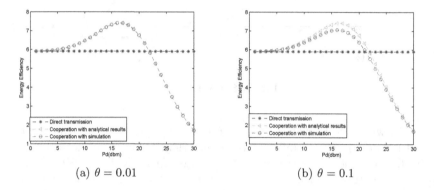

(a) $\theta = 0.01$ (b) $\theta = 0.1$

Fig. 4. ξ_{EE} versus P_d

Figure 4 plots the achievable energy efficiency versus D2D transmission power. When the traffic density is low, which can be seen in Fig. 4(a), ignoring the interference will not cause inaccuracy in Sect. 3. When P_d is small and close to zero, it is hard to decode successfully. Thus the system is nearly equal to direct transmission. For a wide range of P_d, by using D2D technology the VUE pair can decode successfully and performance can obtain a better performance. When P_d becomes larger, the probability that VUEs decode successfully is equal to one and increasing P_d leads to waste of energy. So, the energy efficiency decreases. In Fig. 4(b), when traffic is high, ignoring inter-vehicle interference may become serious and hence affect the system performance.

5 Conclusion

We have proposed a D2D-based cooperative transmission scheme to improve the uplink transmission energy efficiency for mobile users locating inside public transportation vehicles. Specifically, before the scheduled uplink transmissions, two such VUEs are permitted to share their messages via D2D transmission. If the message sharing is successful, then they adopt the Alamouti space-time coding to carry out uplink transmissions. To improve system channel usage efficiency, we have considered the case that the D2D transmissions can reuse the uplink channel of a regular cellular user. We have provided the method to analyze the system achievable energy efficiency, and used numerical results to clearly exhibit the advantages of the proposed scheme over conventional direct uplink transmissions.

Acknowledgement. This work was supported by the National Natural Science Foundation of China (61331009), the Fundamental Research Funds for the Central Universities (1709219004), the National Science and Technology Major Project of MIIT under Grant 2015ZX03002009-003, and the EU FP7 QUICK project (PIRSESGA-2013-612652).

References

1. Sui, Y., Guvenc, I., Svensson, T.: Interference management for moving networks in ultra-dense urban scenarios. EURASIP J. Wirel. Commun. Netw. **2015**(1), 1–32 (2015)
2. Sui, Y., Svensson, T.: Uplink enhancement of vehicular users by using D2D communications. In: 2013 IEEE GLOBECOM, pp. 649–653 (2013)
3. Aziz, D., Kusume, K., Queseth, O., Tullberg, H.: ICT-317669-METIS/D8.4 METIS final report. https://www.metis2020.com
4. Benkhelifa, F., Tall, A., Rezki, Z., Alouini, M.-S.: On the low SNR capacity of MIMO fading channels with imperfect channel state information. In: Modeling and Optimization in Mobile, Ad Hoc and Wireless Networks (WiOpt), pp: 303–310 (2001)
5. Phan-Huy, D.T., Sternad, M., Svensson, T.: Making 5G adaptive antennas work for very fast moving vehicles. IEEE Intell. Transp. Syst. Mag. **7**, 71–84 (2015)
6. Caire, G., Shamai, S.: On the capacity of some channels with channel state information. IEEE Trans. Inf. Theory **45**, 2007–2019 (1999)
7. Alamouti, S.: A simple transmit diversity technique for wireless communications. IEEE J. Sel. Areas Commun. **16**, 1451–1458 (1998)
8. Papoulis, A., Pillai, S.U.: Probability, Random Variables, and Stochastic Processes. Tata McGraw-Hill Education, New York City (2002)
9. Ni, Y., Jin, S., Tian, R., Wong, K.-K., Zhu, H., Shao, S.: Outage analysis for device-to-device communication assisted by two-way decode-and-forward relaying. In: 2013 International Conference On Wireless Communications & Signal Processing (WCSP), pp. 1–6 (2013)
10. GT. 36814: Further advancements for E-UTRA physical layer aspects. Technical report, March 2010

Research on Data Transmission Protocol Performance of DTN Relay Channel Based on CFDP

Dezhi Li[✉], Yaoqing Ni, Qun Wu, Zhenyong Wang,
and Deyang Kong

School of Harbin Institute of Technology, Harbin 150000, China
{lidezhi,qwu}@hit.edu.cn, niyaoqing1993@yeah.net

Abstract. In this paper, we explore the transmission linking delay technology brings which improvement to DTN network based on CFDP in both theory and simulation. In past research, there are mainly focused on the performance of TCP and UDP as the transmission protocol in DTN network while they paid little attention to CFDP protocol as a transmission layer protocol which is below bundle layer. By a new data automatic retransmission mechanism, the performance of transmission link is strengthened in situation of high error reliable transmission rate. In the simulation, we compared the difference of file transferring delay, proportion of data which successfully reached, the effective data rate and date rate between normal mode and forwarding mode based on CFDP in DTN relay link.

Keywords: DTN · CFDP · Automatic retransmission · High error rate · High transmission delay · Bundle layer

1 Introduction

In deep space exploration, the traditional point to point communication technology has developed to a bottleneck. There is a trend to use the transmission linking relay technology in the future deep-space communication systems. In recent years, with the development of network technology and the increasing demand for people to use, emerges a class of new network. In such networks, the capacity of nodes is low, and the networks are often disconnected and the delay of data round-trip is long. In this context, tolerate network delays (Delay/Disrupted Tolerant Networks, DTN) came into being. Studying this kind of restricted network and data link systems thinking are important.

The propagation delay of the ground and space DTN rises from tens to hundreds of milliseconds to tens of milliseconds to tens of seconds or even longer. In terms of routing, buffer, congestion avoidance and control problems, the TCP protocol for the IP network on the ground is not suitable for the asymmetric DTN network and the uplink/downlink. Due to the congestion of the reverse response link, such as slow start, TCP is widely verified not suitable for the link whose asymmetric rates over 50:1. DTN network hopes the transmission protocol can transmit data with different network

© ICST Institute for Computer Sciences, Social Informatics and Telecommunications Engineering 2017
X.-L. Huang (Ed.): MLICOM 2016, LNICST 183, pp. 86–96, 2017.
DOI: 10.1007/978-3-319-52730-7_9

coverage. Overlay may be the most beneficial to the intermittent and short time of the network connection, so that the DTN nodes can transfer a large number of data as much as possible.

This paper is based on the relevant ideas of the following protocols, the interstellar transmission protocol (TP-Planet) is one of the early ideas for the reliable data transmission of deep space links. The main function of Planet - TP is the control way of congestion detection and processing. In addition, Planet - TP is used to deal with power outage, delay of SACK policy and the bandwidth asymmetry. Based on the rate of the increase in the product type (AIMD algorithm) congestion control, its operation depends on the congestion detection mechanism. However, at least in the current, deep space communications in the prearranged management procedures in the static operation, congestion control is not really needed, covering deep space connection of the flow multiplexing technology does not exist.

Space communication protocol standard transport protocol (TP-SCPS) is proposed by the Spatial Data System Advisory Committee (CCSDS) for the development of space communications. TP - SCPS is based on the widely used transmission control protocol (TCP) and a number of modifications and extensions to the deep space communication link constraints set.

Interstellar reliable communication protocol (RCP – Planet). RCP – Planet's detection rate control and data packet forward error correction resolve link congestion and error rate together. RCP – Planet also deployed a blackout status program and ACK for FEC, to detect the asymmetry of the link. RCP – Planet's main target is to transport the real-time application of data to ground, satellite or spacecraft and other nodes.

Deep space transmission protocol (DS-TP) is a transport layer protocol, this protocol is based on rate, hybrid quick response strategy and double automatic retransmission, the transmission efficiency is 2 times faster than the traditional protocol. DS – TP transmits data in the case of predictable and scheduled line speeds, using the principle of one hop, store and forward mail to control the current space communications and ease congestion avoidance and control requirements.

LTP protocol is a point to point protocol for the application of DTN overlay. LTP can send anonymous data blocks, and introduced through each data block is divided into two parts of the local reliability point of view, the "red" section is reliable and the "green" is unreliable. Besides, while the receiving terminal receives explicit request of the red section of data, it sends concise receive reply Report.

2 Simulation Scenarios

As the Fig. 1 showed, the deep space communication simulation scene can be simplified to the communication between the three nodes, the transmitting node, relay node and receiving node. Traffic in the network is transmitted to the receiving node by the sending node, and it is not a hop. By the experiment platform based on C language in Linux, we design node model and process model to achieve different forward behaviors. On the basis of this, two kinds of transmission performance are compared.

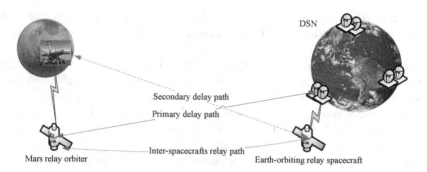

Fig. 1. Simulation scenarios.

3   The Transmission Model Based on CFDP

The transport mechanism for storage and automatic forwarding of redundant data with the new automatic retransmission mechanism is based on CFDP. The source node of the file to be transmitted is divided into several sections, each section forming a protocol data unit. First send metadata package, containing the file name, file size, source and destination ID and other information. But it and most of the data unit, there is no confirmation, that the sender does not need to wait for the receiver to return confirmation information, whether metadata packet is successfully received or not, the sender will be issued after the metadata packet data continues to transfer files unit. That is the sender and the recipient does not need a handshake can start transferring files, which will save a lot of time in deep space. Each data unit header are marked active ID. If the recipient receives a file labeled with the new ID data unit represents a new file transfer begins. Each data unit also contains a special field, which indicated the data unit carried by the contents of the file start and stop bits, the data unit sequence by checking the recipient has received it can determine which data units sent failed. Simultaneous retransmission ensure the reliability of the information (Fig. 2).

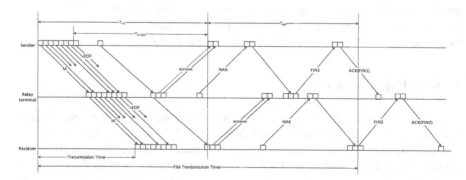

Fig. 2. Schematic view of the transport mechanism

The forwarding node could storage data elements at first by the transmission of data elements from the sender to the relay terminal. Feedback EOF and the corresponding lost package NAK which marked the sequence information of retransmission data elements by the specific domain will be transmitted when the relay terminal received the sender file confirmation EOF, but when there is no data elements missing data, it is empty. Then the interaction between the sender and the relay terminal will begin to work for the purpose of data retransmission from the sender to relay terminal. Relay terminal keep sending after receiving data elements. When relay terminal accept EOF ACK sent to the receiver, relay terminal begin with the data elements interaction, start the second phase of data retransmission work. Application of automatic retransmission mechanism, when transmitting data error will send negative confirmation. Each NAK information indicates the start bit file area should the end bit of the other area that before NAK information indicates.

First, it needs some instructions on variables to estimate two rounds of interaction time of this model. Assumptions for file transfer in the of data elements (PDU) number (including the MPDU) is N, L_{PDU} is the length of each PDU, P_e is link error rate, L_{EOF} is the length of the EOF PDU, T_{tran} represents a single link transmission delay, T_{PDU} represents the time needed for transmitting PDU, T_{ACK} represents the time needed for transmitting NAK, T_{EOF} represents the time needed for single link transmitting EOF, $T_{ACK-EOF}$ represents the time needed for single link transmitting ACK-EOF, T_{D-EOF} is the transmission time of EOF, P_{ePDU} represents the two links PDU error probability, P_{eEOF} represents the probability of transmission error of EOF, P_{eNAK} represents the probability of transmission error of NAK. The length and the BER of all PDU are same. Also the length and the BER of all NAK are same. So the probability of errors that single link PDU occurs P_{ePDU} and the relationship between P_{eEOF} and bit error rate P_e is

$$P_{ePDU} = 1 - (1 - P_e)^{L_{PDU}}, \; P_{eEOF} = 1 - (1 - P_e)^{L_{EOF}} \tag{3-1}$$

Expectations for the T_{inc} is

$$E(T_{inc}) = 2N \times T_{PDU} + E(T_{D-EOF}) \tag{3-2}$$

Set $E(\eta_{EOF})$ is the number of the sender sends a total EOF when receiver successfully received EOF, so that we can know the expectation for $E(\eta_{EOF})$ is

$$E(\eta_{EOF}) = 2\sum_{i=1}^{\infty}(1 - P_{eEOF})iP_{eEOF}^{i-1} = 2(1 - P_{eEOF})\sum_{i=1}^{\infty}iP_{eEOF}^{i-1} = \frac{2}{1 - P_{eEOF}} \tag{3-3}$$

Suppose $T_{timeout-EOF}$ is a retransmission time for EOD PDU, so

$$T_{timeout-EOF} = 2T_{tran} + T_{ACK-EOF} \tag{3-4}$$

So the expectation for the transmission time of EOF can represented as

$$E(T_{inc}) = \frac{(2 + 2P_{eEOF})T_{tran} + 2T_{EOF} + 2T_{ACK-EOF}}{1 - 2P_{eEOF}} \tag{3-5}$$

The expectation of T_{inc} can be obtained to get (3-5) into the Eq. (3-2)

$$E(T_{inc}) = 2N \times T_{PDU} + \frac{(2 + 2P_{eEOF})T_{tran} + 2T_{EOF} + 2T_{ACK-EOF}}{1 - 2P_{eEOF}} \tag{3-6}$$

Suppose the transmission number of entire T_{def} period in two links are $M1$ and $M2$. Then, after N PDUs data packets through the T_{inc} stage, the expectation for $2N'$ which is the number of PDU that still need transmission interaction is

$$E\left(2N'\right) = \sum_{i=0}^{\infty} C_N^i (1 - P_{ePDU})^{N-i} P_{ePDU}^2 i + \sum_{i=0}^{\infty} C_{NP_{ePDU}}^i (1 - P_{ePDU})^{NP_{ePDU} - i} P_{ePDU}^2 i$$

$$= N \times P_{ePDU} + N \times P_{ePDU} + N \times P_{ePDU}^2 = N1' + N2' \tag{3-7}$$

After entering T_{def} stage, we had to consider the probability of error of NAK P_{eNAK}, therefore, the single link failure PDU interaction probability at T_{def} stage is

$$P_{ef} = 1 - (1 - P_{ePDU})(1 - P_{eNAK}) \tag{3-8}$$

The total number of packets required for the interaction at T_{def} stage is $2N'$, From the above formula, we need to know the expectation for retransmission number of interactions when $N1' + N2'$ packets interaction is complete.

Suppose N_i' is the expectation for the number of packets after i times retransmission interaction still remaining retransmission, there is

$$N1_i' = N1' \times P_{ef}^i \qquad N2_i' = N2' \times P_{ef}^i \tag{3-9}$$

Suppose after $M1_1$ times transmission interaction, $N1''$ is the expectation for the number of remaining packets after times retransmission interaction. $M1_1$ satisfies the following two relationship

$$N1'' = N1' \times P_{ef}^{M1_1} \leq 1 \qquad N1' \times P_{ef}^{(M1_1 - 1)} \geq 1 \tag{3-10}$$

Then, the expectation for the remaining packet number of transmission finished $M1_2$ is

$$M1_2 = N1'' \times \frac{1}{1 - P_{ef}} \tag{3-11}$$

So you can find the expectation for the number of retransmission interaction that $N1'$ packets retransmit for the entire T_{def} period.

Known the number of retransmission interactions in the first phase of the link T_{def} period M. In every single link exchange process, the interaction phase delay of the first link i times interaction is

$$T_{spurt}(i) = T_{NAK} + N_i \times T_{PDU} + 2T_{tran} \tag{3-12}$$

The delay of entire retransmission stage consist of M times interaction delay and $T_{ACK-EOF}$, there is

$$T_{def} = T_{ACK-EOF} + \sum_{i=1}^{M} T_{spurt}(i) = T_{ACK-EOF} + M(T_{NAK} + 2T_{tran}) + (\sum_{i=1}^{\infty} N_i)T_{PDU}$$

The number of packets which need to retransmit at T_{def} stage is N', the N' PDUs come from the packets which resent in the interactive stage and passed the channel which error rate is P_{ePDU}, there is

$$E(\sum_{i=1}^{M} N_i)(1 - P_{ePDU}) = N' \tag{3-13}$$

Therefore the expectation for the time of retransmission stage is

$$E(T_{def}) = T_{ACK-EOF} + E(M)(T_{NAK} + 2T_{tran}) + E(\sum_{i=1}^{M} N_i)T_{PDU} \tag{3-14}$$

The expectation for the entire time T_{file} is

$$E(T_{file}) = E(T_{inc}) + E(T_{def}) \tag{3-15}$$

4 Simulation and Analysis

4.1 The Relationship Between File Transmission Delay and Packet Loss Rate

Form the Fig. 3 we can see that the network communications situation of two models is similar when the packet loss rate is low. And in this case the file transfer delay of store and forward mode and non-store and forward mode is basically same. When loss rate of channel a and channel b increases, that is when the communication conditions getting closer to of deep space communication, the gap of two modes delay in increasing. So the store and forward mechanism has a strong research value when the packet loss rate as high as deep space is.

Next, we will do a comprehensive analysis of total transfer delay from the theoretical and practical measured data (Tables 1 and 2).

The situation of the relay node to receiving node is same to sending node to relay node. Therefore, the total delay of store-and-forward mode is

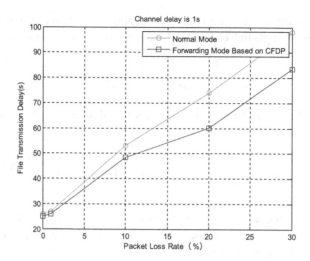

Fig. 3. The file transfer delay of two models in different packet loss rate

Table 1. Retransmission times (n) and the arrival probability (p)

n	0	1	2
p	0.64	$0.36 \times 0.64 = 0.2034$	$0.36^2 \times 0.64 = 0.0829$
n	3	4	5
p	$0.36^3 \times 0.64 = 0.0299$	$0.36^4 \times 0.64 = 0.0107$	$0.36^5 \times 0.64 = 0.0037$

Table 2. The probability of retransmission from sending node to relay node

n	1	2	3	4
p	0.16	0.032	0.0064	0.00128

$$T_{file} - T_{file0} = 36.4675s$$

In theory, $T_{file} - T_{file0}$ is 35.8457 s. The output of simulation is equal to theoretical analysis. The total delay of Store-and-forward mode is shorter than the delay of non-store-and-forward mode.

4.2 The Relationship Between File Transmission Delay and Channel Delay

When the channel delay of link a and link b is different, we measured the total delay of sending data packets of the platform. Each data was measured ten times and then we get the average of these data. The Fig. 4 is the result of the measurement.

When the channel delay is 0 s, two modes are same. When the channel delay reaches 1 ~ 2 s, the differences between the two models become clear. With the delay further increasing, the total delay of forwarding mode is better than normal mode.

We set the channel delay is 10 s and do a theoretical analysis.
Non-store and forward

$$T'_{file} - T'_{file0} = 50.3582s$$

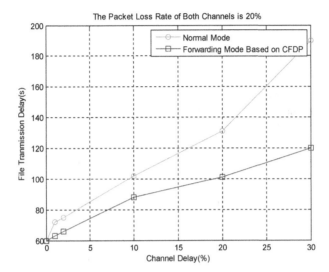

Fig. 4. The total delay of two models

Store and forward

$$T_{file} - T_{file0} = 29.4839s$$

The output of simulation is equal to the result of theoretical analysis.
In the Table 3, the packet loss rate is 20%.

4.3 The Relationship Between Data Transferring Rate and Packer Loss Rate

When the packet loss rate of link a and b is different, we measured the total rate of sending data packets of the platform. Each data was measured five times and then we get the average of these data. The Fig. 10 and Fig. 11 is the result of the measurement.

From the Fig. 5 we can see that, when the channel delay time is 1 s and packet loss rate is 0%, the total data transfer rate of store and forward mode is slightly higher than the normal mode. With the packet loss rate gradually increasing, the data transfer rate of store and forward mode total declined slightly, but remained equal to the normal model. When the channel delay time is 1 s and packet loss rate is 0%, the effective transfer rate of store and forward mode is consistent with the normal model. With the

Table 3. Retransmission times and remaining packets

Retransmission times	Sent packets (normal mode, packet loss data is 20%)	Remaining packets (packet loss data is 20%)
0	384	200
1	138	40
2	50	8
3	18	2
4	6	0
5	3	0
6	0	0

packet loss rate gradually increasing, forwarding mode based on CFDP has a better performance than the normal mode (Table 4).

When the packet loss rate of link a and link b is 20%, the number of the data package which had been transferred in normal model is 1199, while in the store and forward mode is 850. In 4.1, the transfer delay is analyzed when the packet loss rate is 20%. So,

$$V_N = 15234 \, byte/s$$
$$V_S = 14516 \, byte/s$$

The actually measured data is basically same to theoretical analysis. The rate of data transmission of two model is the same.

In the same case, the number of valid data packet which had been transmitted of both in normal model and forwarding mode based on CFDP is 600. Then we can get

$$V_N = 10169 \, byte/s$$
$$V_S = 8180 \, byte/s$$

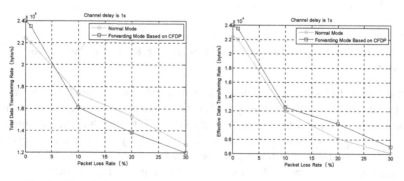

Fig. 5. The relationship between data transferring rate and packer loss rate

Table 4. Retransmission times and remaining packets

Retransmission times	Sent packets (normal mode, packet loss data is 20%)	Remaining packets (packet loss data is 20%)
0	600	600
1	984	800
2	1122	840
3	1172	848
4	1190	850
5	1196	850
6	1199	850

Table 5. The comparison between the two modes

	Transmission Delay (Same channel delay)	Transmission Delay (Same packet loss rate)	Data Arrival Proportion
Normal Mode	high	high	low
Forwarding Mode based on CFDP	low	low	high
	Total Data Transferring Rate	Effective Data Transmission Rate	
Normal Mode	slightly high	low	
Forwarding Mode based on CFDP	slightly low	high	

5 Conclusion

In this paper, we explore what improvement that link forwarding technology brings to transmission in DTN network based on CFDP. By using new redundant data automatic retransmission mechanism, the effectiveness of reliable transmission in links of high error rate has strengthened. The comparison between the two modes is as shown in Table 5.

In bundle layer, the forwarding node was changing toward destination node by reliable protocol of transmission layer and store and forward technology. The moving of forwarding node minimize the potential of forwarding hops, the additional load on the network caused by retransmission of information and the total time of a reliable transmission bundle to its destination. This improvement has enhanced the link which has a long delay or decay. In the links which have a large decay, forwarding node retransmission needed less retransmission than sending node retransmission (linear growth vs exponential growth associated with the number of hops). When the link is unreliable, store and auto-forwarding mode has better reliability and validity than normal. This advantage is mainly because that the responsibility of retransmission moves from the source node to the destination node. This model divides the long length of the propagation path into short length path and long delay path into short delay path by hops. And this model assigned the responsibility to whole forwarding nodes instead of source node and destination node. Above all, this kind of transmission node based on CFDP has a better transferring performance in DTN network.

References

1. Wang, R., Wei, Z., Zhang, Q., Hou, J.: LTP aggregation of DTN bundles in space communications. IEEE Trans. Aerosp. Electron. Syst. **49**(3), 1677–1691 (2013)
2. Andrew, J., Sebastian, K., Kevin, K.: Delay/Disruption-Tolerant networking: flight test results from the international space station. In: 2010 IEEE Aerospace Conference, pp. 1–8 (2010)
3. Nikolaos, B., Vassilis, T.: Packet size and DTN transport service: evaluation on a DTN testbed. In: ICUMT, pp. 1198–1205 (2010)
4. De Sanctis, M.: Space system architectures for interplanetary internet. In: 2010 IEEE Aerospace Conference, pp. 1–8 (2010)
5. Park, I., Ida, P.: Energy efficient expanding ring search. In: The First Asia International Conference on Modelling & Simulation, no. 3, pp. 198–199 (2007)
6. Choo, F.C., Seshadri, P.V., Chan, M.C.: Application-aware disruption tolerant network. In: 2011 IEEE 8th International Conference on Mobile Adhoc and Sensor Systems (MASS), pp. 1–6. IEEE (2011)

Multiple Relay Selection Scheme for Underwater Acoustic Cooperative Communication Based on Steady-State Mean-Square-Error Threshold

Zhiyong Liu$^{(\boxtimes)}$, Yinghua Wang, and Baoqi Ding

School of Information and Electrical Engineering, Harbin Institute of Technology
(Weihai), Weihai 264209, People's Republic of China
lzyhit@aliyun.com, 952684330@qq.com, 453873486@qq.com

Abstract. A multiple relay selection scheme for underwater acoustic cooperative communication is proposed. In the scheme, the steady-state mean-square-error (SMSE) of each relayed path is used to order the relays, and then the relay with smaller steady-state mean-square-error (SMSE) value will be more preferential to participate in cooperation. Simulation results demonstrate that the proposed scheme can adaptively select the number of relay nodes to cooperate by the threshold, and it has a lower bit error rate (BER) compared with existing counterparts.

Keywords: Underwater acoustic cooperative communication · Multiple relay selection · Steady-state mean-square-error

1 Introduction

In the past few years, underwater acoustic communication (UAC) has received extensive attention due to emerging applications, including seafloor resource exploration, marine observation, offshore oilfield monitoring, submarine communication, marine data collection, pollution monitoring, seismic observation, marine traffic and transport, tactical surveillance applications, and port safety in many others ocean observation [1]. However, distinct characteristics of UAC, like, large propagation delay, limited bandwidth, highly dynamic topology, and serious multipath spread introduce new challenges to design reliable and efficient communication protocols [2, 3].

Cooperative communication, which can promise significant performance gains with respect to the capacity of system, communication reliability and spectrum utilization, have attracted growing interest from UAC researchers. The concept of the cooperative

Z. Liu—The work was supported by National Natural Science Foundation of China (NO. 61201145), Subject Construction Guiding Foundation of HITWH (HITWH2016), the Graduate Education and Teaching Reform Research Project in Harbin Institute of Technology (JGYJ-201625), and the Foundation of Key Laboratory of Communication Network Information Transmission and Dissemination.

© ICST Institute for Computer Sciences, Social Informatics and Telecommunications Engineering 2017
X.-L. Huang (Ed.): MLICOM 2016, LNICST 183, pp. 97–104, 2017.
DOI: 10.1007/978-3-319-52730-7_10

communication has been employed to UAC in some recent papers [4–11]. It will not only increase the complexity of the system, but also affect the overall performance if all the relay nodes are used to cooperate. Thus the relay selection is very important in the research of underwater acoustic cooperative communication (UACC). An asynchronous relaying protocol tailored for UAC is proposed in [9], this method only selects the relay with the maximal effective SNR to forward signal while other relay nodes keep quiet. In [10], relay selection method according to propagation delay for UACC has been discussed. First, the scheme evaluates the propagation delay of each path, and then the relay with the minimum-delay-difference will be selected by comparing the propagation delay of relays with direct path. In [11], two kinds of relay selection schemes are considered. One uses the SNR to select the relays. In this scheme, the relayed path with maximum SNR will be given priority to cooperate. And the other is based on the minimum probability of error (PoE). The aforementioned schemes [9–11] require that the channel state information (CSI) is known. However, the CSI of underwater acoustic channel is difficult to be obtained. Furthermore, these schemes are all single relay selection (SRS) schemes, so the diversity gain they provide is limited.

In this paper, we put forward a multiple relay selection (MRS) strategy for UACC based on steady-state mean-square-error threshold (SMT), which does not need the channel state information. In the proposed MRS-SMT scheme, the steady-state mean-square-error (SMSE) is used to order the relays. One or multiple relay can be sequentially selected out from L relays according to the relay ordering. The proposed scheme can meet the performance of system conveniently by set a proper threshold.

2   System Model

The structure of the underwater acoustic cooperative communication system is shown in Fig. 1. It contains a source node S, L amplify-and-forward (AF) relay nodes $R_i|_{i=1}^{L}$ and a destination node D. We assume that all relays are half-duplex. That means they cannot transmit and receive simultaneously in the same frequency band. As usual, the communication between source and destination occur in two phases. In the first phase, S broadcasts to relay nodes and the destination node D, we call this the broadcast phase. In the second phase, the selected $L_c(0 \leq L_c \leq L)$ relay nodes sequentially transmit the amplified signal of the source node to D, this is usually referred to as the relaying phase.

We consider that the system works in shallow sea, every distinct may include a dominant component and a number of random sub-eigenpath components, so the channel between each node can be modeled as Rice fading channel as follows [12–15]:

$$h_i(n) = \sum_{k=1}^{M_i} A_{i,k} \delta(n - \tau_{i,k}) \tag{1}$$

where $i \in \{SD, SR_1, \cdots, SR_L, R_1D, \cdots, R_LD\}$, $\tau_{i,k}$ is the path delay, $A_{i,k}$ is the normalized amplitude of the signal in the propagation path of k, and $A_{i,k}$ obey the Rice distribution. Only a small amount of $A_{i,k}$ is not equal to zero in the light of the sparse characteristic of the underwater acoustic channel.

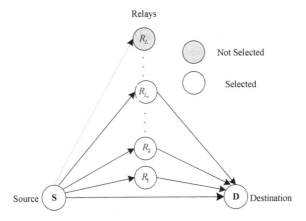

Fig. 1. UACC transmission system

During the broadcasting phase, the signals which obtained at the relay and the destination terminal are expressed by

$$r_{SR_i} = \sqrt{P_S} \cdot h_{SR_i} \cdot x + n_{SR_i} \tag{2}$$

$$r_{SD} = \sqrt{P_S} \cdot h_{SD} \cdot x + n_{SD} \tag{3}$$

where $i \in \{1, \cdots, L\}$, P_S is the signal transmission power, h_{SD}, h_{SR_i} are parameters of the channels $S \to D$ and $S \to R_i$, respectively. x is a transmitted signal with unit energy, $n_{SR_i}(t)$, $n_{SD}(t)$ are independent zero mean circularly symmetric additive white Gaussian noise with variance σ_j^2 for the channel $S \to R_i$ and the channel $S \to D$, respectively.

During the relaying phase, the amplified signal which is transmitted by relay node R_i at the destination terminal is given by:

$$r_{R_iD} = \sqrt{P_{R_i}} \cdot h_{R_iD} \cdot \beta_{R_i} \cdot r_{SR_i} + n_{R_iD} \tag{4}$$

where P_{R_i} is the transmission power of relays, n_{R_iD} is additive Gaussian noises, β_R is the amplifying factor of R_i. β_R can be defined as follows:

$$\beta_{R_i} = \sqrt{\frac{P_{R_i}}{|h_{SR_i}|^2 P_S + \sigma_j^2}} \tag{5}$$

Frequency-domain equalization (FDE) has been used for received signal of each path, so the SMSE of each path can be calculated. The specific receiver structure of FDE is shown in Fig. 2. The SMSE of each path is expressed by

$$SMSE_i = E\left[|d_n - y_n|^2\right] = \frac{\sum_{l=1}^n e_l^2}{n} \tag{6}$$

Fig. 2. The structure of FDE

where d_n is the desired signal, $e_n = d_n - y_n$ is error signal, i denotes i^{th} branch, $i \in \{SD, R_1D, R_2D, \cdots, R_LD\}$.

3   Proposed Multiple Relay Selection Scheme

The multiple relay selection strategy for UACC based on SMSE is described in this section. In the implementation process, the role of the direct path is considered. In addition, the SMSE of each relayed path is ordered, and the relay with minor SMSE value will be given priority to participating in the cooperation. The first $L_c (1 \le L_c \le L)$ relays will be selected when the combined SMSE performance index of the direct path and the L_c relayed paths exceeds a preset threshold Γ_{th} for the first time. The threshold Γ_{th} can be chosen according to the requirement of system. The combined SMSE performance index Γ_c is given by:

$$\Gamma_c = \Gamma_{sd} + \sum_{i=1}^{L_c} \Gamma_{R_i} = \frac{1}{SMSE_{sd}} + \sum_{i=1}^{L_c} \frac{1}{SMSE_i} \qquad (7)$$

where $\Gamma_{sd} = 1/SMSE_{sd}$, $\Gamma_{R_i} = 1/SMSE_i$, $SMSE_{sd}$ and $SMSE_i$ is the SMSE of the direct path and R_i relayed path, respectively.

Figure 3 shows the process of MRS scheme based on SMSE. First, we set a preset threshold Γ_{th}, and the destination D receives the signal sent by the source node S. Next, all relays are listed in descending order by Γ_{R_i}, the first relay (denoted as R_1) is chosen to participate in cooperation in the first time slot of the relaying phase. Then, the combined SMSE performance index of the first relayed branch and the direct branch is calculated. If the combined SMSE performance index exceeds threshold Γ_{th}, i.e., the communication quality can meet the requirement of system, no more relays are chosen. Otherwise, the scheme selects remaining relays to cooperate in subsequent time slots one by one until the cumulative SMSE performance index exceeds Γ_{th}. The worst case is that all L relay nodes are chosen. This strategy can also be modeled as follow:

$$\Gamma_c = \begin{cases} \Gamma_{sd} + \Gamma_{R_1}, & \Gamma_{sd} + \Gamma_{R_1} \ge \Gamma_{th} \\ \Gamma_{sd} + \sum\limits_{i=1}^{L_c} \Gamma_{R_i}, & \Gamma_{sd} + \sum\limits_{i=1}^{L_c} \Gamma_{R_i} \ge \Gamma_{th} \ and \ \Gamma_{sd} + \sum\limits_{i=1}^{L_c-1} \Gamma_{R_i} < \Gamma_{th} \\ \Gamma_{sd} + \sum\limits_{i=1}^{L} \Gamma_{R_i}, & otherwise \end{cases} \qquad (8)$$

The MRS-SMT scheme can adaptively choose the number of relays according to the threshold. And the average number of chosen relay nodes $\bar{L}_c$ is expressed as:

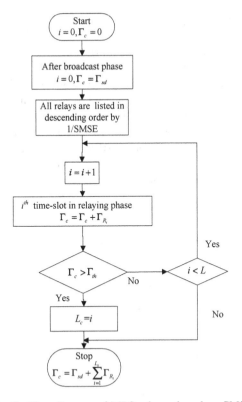

Fig. 3. Flow diagram of MRS scheme based on SMSE

$$\bar{L}_c = \sum_{i=1}^{L} i \ \Pr(L_c = i) \qquad (9)$$

where $\Pr(L_c = i)$ is given as:

$$\Pr(L_c = i) = \begin{cases} \Pr(\Gamma_{sd} + \Gamma_{R_i} \geq \Gamma_{th}), & i = 1 \\ \Pr(\Gamma_{sd} + \sum_{i=1}^{L_c} \Gamma_{R_i} \geq \Gamma_{th}, \\ \quad \cap \left[\Gamma_{sd} + \sum_{i=1}^{L_c} \Gamma_{R_i} < \Gamma_{th} \right]), & i \in \{2, \ldots, L-1\} \\ \Pr\left(\Gamma_{sd} + \sum_{i=1}^{L} \Gamma_{R_i} < \Gamma_{th} \right), & i = L \end{cases} \qquad (10)$$

4   Simulation Results

In order to evaluate the capability of the MRS-SMT scheme, the emulation results have been given and analyzed in this section. QPSK is used as the modulation mode in our simulation study. The same power is assumed to be given to all nodes. In addition, frequency domain adaptive equalization based on LMS algorithm is used for each path. The bit error rate (BER) performance of different thresholds for UACC system can be seen in Fig. 4. Threshold 1 and threshold 2 are 10^3 and 5×10^3, respectively. It is clear that the higher the threshold is, the lower the BER performance becomes. Meanwhile, the performance of the system with higher threshold is better as we expect. It demonstrates that the performance of system can conveniently be met by setting a proper threshold.

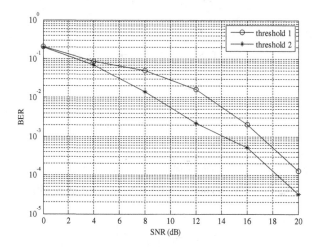

Fig. 4. The BER performance of different thresholds

Under the condition of SNR = 10 dB, Fig. 5 describes the effects of different threshold on the average number of chosen relay nodes. $L = 4, 7, 10$ mean that the number of potential nodes in the system are 4, 7 and 10, respectively. With the increase of threshold, the requirement of the system goes higher, and the average number of relay nodes increases. The number of relay nodes is no longer increased when the threshold value is high enough, meanwhile, all relay nodes have been selected. That is, the number of relay nodes can be chosen adaptively by the threshold.

Figure 6 shows the BER performance of different strategies. As seen from the figure, the BER performance of the MRS scheme and all SRS schemes are better than the no-cooperation scheme. This is because the MRS scheme and all SRS schemes get the diversity gain through the cooperation of relays. In addition, the BER performance of the SRS based on the minimization of SMSE is close to the SRS schemes based on the minimization of delay, the maximization of SNR and the minimization of probability of error (PoE). It is also obvious that the performance of the scheme we proposed is better than the existing SRS schemes. This is because the scheme we proposed can improve the diversity gain.

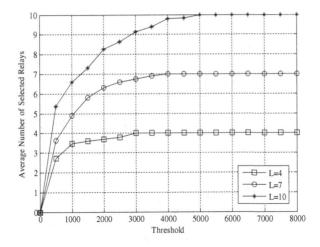

Fig. 5. The effects of different threshold on the average number of chosen relays

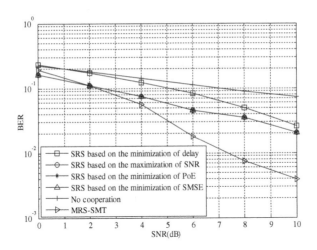

Fig. 6. BER performance of no cooperation, the proposed scheme and existing methods

5 Conclusions

This paper presented a MRS scheme for underwater acoustic cooperative communication based on SMSE. The effects of the threshold and the BER performance have been analyzed respectively. This scheme can meet the demand of system conveniently by a proper selection of the threshold. The relay nodes can adaptively be selected according to the SMSE of each path. Furthermore, it is not required to assume that the perfect and complete channel state information is known. Simulation results have shown that the MRS strategy we proposed is effectiveness and feasibility, and it can obtain better performance than existing methods.

References

1. Al-Dharrab1, S., Uysal, M.: Information theoretic performance of cooperative underwater acoustic communications. In: IEEE 22nd International Symposium on Personal, Indoor and Mobile Radio Communications, pp. 1562–1566, Toronto (2011)
2. Umar, A., Akbar, M., Iqbal, Z., Khan, Z.A.: Cooperative partner nodes selection criteria for cooperative routing in underwater WSNs. In: 2015 5th National Symposium on Information Technology: Towards New Smart World (NSITNSW), pp. 1–7, Riyadh (2015)
3. Zhan, C., Xu, F., Xie, Y., Wang, D.: BER performance of underwater multiuser cooperative communication based on MMSE-DFD algorithm. In: Oceans-Taipei, pp. 1–6. IEEE Press, Taipei (2014)
4. Yang, H., Ren, F., Lin, C., Liu, B.: Energy efficient cooperation in underwater sensor networks. In: 2010 18th International Workshop on Quality of Service, pp. 1–9. IEEE (2010)
5. Carbonelli, C., Mitra, U.: Cooperative multihop communication for underwater acoustic networks. In: Proceedings of the 1st ACM International Workshop on Underwater Networks, ser. WUWNet 2006, pp. 97–100. ACM, New York (2006)
6. Han, Z., Sun, Y., Shi, H.: Cooperative transmission for underwater acoustic communications. In: IEEE International Conference on Communications, pp. 2028–2032 (2008)
7. Vajapeyam, M., Mitra, U., Preisig, J., Stojanovic, M.: Distributed space-time cooperative schemes for underwater acoustic communications. In: Proceedings of OCEANS 2006-Asia Pacific, pp. 1–8 (2006)
8. Carbonelli, C., Chen, S.-H., Mitra, U.: Error propagation analysis for underwater cooperative multi-hop communications. Ad Hoc Netw. 7(4), 759–769 (2009)
9. Cheng, X., Cao, R., Qu, F., et al.: Relay-aided cooperative underwater acoustic communications: selective relaying. In: OCEANS 2012-Yeosu, pp. 1–7. IEEE Press, Yeosu (2012)
10. Gao, C., Liu, Z., Cao, B., et al.: Relay selection scheme based on propagation delay for cooperative underwater acoustic network. In: 2013 International Conference on Wireless Communications & Signal Processing (WCSP), pp. 1–6. IEEE (2013)
11. Karakaya, B., Hasna, M.O., Uyasl, M., et al.: Relay selection for cooperative underwater acoustic communication system. In: 2012 19th International Conference on Telecommunications (ICT), pp. 1–6. IEEE (2012)
12. Li, B., Zhou, S., Stojanovic, M., Freitag, L., Willett, P.: Multicarrier communication over underwater acoustic channels with nonuniform doppler shifts. IEEE J. Oceanic Eng. 33(2), 198–209 (2008)
13. Geng, X., Zielinski, A.: An eigenpath underwater acoustic communication channel model. In: OCEANS 1995 MTS/IEEE Challenges of Our Changing Global Environment Conference Proceedings, pp. 1189–1196 (1995)
14. Kilfoyle, D.B., Baggeroer, A.B.: The state of the art in underwater acoustic telemetry. IEEE J. Oceanic Eng. 25, 4–27 (2000)
15. Munoz Gutierrez, M.A., Prospero Sanchez, P.L., Do, V.N.J.V.: An eigenpath underwater acoustic communication channel simulation. In: Proceedings of OCEANS 2005 MTS/IEEE, pp. 355–362 (2005)

Intelligent Massive MIMO

A Serial Time-Division-Multiplexing Chip-Level Space-Time Coded Multi-user MIMO System Based on Three Dimensional Complementary Codes

Siyue Sun[1(⊠)], Guang Liang[1], and Kun Wang[2]

[1] Shanghai Engineering Center for Micro-satellites, Shanghai, China
sunsiyue@hit.edu.cn
[2] Huawei Technologies Co., Ltd, Shanghai, China

Abstract. This paper presents a serial time-division-multiplexing multi-user MIMO system for the chip level space-time coding scheme based on three dimensional complementary codes (3DCCs). In such a 3DCC-based MIMO system, the spread signals corresponding to different element sequences of a 3DCC are transmitted in different time slots in a series way. A "Diversity/Multiplexing Controller" is designed to control the transmit spatial diversity and multiplexing gains according to different channel conditions. Both the theoretical analysis and the computer simulation will prove the capability of the proposed system to eliminate multi-path and multi-user interference compared to the traditional multiuser MIMO solutions.

Keywords: Multi-user MIMO · 3D complementary code · Chip-level space-time coding · CDMA

1 Introduction

As an effort to integrate multiple-input multiple-output (MIMO) and code division multiple access (CDMA) techniques, a kind of chip level space-time coding scheme (CLSTC) is proposed [1], which employs direct sequence spreading to serve users for multiple access and antenna separation simultaneously in MIMO systems with the help of special complementary codes (CCs), or three dimensional complementary codes (3DCCs).

3DCCs are an evolutional version of complementary codes [2] to provide the orthogonality not only among users to achieve code division multiple access, but also among different antennas in a MIMO system to achieve space diversity/multiplexing gains, in order to offer a new paradigm integrating multi-user and multi-antenna techniques to facilitate system optimization.

S. Sun and G. Liang—This work was supported by Shanghai Sailing Program (16YF1411000).

© ICST Institute for Computer Sciences, Social Informatics and Telecommunications Engineering 2017
X.-L. Huang (Ed.): MLICOM 2016, LNICST 183, pp. 107–116, 2017.
DOI: 10.1007/978-3-319-52730-7_11

In order to offer orthogonality in time-frequency-spatial three fields, the construction of 3DCCs is an extremely challenging issue and a new attempt. In [3] and [4], two construction methods have been proposed to show the feasibility to construct such codes. The design of CDMA systems based on classic complementary codes have been well studied [5,6]. However, few system design or performance analysis of 3DCC-based MIMO system are present till now. The main contribution made in this paper is to present a serial time-division-multiplexing chip-level space-time coded multi-user MIMO system. In such a 3DCC-based MIMO system, the spread signals corresponding to M element sequences of a 3DCCs are transmitted in different time slots in a series way. Additionally, a "Diversity/Multiplexing Controller" is designed to control the transmit spatial diversity and multiplexing gains according to different channel conditions. Based on attractive correlation properties of 3DCCs, this paper analyzes the elimination of multi-user interference (MUI) elimination and diversity gains of such 3DCC-based MIMO system. Finally, the simulated bit error rate performance comparison with traditional multi-user MIMO system will proof the benefits of the proposed 3DCC-based MIMO system on MUI- and multi-path interference (MPI)-resistant performance.

2   Definitions and Code Construction

2.1   Three Dimensional Complementary Codes

Let $\mathcal{G}(K, A, M, N)$ be a family of 3DCCs, which contains K 3DCCs denoted as $\mathbb{G}^{(k)}$, $k \in \{1, 2, \cdots, K\}$, and K is the family size. Each 3DCC contains A sub-2DCCs, $\mathbf{G}^{(k,a)}$, with the same flock size M and code length N, where A is the number of transmit antennas used by the transmitter. Each sub-2DCC $\mathbf{G}^{(k,a)}$ contains M element sequences $\mathbf{g}_m^{(k,a)}$ with the same code length N, $m \in \{1, 2, \cdots, M\}$, and M is the flock size (which determines the number of element codes used by the same user). Therefore, $\mathbb{G}^{(k)}$ can be viewed as a three-dimensional code array, and we have $\mathbb{G}^{(k)}(:, :, a) = \mathbf{G}^{(k,a)}$, $\mathbf{G}^{(k,a)}(m, :) = \mathbf{g}_m^{(k,a)}$, and $\mathbf{g}_m^{(k,a)}(n) = g_{m,n}^{(k,a)}$, where $k \in \{1, 2, \cdots, K\}$, $a \in \{1, 2, \cdots, A\}$, $m \in \{1, 2, \cdots, M\}$, $n \in \{1, 2, \cdots, N\}$ and $g_{m,n}^{(k,a)} \in \{1, -1\}$.

Generally, A determines the number of transmit antennas of a MIMO system and the family size K determines the user capacity of such a system. $\mathbb{G}^{(k)}$ is assigned to user k as its signature code and space-time code. The flock size M determines the number of independent sub-channels required to implement a 3DCCs-base MIMO-CDMA system, and the independent sub-channels can be separated by different sub-carriers or time slots.

2.2   Complementary Correlation and Perfect 3DCCs

Aperiodic correlation, also called partial correlation, will be considered in this paper, because the aperiodic correlation property of spreading codes is more general in system performance evaluation than the periodic correlation property, due to the fact that a periodic correlation function can always be decomposed into two partial correction functions. Therefore, if we have ideal partial correction functions for a code, we can always ensure ideal periodic correlation functions for the same code as shown later.

For any two sequences of length N, $\mathbf{a} = \{a_1, a_2, \cdots, a_N\}$ and $\mathbf{b} = \{b_1, b_2, \cdots, b_N\}$, the aperiodic correlation function $\psi(\mathbf{a}, \mathbf{b}; \tau)$ with positive relative delay τ is defined as

$$\psi(\mathbf{a}, \mathbf{b}; \tau) = \sum_{n=0}^{N-1-\tau} a_n b_{n+\tau}, \quad 0 \leq \tau \leq N - 1. \tag{1}$$

The correlation properties of 3DCCs are characterized by the complementary aperiodic correlation function, which is calculated as the sum of the aperiodic correlation functions of all element codes with the same delay τ, or

$$\rho(\mathbf{G}^{(k_1, a_1)}, \mathbf{G}^{(k_2, a_2)}; \tau) = \sum_{m=0}^{M-1} \psi(\mathbf{g}_m^{(k_1, a_1)}, \mathbf{g}_m^{(k_2, a_2)}; \tau)$$

$$= \sum_{m=0}^{M-1} \sum_{n=0}^{N-1-\tau} g_{m,n}^{(k_1, a_1)} g_{m,n+\tau}^{(k_2, a_2)}, \quad 0 \leq \tau \leq N - 1, \tag{2}$$

where $k_1, k_2 \in \{1, 2, \cdots, K\}$ and $a_1, a_2 \in \{1, 2, \cdots, A\}$.

A family of 3DCCs should provide the orthogonality among the signals from both different users and different antennas in order to achieve spatial diversity and/or spatial multiplexing along with orthogonal multiple access in multipath asynchronous communications. Therefore a perfect family of 3DCCs should satisfy the following three constrains:

1. Ideal auto-correlation property to eliminate MPI, i.e.

$$\text{ACF} = \rho(\mathbf{G}^{(k,a)}, \mathbf{G}^{(k,a)}; \delta) = \begin{cases} MN, & \delta = 0 \\ 0, & \delta \neq 0 \end{cases} \tag{3}$$

2. Ideal cross-correlation property for the sub-2DCCs belonging to the same 3DCCs to get orthogonality among the signals from different antennas of the same user. We name it as Cross-Correlation Function among Antennas (CCF-A), i.e.

$$\text{CCF-A} = \rho(\mathbf{G}^{(k,a)}, \mathbf{G}^{(k,b)}; \delta) = 0, \quad a \neq b \tag{4}$$

3. Ideal cross-correlation property for the sub-2DCCs belonging to different 3DCCs to get orthogonality among the signals from different users. We name it as Cross-Correlation Function among Users (CCF-U), i.e.

$$\text{CCF-U} = \rho(\mathbf{G}^{(k,a)}, \mathbf{G}^{(g,b)}; \delta) = 0, \quad k \neq g \tag{5}$$

The construct of a perfect family of 3DCCs can be found in, and in this paper a system architecture of a MIMO-CDMA system based on such perfect 3DCCs will be discussed.

3 A Serial Time-Division-Multiplexing Multi-user MIMO System Based on 3DCCs

3.1 System Models

In 3DCCs based multi-user system, each user will be allocated a particular 3DCCs from a code set as both its signature code and space-time code. A user should spread its data with M element sequences of 3DCCs, respectively. Transmitted over wireless channels, M streams of spread signals are required to be separated at a receiver, because there is no correlation constraint on the correlation properties between different element sequences. Therefore, M streams of spread signals are normally transmitted in M independent subchannels. Considering the compatibility of the existing, this paper proposed a serial time-division-multiplexing (TDM) multi-user MIMO system based on 3DCCs mentioned above, as shown in Figs. 1 and 2.

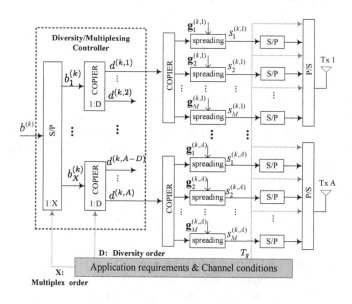

Fig. 1. The structure of the transmitter of user k.

Let us consider an $A \times X$ MIMO system with K users, where the corresponding transmitter (taking user k as an example) and receiver (e.g., user g) are shown in Figs. 1 and 2. Let $b^{(k)}$ represent the polarized binary source data from user k, and it is mapped to A streams of signals, $\{d^{(k,a)}\}_{a=1}^{A}$, by "Diversity/Multiplexing Controller" module. Owing to the orthogonality of the signals among different antennas ensured by 3DCCs, both space diversity and multiplexing are supported by the 3DCC-based multiuser MIMO system, which is controlled by the "Diversity/Multiplexing Controller" module according to

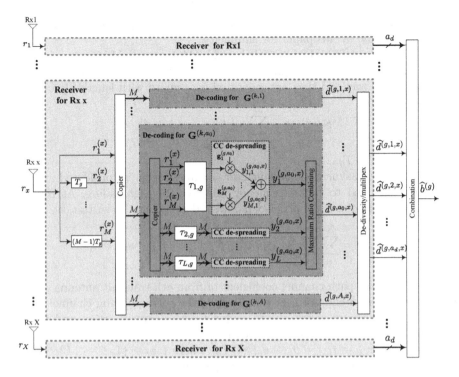

Fig. 2. The structure of the receiver of user g.

channel conditions and performance requirements, i.e., (1) if the diversity mode is employed to improve the error probability, $\{d^{(k,a)}\}_{a=1}^{A}$ are A copies from the source signal $b^{(k)}$; (2) if the multiplexing mode is employed to enhance data rate, they are generated by serial to parallel operation (or S/P) from $b^{(k)}$; (3) if $A > 2$, a hybrid diversity and multiplexing mode can also be employed to achieve both diversity gain and multiplex capability.

Then, A streams of signals $\{d^{(k,a)}\}_{a=1}^{A}$ of user k are spread by M element sequences of ath of the 3DCC $\mathbb{G}^{(k)}$. Take ath antenna and mth element sequence as an example. The mathematical expressions are given as follows:

$$s_m^{(k,a)}(t) = \sqrt{p_t} \sum_{i=0}^{B-1} d^{(k,a)}(i) G_m^{(k,a)}(t - iT) \qquad (6)$$

where B is the length of a data block and p_t is the transmit power. $T = NT_c$ and T_c is a chip duration $G_m^{(k,a)}(t)$ is the spreading chip waveform of the mth element sequence of the ath sub-2DCC of $\mathbb{G}^{(k)}$, or

$$G_m^{(k,a)}(t) = \sum_{n=1}^{N} g_{m,n}^{(k,a)} q(t - nT_c + T_c) \qquad (7)$$

where $q(t)$ is the impulse response of chip waveform-shaping filter, which is a rectangular shape in this paper for simplicity.

In the proposed The TDM 3DCCs-based MIMO system, M streams of spread signals corresponding to M element sequences are transmitted in different time slots in a series way and they are separated by a guard with length T_g, or

$$S^{(k,a)}(t) = \sum_{m=1}^{M} s_m^{(k,a)}(t - m\Delta + \Delta) \tag{8}$$

where $\Delta = T_c N + T_g$.

In this paper, asynchronous multiuser communication is considered and the channel is assumed to suffer frequency selective Rayleigh fading. At the xth receiver antenna of user g, carrier-demodulated signal can be written as

$$r^{(x)}(t) = \sum_{k=1}^{K} \sum_{a=1}^{A} \sum_{l=1}^{L} \sqrt{p_t} h_{k,l}^{(a,x)} S^{(k,a)}(t - \tau_{l,k} - \theta_k) + n(t) \tag{9}$$

where $h_{k,l}^{(a,x)}$ is lth path channel coefficient of from ath transmit antenna of user k to xth receive antenna of user g and $\tau_{l,k}$ is the corresponding channel delay. θ_k is the delay among users due to asynchronous communication and $n(t)$ is Gaussian noise with power spectrum density N_0.

Assuming $|\tau_{l,k} - \tau_{z,g} + \theta_k - \theta_g| \leq T_g$, where $l, z \in \{1, 2, \cdots, L\}$, $k, g \in \{1, 2, \cdots, K\}$. After time-division de-multiplexing, we get

$$r_m^{(x)}(t) = \sum_{k=1}^{K} \sum_{a=1}^{A} \sum_{l=1}^{L} \sqrt{p_t} h_{k,l}^{(a,x)} s_m^{(k,a)}(t - \tau_{l,k} - \theta_k) + n(t) \tag{10}$$

Then the M signal streams $\{r_m^{(x)}\}_{m=1}^{M}$ are de-coded by the 3DCCs $\mathbb{G}^{(g)}$, as the following three steps.

Step 1. CC de-spreading for a_0th antenna of user g for l_0 path, $a_0 = \{1, 2, \cdots, A\}$ and $l_0 = \{1, 2, \cdots, L\}$.

$$\begin{aligned}
y_{l_0}^{(g,a_0,x)} &= \sum_{m=1}^{M} \int_0^{NT_c} r_m^{(x)}\left(t + \tau_{l_0,g}\right) G_m^{(g,a_0)}(t) dt \\
&= \sum_{m=1}^{M} \sum_{k=1}^{K} \sum_{a=1}^{A} \sum_{l=1}^{L} \sqrt{p_t} h_{k,l}^{(a,x)} d^{(k,a)} \int_0^{NT_c} G_m^{(k,a)}(t - \tau_{l,k} - \theta_k + \tau_{l_0,g}) G_m^{(g,a_0)}(t) dt + \omega \\
&= \frac{1}{MN} \sum_{k=1}^{K} \sum_{a=1}^{A} \sum_{l=1}^{L} \sqrt{p_t} h_{k,l}^{(a,x)} d^{(k,a)} \underbrace{\sum_{m=1}^{M} \sum_{n=1}^{N-\delta} g_{m,n}^{(k,a)} g_{m,n+\delta}^{(g,a_0)}}_{\rho(\mathbf{G}^{(k,a)}, \mathbf{G}^{(g,a_0)};\delta)} + \omega
\end{aligned} \tag{11}$$

where $\delta = (\tau_{l,k} + \theta_k - \tau_{l_0,g})/T_c$, $\omega = \sum_{m=1}^{M} \int_0^{NT_c} n(t) G_m^{(g,a_0)}(t) dt$ is sampled additive white noise. According to the ideal CCF-U of 3DCCs in (5), we get

$$y_{l_0}^{(g,a_0,x)} = \sqrt{p_t} h_{g,l_0}^{(a_0,x)} d^{(g,a_0)} + \omega \tag{12}$$

Step 2. Combine L detected data of a_0 sub-2DCCs of user g by maximum ratio combining (MRC), or

$$\widehat{d}^{(g,a_0,x)} = \sum_{l=1}^{L} \left(h_{g,l}^{(a_0,x)}\right)^* y_l^{(g,a_0,x)} = \sqrt{p_t} \sum_{l=1}^{L} |h_{g,l}^{(a_0,x)}|^2 d^{(g,a_0)} + \sum_{l=1}^{L} \left(h_{g,l}^{(a_0,x)}\right)^* \omega$$

Step 3. De-diversity/multiplex the detected A according to the diversity/multiplexing modes employed by the transmitter of user g. (1) to recover the multiplexed streams through parallel to serial operation (or P/S); (2) to combine the streams for space diversity gain.

Step 4. Combine the detected data from X receive antennas of user g, or

$$\widehat{b}^{(g)} = \sqrt{p_t} \sum_{x=1}^{X} \sum_{a=a_1}^{a_d} \sum_{l=1}^{L} |h_{g,l}^{(a,x)}|^2 b^{(g)} + \sum_{x=1}^{X} \sum_{a=a_1}^{a_d} \sum_{l=1}^{L} \left(h_{g,l}^{(a,x)}\right)^* \omega \qquad (13)$$

3.2 Analysis on MUI, MPI and Diversity Gains

In (11), the detected signal $y_{l_0}^{(g,a_0)}$ for l_0th path a_0 transmit antenna of user g contains useful signal U, multipath interference I_{MP}, multiuser interference I_U, interference among antennas I_A and noise ω, or

$$y_{l_0}^{(g,a_0)} = \frac{1}{MN} \sum_{k=1}^{K} \sum_{a=1}^{A} \sum_{l=1}^{L} \sqrt{p_t} h_{k,l}^{(a,x)} d^{(k,a)} \rho(\mathbf{G}^{(k,a)}, \mathbf{G}^{(g,a_0)}; \delta) + \omega$$

$$= U + I_{MP} + I_A + I_U + \omega \qquad (14)$$

According to the ideal correlation properties of 3DCCs in (3) (5), we get

$$\begin{cases} U &= \frac{1}{MN} \sqrt{p_t} h_{g,l_0}^{(a_0,x)} d^{(g,a_0)} \rho(\mathbf{G}^{(g,a_0)}, \mathbf{G}^{(g,a_0)}; 0) = \sqrt{p_t} h_{g,l_0}^{(a_0,x)} d^{(g,a_0)} \\ I_{MP} &= \frac{1}{MN} \sum_{l=1,l\neq l_0}^{L} \sqrt{p_t} h_{g,l}^{(a_0,x)} d^{(g,a_0)} \rho(\mathbf{G}^{(g,a_0)}, \mathbf{G}^{(g,a_0)}; \frac{\tau_{l,g} - \tau_{l_0,g}}{T_c}) = 0 \\ I_A &= \frac{1}{MN} \sum_{a=1}^{A} \sum_{l=1}^{L} \sqrt{p_t} h_{g,l}^{(a,x)} d^{(g,a)} \rho(\mathbf{G}^{(g,a)}, \mathbf{G}^{(g,a_0)}; \frac{\tau_{l,g} - \tau_{l_0,g}}{T_c}) = 0 \\ I_U &= \frac{1}{MN} \sum_{k=1,k\neq g}^{K} \sum_{a=1}^{A} \sum_{l=1}^{L} \sqrt{p_t} h_{k,l}^{(a,x)} d^{(k,a)} \rho(\mathbf{G}^{(k,a)}, \mathbf{G}^{(g,a_0)}; \frac{\tau_{l,k} - \theta_k - \tau_{l_0,g}}{T_c}) = 0 \end{cases}$$

Therefore, both the multipath interference and multiuser interference are eliminated in the proposed system. Now we deduce the bit error rate (BER) and diversity gain of the above system with assumption $L = 1$ for simplicity.

$$\widehat{b}^{(g)} = \sqrt{p_t} \sum_{x=1}^{X} \sum_{a=a_1}^{a_d} |h_g^{(a,x)}|^2 b^{(g)} + \sum_{x=1}^{X} \sum_{a=a_1}^{a_d} \left(h_g^{(a,x)}\right)^* \omega \qquad (15)$$

where A_d is the number of antennas of user g transmit the same data, i.e. using diversity mode. Assuming $p_t = \frac{E_b}{A_d X M N T_c}$, the instantaneous signal-noise ratio (SNR) before decision is:

$$\gamma_b = \frac{1}{A_d X} \sum_{x=1}^{X} \sum_{a=a_1}^{a_d} |h_g^{(a,x)}|^2 \frac{E_b}{N_0} \qquad (16)$$

Assuming the channels between any antennas are suffer independent Rayleigh fading, we get the probability density of γ_b is

$$p(\gamma_b) = \frac{1}{(A_d X - 1)! \gamma_b^{A_d X}} \overline{\gamma}^{A_d X - 1} e^{-\gamma_b / \overline{\gamma}} \tag{17}$$

where $\overline{\gamma} = \frac{E_b}{A_d X N_0}$. Then we get the BER as

$$P(\gamma_b) = \int_0^\infty Q \sqrt{2 \gamma_b} p(\gamma_b) d\gamma_b$$

$$= \left[\frac{1}{2} \left(1 - \sqrt{\frac{\overline{\gamma}}{1 + \overline{\gamma}}} \right) \right]^{A_d X} \sum_{x=0}^{A_d X - 1} \binom{A_d X - 1 + x}{x} \left[\frac{1}{2} \left(1 + \sqrt{\frac{\overline{\gamma}}{1 + \overline{\gamma}}} \right) \right]^x \tag{18}$$

If $\overline{\gamma} \gg 1$, we get

$$P_2(\gamma_b) \approx (4\gamma)^{-A_d X} \binom{2 A_d X - 1}{A_d X} \tag{19}$$

Finally, we get the diversity gain of the above system as

$$g_d(\gamma) = -\lim_{\gamma \to \infty} \frac{\log \left[(4\gamma)^{-A_d X} \binom{2 A_d X - 1}{A_d X} \right]}{\log(\gamma)} = A_d X \tag{20}$$

4   Simulation Results and Discussion

BERs of five kinds of 2×1 multiuser MIMO systems over either a Rayleigh flat fading channel or a multi-path channel are shown in Figs. 3 and 4, respectively. In the simulations, a family of 3DCCs $\mathcal{G}(8, 2, 8, 8)$ was employed in the proposed S-TDM MIMO system. "STBC+Gold" denotes a MIMO system using space-time block codes (STBCs) [7] as space-time codes and Gold sequences ($N = 63$) as spreading codes. "STS+Walsh" means a MIMO system using space-time-spreading (STS) [8] as space-time and spreading codes and Walsh sequences ($N = 64$) as spreading codes. In the simulations, $T_c = 0.025$ μs and an uncoded BPSK modulation were employed. The multi-path channel is a three-path tapped-delay line model with its normalized path gain coefficients vector as $[-1.92, -5.92, -9.92]$ dB and delay vector as $[0\ 0.025\ 0.075]$ μs.

As seen from the results in Fig. 3, in the flat fading channel, both 3DCC-based and "STS+Walsh" schemes achieve MUI-free, while the BER performance of "STBC+Gold" scheme deteriorates significantly in multiuser scenario due to the bad cross-correlation property of Gold sequences.

In the multi-path fading channel, only the proposed system achieves MUI-free multiuser communications, while the BER performance of "STBC+Gold" degrades significantly in multiuser scenario due to the bad cross-correlation

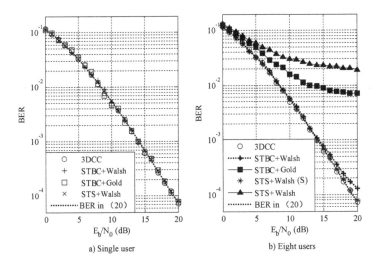

Fig. 3. BER performance of 2 × 1 3DCC-based MIMO systems with different space-time and spreading codes over a flat fading channel.

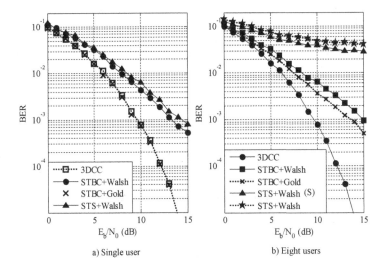

Fig. 4. BER performance of 2 × 1 3DCC-based MIMO systems with different space-time and spreading codes over a multi-path channel.

property of Gold sequences. The BER performance of "STS+Walsh" deteriorates significantly in both single-user and multiuser scenarios due to the bad auto-correlation property of Walsh sequences. Additionally, the proposed system achieves a better BER in the multi-path fading channel than it in a flat fading channel. This result not only proves the capability of such a system to eliminate both MPI and MUI, but also shows its superior capability to achieve multi-path diversity gains.

5 Conclusions

In this paper, we presented a serial time-division-multiplexing chip-level space-time coded multi-user MIMO system based on such 3DCCs. Both the theoretical analysis and the computer simulation prove the capability of the proposed system to eliminate MPI and MUI in multiuser MIMO communications. Additionally, providing both diversity and multiplex is a salient feature of the proposed system, which make it in particular well suited for futuristic wireless communication systems to fit to varying channel conditions and application requirements.

References

1. Sun, S.-Y., Yu, Q.-Y., Meng, W.-X., Chen, H.-H.: Evolution from symbol-level space-time coded MIMO to chip-level space-time coded MIMO. Wireless Commun. Mob. Comput. **14**(13), 1219–1230 (2014)
2. Tseng, C.-C., Liu, C.L.: Complementary sets of sequences. IEEE Trans. Inf. Theory **18**(5), 644–652 (1972)
3. Sun, S.-Y., Chen, C.-W., Chen, H.-H., Meng, W.-X.: Multi-user interference free space-time spreading MIMO systems based on three dimensional complementary codes. IEEE Syst. J. **9**(1), 45–57 (2015)
4. Sun, S.-Y., Chen, H.-H., Meng, W.: A framework to construct 3-dimensional complementary codes for multiuser MIMO systems. IEEE Trans. Veh. Technol. **64**(7), 2861–2874 (2015)
5. Chen, H.-H., et al.: Design of next-generation CDMA using orthogonal complementary codes and offset stacked spreading. IEEE Wireless Commun. **14**(3), 61–69 (2007)
6. Sun, S.-Y., Chen, H.-H., Meng, W.-X.: A survey on complementary coded MIMO CDMA wireless communications. Submitted to IEEE Survey and Tutorial on Communications. Sun, S-Y., Chen, H-H., Meng, W.: A survey of complementary coded wireless communications. IEEE Commun. Surv. Tutor. **17**(1), 52–69 (2015)
7. Tarokh, V., Jafarkhani, H., Calderbank, A.: Space-time block codes from orthogonal designs. IEEE Trans. Inf. Theory **45**(5), 1456–1467 (1999)
8. Hochwald, B., Marzetta, T.L., Papadias, C.B.: A transmitter diversity scheme for wideband CDMA systems based on space-time spreading. IEEE J. Sel. Areas Commun. **19**(1), 48–60 (2001)
9. Leppanen, P.A., Pirinen, P.O.: A hybrid TDMA/CDMA mobile cellular system using complementary code sets as multiple access codes. In: IEEE Proceedings of Personal Wireless Communications, pp. 419–423 (1997)

A Spectrum Access Scheme for MIMO Cognitive Networks with Beamforming Design

Yanbing Wang[1(✉)], Weidang Lu[1], Hong Peng[1], Zhijiang Xu[1], and Xin Liu[2]

[1] College of Information Engineering, Zhejiang University of Technology, Hangzhou 310014, Zhejiang, China
wangyanbing2016@gmail.com
[2] School of Information and Communication Engineering, Dalian University of Technology, Dalian 116024, China

Abstract. This paper studies a spectrum access scheme to increase the spectral utilization for the multi-antenna cognitive radio (CR) network. The network comprises of one transmitter-receiver pair for primary users (PUs) and another pair for secondary users (SUs) with secondary transmitter equipped with multi-antenna. We divide the transmission time into two equal time phases, the first for PUs to transmit signals and the second for SUs to relay and transmit self-information. We aim to maximize the data rate of the secondary system through designing the beamforming vectors and power scaling factors with zero-force beamforming (ZFBF) to eliminate the co-channel interference, under the condition that PUs can achieve the target rate. The simulation results demonstrate a higher spectral utilization and a performance gain with the proposed scheme.

Keywords: Cognitive network · MIMO · Beamforming · Multi-antenna

1 Introduction

With the booming of the wireless radio, spectrum resources are getting more and more scarce. As the conception of cognitive radio (CR) first proposed by Joseph Mitola in 1999 [1], there emerged a new model to improve the utilization of spectrum resources as cognitive radios have a good performance in increasing spectral efficiency [2]. In the common wireless network, once users with licensed spectrum have connection interruptions, there is a waste of spectrum. While in a cooperative cognitive network, the situation is quite different. Users are generally divided into two parts in a cooperative cognitive network, PUs with licensed spectrum and SUs without license. Then SUs sense the spectrum whether it is spare or not. Once it's spare, SUs can access the spectrum after bargaining with PUs.

In a cooperative cognitive network, generally PUs lease the spectrum to SUs and SUs separate part of the power to assist PUs to achieve win-win results [3–5]. In [6], the authors proposed a two-phase transmission protocol aimed to achieve spectrum access in a cooperative cognitive network. In the first phase, only the primary transmitter is allowed to broadcast and in the second phase it turns to secondary system. The

© ICST Institute for Computer Sciences, Social Informatics and Telecommunications Engineering 2017
X.-L. Huang (Ed.): MLICOM 2016, LNICST 183, pp. 117–125, 2017.
DOI: 10.1007/978-3-319-52730-7_12

power of the secondary transmitter is designed to ensure the primary system quality-of-service (QoS). While in [7], in both two phases, PUs and SUs were designed to transmit information at same time with the SUs transmission power limited to satisfy the QoS of PUs. Both of [6, 7] show that two-phase cognitive transmission protocol can significantly decrease the outage probability of the secondary system while ensuring the PUs QoS and improve the spectral utilization at same time.

Multiple-input multiple-output (MIMO) technology was originally conceived in 1970s when Bell Labs engineers tried to break through the bandwidth limitations caused by signal interferences. MIMO uses antenna arrays at the transmitter and receiver [8]. Thus MIMO cognitive networks contain at least one user with multi-antenna. As the MIMO channels can obviously increase the channel capacity [9], cooperative MIMO becomes a hot issue.

There are generally two main aims in cooperative MIMO cognitive networks, larger capacity and lower bit error rate (BER). Recently, a scheme for optimal power allocation to maximize the secondary throughput in a MIMO cognitive network was proposed in [10]. Furthermore, in [11], the paper has studied both bandwidth and power allocation to maximize the sum rate of an overlay CR system assisted with multiple antennas two-way relays in which PUs cooperate with SUs for mutual benefits [11]. Antenna selection is also a popular research area in MIMO cognitive networks. In [12], antenna selection is used to maximize the CR data rate, while it is designed to decrease the BER in [7, 8].

In this paper, we aim at proposing a cooperative spectrum access scheme for MIMO cognitive network to improve the spectral utilization. We apply the two-phase transmission protocol to our scheme. The design object is to maximize the transmission rate of the secondary system on the condition that the primary system can satisfy the rate constraint. On purpose of eliminating the co-channel interference, we apply ZFBF to construct the beamforming vector and power scaling factors. At last, we compare the transmission rate of secondary system with various parameters and estimate the performance with simulation results.

2 System Model

The system model of the MIMO cognitive network we considered is shown in Fig. 1. The whole system consists of two transmitter and receiver pairs which are PUs (PT and PR) and SUs (ST and SR). The PUs and the SUs share with the same spectrum and we assume the bandwidth is W. We assume that the secondary transmitter ST equipped with M ($M \geq 2$) antennas and other users equipped with a single antenna. The channels are flat Rayleigh fading channels and the Channel State Information (CSI) is perfectly obtained. In this model, let $\mathbf{h}_1$, $\mathbf{h}_2$, $\mathbf{h}_3$, h_d ($\mathbf{h}_1, \mathbf{h}_2 \in \mathbb{C}^{1 \times M}, \mathbf{h}_3 \in \mathbb{C}^{1 \times M}$) represent the channel coefficients of ST $\rightarrow$ PR, ST $\rightarrow$ SR, PT $\rightarrow$ ST, PT $\rightarrow$ PR, respectively.

When there are no SUs, PT sends information directly to PR, the signal received by PR can be written as

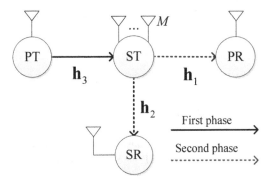

Fig. 1. The system model

$$r_d = \sqrt{P_p}h_d x_1 + n_d \tag{1}$$

where P_P is the transmission power of PT, x_1 is the transmitted symbol from PT, n_d is the complex addictive white Gaussian noise (AWGN) whose means is zero and variance is σ_d^2. Then the transmission rate of PT $\rightarrow$ PR can be written as

$$R_d = W \log_2(1 + \frac{P_p h_d^2}{\sigma_d^2}) \tag{2}$$

We assume that the secondary system gets a chance to access the spectrum when the transmission rate of primary system cannot reach the target rate R_T. But only when the secondary system can assist primary system to reach the target rate, it can really have access to the spectrum. In this paper, we divide the transmission time into two equal time phases. In the first phase, PT transmits signals to ST, the signal received by ST can be written as

$$\mathbf{r}_3 = \sqrt{P_p}\mathbf{h}_3 x_1 + \mathbf{n}_3 \tag{3}$$

where $\mathbf{n}_3$ is the complex addictive white Gaussian noise and $E(\mathbf{n}_3, \mathbf{n}_3^H) = \sigma_3^2\mathbf{I_M}$. Then the transmission rate becomes

$$R_3 = \frac{1}{2} W \log_2(1 + \frac{||\mathbf{h}_3||^2 P_p}{\sigma_3^2}) \tag{4}$$

There is a coefficient of 1/2 because the first phase only possesses half of the transmission time.

In the second phase, as the dotted lines showed in Fig. 1, ST makes use of multi-antenna, on the one hand, forward signals to PR (assume that the received signal is perfectly decoded) and on the other hand, send self-information to ST at same time. Let $x_1, x_2, \mathbf{f}_1, \mathbf{f}_2$ ($\mathbf{f}_1, \mathbf{f}_2 \in \mathbb{C}^{M \times 1}$), and P_1, P_2 be the data symbols, beamforming weight vectors, and transmission power scaling factors respectively (ST $\rightarrow$ PR, ST $\rightarrow$ SR).

Define $\mathbf{x} = [x_1 \ x_2]^T$, $\mathbf{F} = [\mathbf{f}_1 \ \mathbf{f}_2]$, $\mathbf{P} = diag\{p_1, p_2\}$, so that the transmitted signal $\mathbf{s}$ can be written as $\mathbf{s} = \mathbf{Fx}$, as a result, the received signals of PR and SR can be written respectively as

$$\mathbf{r}_{sp} = (\sqrt{P_1}\mathbf{h_1f_1})x_1 + (\sqrt{P_2}\mathbf{h_1f_2})x_2 + \mathbf{n}_1 \tag{5}$$

$$\mathbf{r}_{ss} = (\sqrt{P_2}\mathbf{h_2f_2})x_2 + (\sqrt{P_1}\mathbf{h_2f_1})x_1 + \mathbf{n}_2 \tag{6}$$

where $\mathbf{n}_i$ is the complex addictive white Gaussian noise and $E(\mathbf{n}_i, \mathbf{n}_i^H) = \sigma_i^2 \mathbf{I}_M$. Hence the transmission rate of ST $\rightarrow$ PR and ST $\rightarrow$ SR can be written respectively as

$$R_1 = \frac{1}{2}W \log_2(1 + \frac{P_1|\mathbf{h_1f_1}|^2}{P_2|\mathbf{h_1f_2}|^2 + \sigma_1^2}) \tag{7}$$

$$R_S = R_2 = \frac{1}{2}W \log_2(1 + \frac{P_2|\mathbf{h_2f_2}|^2}{P_1|\mathbf{h_2f_1}|^2 + \sigma_2^2}) \tag{8}$$

subject to the power constraint $P_1||\mathbf{f_1}||^2 + P_2||\mathbf{f_2}||^2 \leq P_S$. Where R_S is the transmission rate of secondary system, after the two phases, the transmission rate of primary system can be written as

$$R_P = \min\{R_1, R_3\} \tag{9}$$

The reason of taking the minimum is that the capacity of the primary system is limited to the worse link between PT $\rightarrow$ ST and ST $\rightarrow$ PR.

Finally, the problem can be concluded that design the beamforming vector $\mathbf{F}$ and the transmission power scaling factors P_1 and P_2 to maximize the transmission rate of secondary system after ST assists the primary system to reach the target rate.

3   Parameters Design Based on ZFBF

We make the beamforming weight vector of one user i be orthogonal to the channel vector of any other user k according to the design of ZFBF to eliminate the interference of other users in the same channels, that is, we select the beamforming weight vector satisfied the condition $\mathbf{h}_i\mathbf{j}_k = 0$, $\forall i \neq k$ [15]. Then we let $\mathbf{H} = [\mathbf{h_1^H} \ \mathbf{h_2^H}]^H$, one easy choice of the beamforming weight matrix $\mathbf{F}$ is the pseudoinverse of the $\mathbf{H}$

$$\mathbf{F} = \mathbf{H}^\dagger = \mathbf{H}^H(\mathbf{H}\mathbf{H}^H)^{-1} \tag{10}$$

where $\mathbf{F} = [\mathbf{f}_1 \ \mathbf{f}_2]$. Bring $\mathbf{f}_1, \mathbf{f}_2$ into (7), (8) respectively, we have

$$R_1 = \frac{1}{2}W \log_2(1 + \frac{P_1}{\sigma_1^2}) \tag{11}$$

$$R_S = R_2 = \frac{1}{2} W \log_2(1 + \frac{P_2}{\sigma_2^2}) \tag{12}$$

subject to $P_1||\mathbf{f}_1||^2 + P_2||\mathbf{f}_2||^2 \leq P_S$.

Next we just design the power scaling factors to maximize the transmission rate of secondary system satisfied the condition that the transmission rate of primary system reach the target. That is, we have

$$\max_{P_1} R_S \tag{13}$$

subject to

$$\begin{cases} R_1 \geq R_T \\ R_3 \geq R_T \\ P_1||\mathbf{f}_1||^2 + P_2||\mathbf{f}_2||^2 = P_S \end{cases} \tag{14}$$

Bring $P_1 = \frac{P_S - P_2||\mathbf{f}_2||^2}{||\mathbf{f}_1||^2}$ into R_1, we can get

$$\frac{1}{2} W \log_2(1 + \frac{P_S - P_2||\mathbf{f}_2||^2}{||\mathbf{f}_1||^2 \sigma_1^2}) \geq R_T \tag{15}$$

$$P_2 \leq \frac{P_S - (2^{\frac{2R_T}{W}} - 1)||\mathbf{f}_1||^2 \sigma_1^2}{||\mathbf{f}_2||^2} \tag{16}$$

It's obvious that R_S monotonically increases with the increase of P_2 according to (14). Thus we choose the maximum value of P_2 as the optimal solution which can be written as

$$P_2^* = \frac{P_S - (2^{\frac{2R_T}{W}} - 1)||\mathbf{f}_1||^2 \sigma_1^2}{||\mathbf{f}_2||^2} \tag{17}$$

Finally, we can get the maximum rate of R_S which can be written as

$$R_{smax} = \frac{1}{2} W \log_2(1 + \frac{P_s - \left(2^{\frac{2R_T}{W}} - 1\right)||\mathbf{f}_1||^2 \sigma_1^2}{||\mathbf{f}_2||^2 \sigma_2^2}) \tag{18}$$

4 Simulation Result

In this section, we analyze the result of the proposed scheme with different parameters. For simplicity, we assume that PT, PR, ST and SR are on a same 2-D coordinate diagram. We let PT, PR, SR lie on the same line. PT and PR are located on (0, 0) and

(1, 0) respectively, ST moved between PT and PR. Define that the distance PT → ST on the coordinate diagram is d, that is, the coordinate of ST is $(d, 0)$. Thus the distance ST → PR is $1 - d$. Assume that the distance between ST and SR is half of that between ST and PR. So we set the coordinate of SR to be $(d, 0.5(1 - d))$. In our simulation, we assume that the path-loss exponent v is -3, the licensed spectrum bandwidth W is 1, all noise variance σ^2 are 1, the power of PT P_T is 8 dB and the power of ST P_S is 10 dB.

Figure 2 shows the transmission rate of the secondary system with various R_T versus the different position of ST when ST equipped with 4 antennas. As is shown, $R_s = 0$ when $R_T = 3$bps/Hz and $d < 0.14$. It indicates that when ST is far away from PR, the SNR of the link ST → PR is bad and R_1 cannot reach the target rate R_T. So the secondary system cannot access the spectrum and have no rate. With ST getting close to PR, the SNR of the link ST → PR is getting better and both R_1 and R_3 can achieve the target rate, then the secondary system can access to the spectrum. And the rate of secondary system is increasing for the SNR of the link ST → SR is getting better. When $d > 0.74$, $R_S = 0$ again since the SNR of the link PR → ST is bad and R_3 cannot achieve target rate R_T. The Fig. 2 also indicates that the access range is smaller when R_T increases to 3.5bps/Hz. That's because it makes high demands of the SNR for both of the link PT → ST and ST → PR. Also the power scaling factor of P_1 increases with the increase of R_T. Obviously, another power scaling factor P_2 will decrease which makes the rate of secondary system decrease correspondingly.

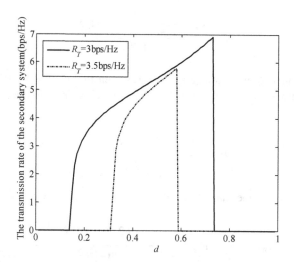

Fig. 2. The rate of secondary system with various R_T versus different position of ST

Figure 3 describes the transmission rate of the secondary system with various M versus different position of ST when $R_T = 3$ bps/Hz. As we can observe from the figure, both of the rate and the access range of the secondary system are increasing with the increase number of ST's antennas. It's because increasing the number of antennas can improve the channel capacity while keeping the SNR and the bandwidth unchanged. That is, the secondary system can reach the R_T with large number of antennas even when the SNR is worse. Then the access range increases to $0.14 < d < 0.74$ and $0 < d < 0.94$ with $M = 4$ and $M = 8$ respectively. Obviously, the rate of the secondary system increases with the capacity of the whole system when the R_T is unchanged.

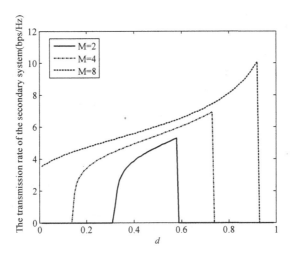

Fig. 3. The rate of secondary system with various M versus different position of ST

In Fig. 4, we set $R_T = 3$bps/Hz and fix ST in the middle of PT and PR, i.e. $d = 0.5$. Figure 4 shows the transmission rate of the secondary system versus the P_S increasing from 5 dB to 25 dB, respectively, with various M. As we can observe, the rate

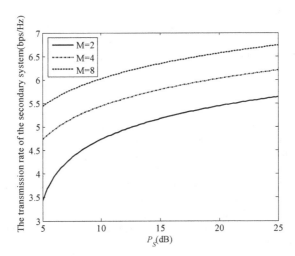

Fig. 4. The rate of secondary system with various M versus different P_S

increases with the increase of P_S. For the position of ST and the value of R_T are fixed, the channel coefficient matrix is considered to be unchanged in our model. Thus the power scaling factor P_1 is unchanged and P_2 increases with the P_S relatively. Also the curve trend corresponds to the relationship between SNR and the channel capacity. Obviously, the rate increase with the M same as Fig. 3 shows.

5   Conclusion

To enhance the spectral utilization, we proposed a cooperative spectrum access scheme in flat fading channel for MIMO cognitive networks where the secondary transmitter ST equipped with multi-antenna. We derived the two-phase transmission protocol in our model to guarantee the service of PUs and the ZFBF to eliminate the co-channel interference between ST $\rightarrow$ PR and ST $\rightarrow$ SR. The results show that the scheme evidently improves the spectral utilization and achieves a win-win result. Note that increasing the antennas can obtain more gain.

Acknowledgment. This work was supported by China National Science Foundation under Grand No. 61402416 and 61303235, Natural Science Foundation of Zhejiang Province under Grant No. LQ14F010003 and LQ14F020005, NSFC-Zhejiang Joint Fund for the Integration of Industrialization and Informatization under grant No. U1509219, Natural Science Foundation of Jiangsu Province under Grant No. BK20140828, the Fundamental Research Funds for the Central Universities under Grant No. DUT16RC(3)045 and the Scientific Foundation for the Returned Overseas Chinese Scholars of State Education Ministry.

References

1. Mitola, J., Maguire Jr., G.: Cognitive radio: making software radios more personal. IEEE Pers. Commun. **6**, 13–18 (1999)
2. Goldsmith, A., Jafar, S.A., Maric, I., Srinivasa, S.: Breaking spectrum gridlock with cognitive radios: an information theoretic perspective. Proc. IEEE **97**, 894–914 (2009)
3. Zou, Y., Yao, Y.D., Zheng, B.: A cooperative sensing based cognitive relay transmission scheme without a dedicated sensing relay channel in cognitive radio networks. IEEE Trans. Sig. Process. **59**, 854–858 (2011)
4. Huang, L.F., Gao, Z.L., Guo, D., et al.: A sensing policy based on the statistical property of licensed channel in cognitive network. Int. J. Internet Protoc. Technol. **5**, 219–229 (2010)
5. Li, Q., Ting, S.H., Pandharipande, A., et al.: Cognitive spectrum sharing with two-way relaying systems. IEEE Trans. Veh. Technol. **60**, 1233–1240 (2011)
6. Han, Y., Pandharipande, A., Ting, S.H.: Cooperative decode-and-forward relaying for secondary spectrum access. IEEE Trans. Wireless Commun. **8**, 4945–4950 (2009)
7. Dai, Z., Liu, J., Long, K.: Cooperative transmissions for secondary spectrum access in cognitive radios. Int. J. Commun. Syst. **27**, 2762–2774 (2014)
8. Lawton, G.: Is MIMO the future of wireless communications? Computer **37**, 20–22 (2004)
9. Goldsmith, A., Jafar, S.A., Jindal, N., et al.: Capacity limits of MIMO channels. IEEE J. Sel. Areas Commun. **21**, 684–702 (2003)

10. Benaya, A.M., Shokair, M., El-Rabaie, E.S., et al.: Optimal power allocation for sensing-based spectrum sharing in MIMO cognitive relay networks. Wireless Pers. Commun. **82**, 2695–2707 (2015)
11. Alsharoa, A., Ghazzai, H., Yaacoub, E., et al.: Joint bandwidth and power allocation for MIMO two-way relays-assisted overlay cognitive radio systems. IEEE Trans. Cogn. Commun. Netw. **1**, 383–393 (2015)
12. Hanif, M.F., Smith, P.J., Taylor, D.P., et al.: MIMO cognitive radios with antenna selection. IEEE Trans. Wireless Commun. **10**, 3688–3699 (2011)
13. Zheng, G., Wong, K.K., Ottersten, B.: Robust cognitive beamforming with bounded channel uncertainties. IEEE Trans. Sig. Process. **57**, 4871–4881 (2009)
14. Kim, H., Kim, J., Yang, S., et al.: An effective MIMO–OFDM system for IEEE 802.22 WRAN channels. IEEE Trans. Circuits Syst. II Express Briefs **55**, 821–825 (2008)
15. Yoo, T., Goldsmith, A.: On the optimality of multiantenna broadcast scheduling using zero-forcing beamforming. IEEE J. Sel. Area. Commun. **24**, 528–541 (2006)

Intelligent Positioning and Navigation Systems

Density-Based Dynamic Revision Path Planning in Urban Area via VANET

Siwei Wu, Demin Li, Guanglin Zhang$^{(\boxtimes)}$, Chang Guo, and Leilei Qi

College of Information Science and Technology,
Engineering Research Center of Digitized Textile and Fashion Technology,
Ministry of Education, Donghua University, Shanghai, China
{wusiwei,guochang}@mail.dhu.edu.cn, {deminli,glzhang}@dhu.edu.cn

Abstract. Dynamic path planning can efficiently enhance the validity and reliability of real-time navigation for vehicles to avoid unexpected traffic congestions and to shorten the whole traveling time. In this paper, we present a dynamic path planning method in which the density of road links is set as the main measurement for traffic status. By analyzing the traveling time and waiting time, we propose a mechanism for vehicles to decide whether to revise the current path. Our method aims at offering a path with smallest time cost for vehicles towards the destination. Furthermore, we give the discussion on time complexity and convergence of our method. Finally, simulation is conducted to evaluate the proposed method.

Keywords: Path planning · Dynamic · Density-based · VANET

1 Introduction

Nowadays, path planning for vehicles plays an essential part in Intelligent Traffic System (ITS). The objective is to provide the drivers with a path leading from source to destination with shortest time or distance. The conventional navigation technology is static, which only takes the length of roads or the historical data of traffic mobility into consideration. This kind of navigational service is launched before the travel starts, so it does not make quick response to traffic congestion and select an alternative route for vehicles. So it is crucial to work out an effective dynamic path planning method for urban transportation scenario.

There are many works on dynamic path planning in traffic scenarios. A partial path planning algorithm is proposed in [1]. Several fixed intermediate destinations were selected between the source and the final destination in order to reduce the re-calculation complexity. The segment route was monitored for vehicles' arrival speed and when necessary, a better route was re-calculated. He et al. presented a skyline for candidate path selection method in paper [3]. It mainly focused on the prediction of traffic condition by a density-speed traffic

© ICST Institute for Computer Sciences, Social Informatics and Telecommunications Engineering 2017
X.-L. Huang (Ed.): MLICOM 2016, LNICST 183, pp. 129–138, 2017.
DOI: 10.1007/978-3-319-52730-7_13

flow model. If the estimated speed for a road link was too low, this road link would be discarded for path planning. Other papers such as [2,4] concerned path planning from a global perspective. They studied how to keep balance of the entire road network to avoid congestion.

All the above works are based on a consumption that the traffic congestion lasts for a rather long time and thus the road segments suffering congestion will not be considered for path planning anymore. In addition, the situation that the alternative route to bypass the congestion may suffer another congestion is not taken into consideration. From a realistic point of view, it makes sense to predict the duration of congestion and the time to bypass the congestion. If the former exceeds the latter, then the vehicle will take the alternative roads, and if the former is less than the latter, the vehicle will just wait in the queue for the smoothness of traffic.

In this paper, we use the real-time traffic information obtained by Vehicular Ad hoc Network (VANET) for dynamic navigation path planning service. We propose a recursive algorithm to continuously revise the route for a vehicle on the purpose of offering a dynamic available path which will take the shortest time to traverse. Moreover, we devise a traveling time and waiting time comparison mechanism, which consider both the congestion duration and alternative candidate path time consumption. This mechanism is in accordance with reality and will improve the reliability for path planning.

The remainder of this paper is organized as follows. System model is presented in Sect. 2. Then the analysis of the proposed problem is illustrated in Sect. 3. The simulation is given in Sect. 4. Finally, Sect. 5 concludes this paper.

2   System Model

2.1   VANET in Urban Area Scenario

VANET in urban area scenario is shown in Fig. 1. Each vehicle is assumed to be equipped with Onboard Unit (ONU). A VANET is a self-constructed network without any host server. All vehicles within one-hop region constitute an independent group. If one vehicle is going to launch communication with one from another group, it should call for a specific routing protocol to transfer data packet in different groups.

To further develop the model of path planning, we first devise a symbolized VANET architecture as (I, R). The set of intersections in the urban road network is denoted as a collection I. Let R be the set of all road links. A road link k is denoted in Fig. 1 as $r_k(t) = \{i_k(x, y), j_k(x, y), v_k(t), \mathbf{d}, w_k(t)\}$, where i and j is the start and end point of the road link. $v_k(t)$ is the traffic flow speed of the road link k. $\mathbf{d}$ is the direction, confining to $\mathbf{d} \in \{i \to j, j \to i\}$. The weight is $w_k(t)$, which is related to the variant density of the road.

The symbols and notations used in this paper is shown in Table 1.

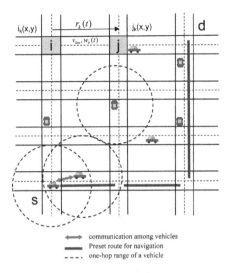

Fig. 1. VANET in urban area scenario

Table 1. Symbols and notations

Symbols	Definitions
$r_k(t)$	A road link k
$|r_k(t)|$	Length of $r_k(t)$
$w_k(t)$	Weight of road link k
$P(t)$	A path in road network
$den_k(t)$	Road density
$N(t)$	The list each vehicle keeps locally
$|N(t)|$	The total number of vehicles
$v_k(t)$	The average spatial velocity of traffic flow on $r_k(t)$
t_{ESTI}	The time candidate path takes to bypass congestion
t_p	The time of congestion to resume smoothless
$p_{cong}(t)$	Set of congested segments

2.2 Dynamic Path Planning Model

In this part, we mainly focus on a dynamic path planning model. Graph Theory is used here to illustrate the model. As is mentioned, I is the set of road network intersections. R is the set of road links. Let $W(t)$ be the set of road link weight, where $W(t) = \{w_k(t)|k \in R\}$. It is related to its real-time traffic density. A navigation path $P(t)$ can be denoted as $P(t) = (r_1(t), r_2(t), \cdots, r_n(t))$ [3], where $r_i \in R$ and $r_n(t) = \{i_n(x,y), j_n(x,y), v_n(t), \mathbf{d}, w_n(t)\}$. Our final objective is to work out a path $P(t)$ along which vehicles can spend the shortest time to the destination.

Referred to the expression in [3], our problem can be formulated as:

Input:

(1) a road network $(I, R, W(t))$
(2) the start point s and the destination point d, where $s, d \in I, s \neq d$

Objective: Find a path $P(t) = (r_1(t), r_2(t), \cdots, r_n(t))$ leading from s to d, which minimize

$$Time = \sum_{i}^{r_i(t) \in P(t)} \frac{|r_i(t)|}{v_i(t)} + \sum_{n} t_n, \tag{1}$$

where $t_n \in Q$, Q is the set of waiting time and $v_i(t)$ is the speed of traffic flow and $|r_i(t)|$ is the length of road link $r_i(t)$.

Subject to: $\sum_{i}^{r_i(t) \in P(t)} |r_i(t)| \leq \Theta$, where Θ is a length constraint.

Our aim is to find a path, which takes the least time to the destination when considering the real time traffic condition. We first find a preset route which consumes minimal time in static situation, which can be obtained by classical algorithms like Dijkstra or A*. In a free traffic situation, a vehicle following the preset route can spend least time to the destination, but if parts of the preset route is congested, our proposed iterative algorithm is used to bypass the congestion.

3   Dynamic Path Planning Solution

3.1   Traffic Information Collection

Since dynamic path planning is based on real-time road condition, it is essential to collect ambient traffic information regularly and efficiently. In vehicular ad-hoc network, this task is accomplished by the vehicle itself. It is assumed that each car maintains a list, in which every entry denotes the parameters of all the other car nodes on the same road link within one-hop region and updates every Δt time. The entry records the vehicle's id, current velocity, road link id it belongs to and time stamp. The standard format is shown in Table 2. As is mentioned above, the list each vehicle keeps can be defined as a set $N(t)$. We define $|N(t)|$ as the total number of vehicles running on the current road link.

Table 2. Beacon message format

Vehicle	Velocity	Road link	Time stamp
ID	$v(t)$	r(i,j)	t

With all these vehicles' information stored, vehicles can get some crucial statistic results such as the mean velocity of vehicles on the road link and its current density. According to the Traffic Flow Theory [5], the space mean speed of traffic flow on road link $r_k(t)$ is denoted as

$$v_k(t) = \frac{|N(t)|}{\sum_{v(t)\in N(t)} \frac{1}{v(t)}}. \tag{2}$$

And density of road link $r_k(t)$ is

$$den_k(t) = \frac{|N(t)|}{|r_k(t)|}. \tag{3}$$

3.2 The Criteria for Revising the Preset Route

When a vehicle launches a request for path planning and navigational service, we assume that with the help of ONU and the static parameters stored, it will first work out an initial path $P_{init} = (r_1, r_2, \cdots, r_n)$ as a preset route leading from the source position to its destination. When the initial route $P_{init} = (r_1, r_2, \cdots, r_n)$ is obtained, it becomes the base to be adjusted and revised with the real-time traffic information. It is assumed that each source node automatically creates a dynamically-updating table, in which there records road links included in P_{init} and the corresponding initial weight and subsequent updated value. The weight of the road links are updated every Δt time. Since road link weight $w(t)$ is a function of the density, it fluctuates all the time. It is impractical to recalculate the route as soon as the weight changes, for in reality a slight fluctuation in road link density may not exert an obvious impact on the real-time traffic condition and thus there is no need to change the current route. So we set a threshold $\alpha(> 0)$ to denote the weight fluctuation. Unless the real-time road link weight fluctuates beyond the threshold, the vehicle keeps moving forward along the preset route. When

$$\frac{w_n(t + \Delta t) - w_n(t)}{w_n(t)} = \frac{w_n(t + \Delta t)}{w_n(t)} - 1 > \alpha, \tag{4}$$

where

$$w_n(t) = \frac{A}{den_k(t)}, \tag{5}$$

A is a non-negative constant, it is necessary to revise the preset route and select a candidate path. By substituting inequality (4) with equation (5), we get

$$\frac{den_k(t)}{den_k(t + \Delta t)} - 1 > \alpha. \tag{6}$$

Since weight $w_n(t)$ is a function of the inverse of density $den_k(t)$, when density increases, its weight should be set lower due to the high possibility of congestion. $w_n(t + \Delta t)$ lower than $w_n(t)$ shows the increased density of the road. When $\frac{w_n(t+\Delta t)}{w_n(t)} - 1 < \alpha$, it means the density of road links fluctuates within a rational range and thus there is no need to launch a path planning service.

3.3   Candidate Path Selection

Once a set of consecutive road links are congested, i.e. the density increases and the corresponding weight declines beyond the threshold α, the candidate path selection algorithm is launched. A general example of the proposed algorithm is explained in Fig. 2(a). When there is a congested segment AB on the preset route $s \rightarrow d$, we just work out a candidate path to bypass the congestion.

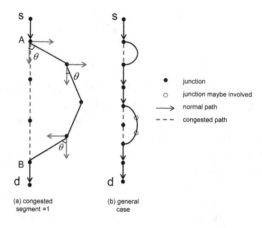

Fig. 2. Example for proposed algorithm

The current route is denoted as $P(t) = \{r_1(t), r_2(t), \cdots, r_n(t)\}$ and the congested road segments is $p_{cong}(t) = \{r_k(t), r_{k+1}(t), \cdots, r_m(t)\}$, which is a subset of $P(t)$. We select the subset $p_{cong}(t)$ as the next miniature of path planning, i.e. $r_k(t)$ is selected as a sub-source and $r_m(t)$ is selected as the sub-destination. A preset route leading from $r_k(t)$ to $r_m(t)$ is first worked out by static methods. The vehicle goes along the preset route and bypass congestion. It is an iterative process and manifests the dynamic feature.

Let $P_{alter}(t) = (r'_k(t), r'_{k+1}(t), \cdots, r'_m(t))$ be the candidate path to bypass the congested roads $p_{cong}(t)$. Then we analyze the time to take the candidate path t_{ESTI} and the time for congestion to consume free t_p.

$$t_{ESTI} = \frac{|r'_k(t)|}{v_{k'}(t)} + \frac{|r'_{k+1}(t)|}{v_{k+1'}(t)} + \cdots + \frac{|r'_m(t)|}{v_{m'}(t)} \tag{7}$$

$$t_p = \frac{|r_k(t)|}{v_k(t)} + \frac{|r_{k+1}(t)|}{v_{k+1}(t)} + \cdots + \frac{|r_m(t)|}{v_m(t)} \tag{8}$$

If $t_{ESTI} < t_p$, it demonstrates that taking the candidate path will help cut the total traveling time towards destination and so the vehicle should take it to bypass the congestion. Otherwise, if $t_{ESTI} > t_p$, the vehicle should wait in the queue for the congestion to resume free. The algorithm is described in pseudocode Algorithms 1 and 2.

Algorithm 1. Dynamic Path Planning

Initialize preset route
$P_{init} = \emptyset$
for $r_n(t_0) \in R$ **do**
  $T_{init}(t_0) = \frac{|r_n(t_0)|}{v_n(t_0)}$
  $P_{init} \leftarrow P_{init} \bigcup \{argmin \sum_i \frac{|r_i(t_0)|}{v_i(t_0)}\}$
  apply Dijkstra to work out P_{init}
end for
procedure DYNAMICREVISING
  $p_{cong}(t) = \emptyset$
  for $i = 0, i <= n, i + +$ **do**

  if $\frac{w_n(t+\Delta t)}{w_n(t)} - 1 > \alpha$ and $w_n(t+\Delta t) < w_n(t)$ **then**
      $p_{cong}(t) \leftarrow p_{cong}(t) \bigcup r_i(t)$
  elsecontinue
  end if
  end for
  launch $FindAlternativePath()$
  compare t_{ESTI} and t_p
  if then$t_{ESTI} > t_p$
    stick to the current path and wait for congestion relieved
  end if
end procedure

In Algorithm 1, a preset route is first worked out by static method. Then the vehicle collects the statistics of road links in the preset route to detect congestion. Once the road weight drops beyond the threshold, Algorithm 2 is launched to find an alternative path. By analyzing the time, whether to take the candidate path or wait in the queue is decided.

Algorithm 2. Find Alternative Path

$source \leftarrow r_k(t), destination \leftarrow r_m(t)$
$P_{init} \leftarrow P_{init} \bigcup \{arg \frac{Q_a(t).length}{|r_a|}\}$
DynamicPathPlanning()
return $P_{alter}(t)$

Algorithm 2 is a subfunction of Algorithm 1. Its function is to find an Alternative path. It also calls Algorithm 2 to form an iterative method.

3.4   Algorithm Analysis

The analysis of time complexity of the algorithm is given as the following theorem.

Theorem 1. The time complexity is in proportion to the number of congested road segments which need revising.

Proof. First we assume that the preset route is free and sound enough to go through without any other effort to revise. In this case the complexity is $O(n)$, where n is the number of road links contained in preset route. Then if there is only one congested segment containing k consecutive road links, the complexity is $O(n) + O(k)$. And further if the there is m congested segments, the complexity is $O(n) + mO(k)$. The deduction shows that the complexity of this algorithm is related to the number of iteration, which is actually the congested road segments to bypass. So the whole time complexity of the algorithm can be denoted as

$$KO(n), \tag{9}$$

where K is the number of iteration and can be regarded as a constant.

Theorem 2. The revision process of a route is convergent as long as the intersection angle between the search direction and the direction pointing from starter to destination or sub-destination is less than 90°.

Proof. Figure 2(b) shows the search direction should be always towards the destination. We define the intersection angle between searching direction and the direction pointing from starter to destination as θ. θ should be less than 90° because based on common sense, drivers never retrace their steps. For further proof, we introduce the theory of gradient descent here to prove the above theorem. Based on the gradient descent, we have the iterative formula $a_{k+1} = a_k + \rho_k s^{-(k)}$, where $s^{-(k)}$ represents the negative gradient direction and ρ_k is the step. In our realistic urban traffic situation, the negative gradient direction $s^{-(k)}$ can be specified as the direction pointing from the starter to the destination or the sub-destination and the step ρ_k end coordinate of one road link for the candidate path search. In addition, a_k stands for the terminal coordinate of road link k and a_{k+1} stands for the successor's terminal coordinate. So the objective function is

$$|a_{k+1} - x_d|, \tag{10}$$

where x_d is the sub-destination's coordinate. During the exploring process, we minimize the objective function, that is, the candidate path is found towards the destination and finally the whole process will be convergent for the destination is locally unique.

4   Numerical Simulation

In this section, we evaluate the performance of the path planning algorithm proposed in Sect. 3. First we construct a virtual urban area road network shown in Fig. 3(a). Fig. 3(b) is the corresponding graph, in which the road length is randomly created by Matlab ranging from 0 to 30.

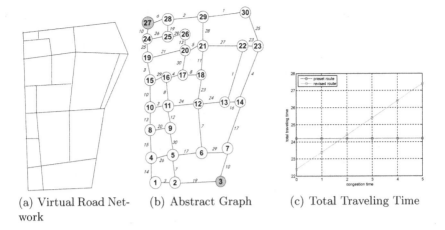

(a) Virtual Road Network

(b) Abstract Graph

(c) Total Traveling Time

Fig. 3. Performance evaluation

We consider the total traveling time from source 3 to destination 27 with the increasing congestion time (recovery time). From the urban area road graph shown in Fig. 3(b), the preset route from node 3 to node 27 is $3 \rightarrow 2 \rightarrow 1 \rightarrow 4 \rightarrow 8 \rightarrow 10 \rightarrow 15 \rightarrow 19 \rightarrow 24 \rightarrow 27$, with the total length of 112. And further we suppose that the velocity of vehicle is 5, so the time to go along the preset route is 22.4. Now the road link $3 \rightarrow 2 \rightarrow 1 \rightarrow 4 \rightarrow 8 \rightarrow 10$ (total length 64) is congested. So we alternatively select $3 \rightarrow 7 \rightarrow 6 \rightarrow 12 \rightarrow 11 \rightarrow 10$ (total length 73) to replace the congested road segments. Results on the total traveling time for vehicles with increasing congestion time is shown in Fig. 3(c).

From Fig. 3(c) we can see that when congestion time is short, sticking to the preset route and wait for congestion recovery is a better choice. When congestion time is long, taking an alternative route to bypass the congestion costs less total traveling time.

5 Conclusion

In this paper, we propose a dynamic path planning mechanism in urban area traffic scenario. Our method is based on the principal of consuming the least time to travel to the destination rather than the shortest distance in the time-sensitive traffic situation. If congestion emerges, the source node would be notified ahead of time and launch alternative path selection algorithm to bypass the congestion. In addition, we devise a time-comparison mechanism to decide whether to take the alternative route or wait for the congestion recovery. Further, we discuss the time complexity and convergence of our proposed algorithm and simulation results are given, which shows that the revision-based dynamic path planning fits the traffic reality better. Future work will concentrate on the study of routing protocols in VANET for this dynamic path planning method.

Acknowledgments. This work is supported by the NSF of China under Grant No. 61301118; the Innovation Program of Shanghai Municipal Education Commission under Grant No. 14YZ130; the International S&T Cooperation Program of Shanghai Science and Technology Commission under Grant No. 15220710600; and the Fundamental Research Funds for the Central Universities under Grant No. 16D210403.

References

1. Xu, J., Guo, L., Ding, Z., Sun, X., Liu, C.: Traffic aware route planning in dynamic road networks. In: Lee, S., Peng, Z., Zhou, X., Moon, Y.-S., Unland, R., Yoo, J. (eds.) DASFAA 2012. LNCS, vol. 7238, pp. 576–591. Springer, Heidelberg (2012). doi:10.1007/978-3-642-29038-1_41
2. Wang, M., Shan, H., Lu, R., et al.: Real-time path planning based on hybrid-VANET-enhanced transportation system. IEEE Trans. Veh. Technol. **64**(5), 1664–1678 (2015)
3. He, Z., Cao, J., Li, T.: MICE: a real-time traffic estimation based vehicular path planning solution using VANETs. In: International Conference on Connected Vehicles and Expo, pp. 172–178. IEEE (2012)
4. Rajabi-Bahaabadi, M., Shariat-Mohaymany, A., Babaei, M., et al.: Multi-objective path finding in stochastic time-dependent road networks using non-dominated sorting genetic algorithm. Expert Syst. Appl. **42**(12), 5056–5064 (2015)
5. Lieu, H.: Traffic-flow theory. Public Roads **62**(1–2), 5–7 (1999)

An Intelligent Mobile Crowdsourcing Information Notification System for Developing Countries

Arun Singh[1], YueXin (Sophia) Li[2], Yu Sun[1(✉)], and Qingquan Sun[3]

[1] Department of Computer Science,
California State Polytechnic University, Pomona, CA 91768, USA
{alsingh,yusun}@cpp.edu
[2] Branksome Hall, Toronto, ON M4W 1N4, Canada
sli2@branksome.on.ca
[3] School of Computer Science and Engineering,
California State University San Bernardino, San Bernardino, CA 92407, USA
qsun@csusb.edu

Abstract. Crowdsourcing is an important computing technique that taps into the collective intelligence of the public at large to complete business-related tasks and solve many real-time problems. It is changing the way we work, hire, research, make and market. Many developing nations are trying to take advantage of crowdsourcing for information notification to make cost effective system, like real-time transit system, disaster notification system and other services which are available to the masses. However, many of them are still not able to completely benefit from it compared to developed nations. In this paper, we have identified a series of limitations of using crowdsourcing for information gathering and providing real-time notification in developing countries due to their unstable electronic communication infrastructure, their lack of contribution, lack of crowdsource (participating people), less exposure to English language, and unawareness of crowdsourcing. We proposed, and demonstrated, a solution to overcome these limitations by developing a prototype which uses SMS as a reliable method for providing real-time notification and information gathering. Our prototype uses prediction algorithms to fill the gaps in real-time notification. It also uses the prediction of a user's behavior to provide a better reward and motivational platform, as well as good usability.

Keywords: Mobile crowdsourcing · Machine learning · Information prediction · Reliable communication

1 Introduction

Notification systems have been a crucial part of any system in recent decades. A notification system is a set of protocols and procedures that can involve both human and computer or mobile components. The purpose of these systems is to generate and send timely messages to a person or group of people. Simple notification systems use a single means of communication, such as an email or text message [1]. More complex

© ICST Institute for Computer Sciences, Social Informatics and Telecommunications Engineering 2017
X.-L. Huang (Ed.): MLICOM 2016, LNICST 183, pp. 139–149, 2017.
DOI: 10.1007/978-3-319-52730-7_14

information notification systems involve the processing of bulk information and providing some meaningful result from processed information. For example, a transit system gets information from different sources to notify user about the real-time schedules. Data reins as the great equalizer and democratizer in an era of the global economies. Vast amounts of data in developed societies are being handled by the leading tech companies such as Google [8] and Facebook [9]. Crowdsourcing can be seen as a catalyst for global innovation and is something businesses should keep an eye on going forward [2, 3]. Figure 1 shows the main participants and abstract dataflow for crowdsourcing. The main participants include: a contributor, which is the crowd, a consumer which could be anyone from a single business to the whole world, an administrator/business that manages all of the activity and an operator who does monitoring and management. One of the great examples is Wikipedia, which has opened a number of possibilities for crowd sourcing learning resources [4].

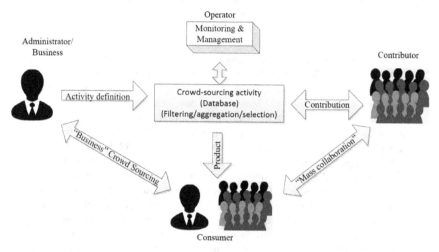

Fig. 1. Crowdsourcing main participant and abstract dataflow (adapted from [25])

There are a number of successful applications where crowdsourcing is used for information notification; one of popular apps is Waze [5]. Its mapping and traffic information are built from 70,000 volunteer map editors and more than 15 million active users. These users contribute their live driving data by default so others can benefit by seeing how fast they are going. Given the increasing number of people who carry smartphones, it seems likely other services could be built leveraging willing users who contribute a small portion of the data from their travels. That data can be mapped and analyzed for the common good [6]. While the west is doing its best to tackle the immense flow of data, the case is very different in developing countries where entire societies are cut off from this transfer of data. Action needs to be taken for the global disparity related to the data that fuels crowdsourcing.

Open Problems: Numerous technical and non-technical barriers exist in developing countries that limit the effective usage of crowdsourcing platforms.

One such barrier is the lack Internet availability, which is typically needed for mobile crowdsourcing. Public Internet services may be available but that not conducive to real-time information acquisition and dissemination. Other barriers include gaps between real time notifications, understanding of user behavior due to technology limitations, the lack of skill in understand complex English instructions, minimum awareness of crowdsourcing, and motivations to use crowdsourcing platform.

Proposed Solution → An intelligent mobile-based information notification system applicable for developing countries. In order to address the problem, our proposed solutions are using SMS as a reliable method for providing real-time notification and information gathering, using prediction algorithms to fill the gaps in real-time notification, using the prediction of user's behavior to provide better a reward and motivational platform.

The solution presented in this paper focuses on stable communication protocol between mobile devices and server, prediction algorithm to encourage crowd to use crowdsourcing applications in developing countries. The main contribution of this paper is to provide the practical crowdsourcing solution for developing countries with a completely intelligent mobile-based crowdsourcing platform.

The remainder of this paper is organized as follows: in Sect. 2, we provide a motivating example for this paper and in Sect. 3 we list the challenges that are faced to institute a crowdsourcing application in developing countries. In Sect. 4, we present our solution to address these challenges. We analyze the related work in Sect. 5 and conclude this paper as well as provide some insight into future work and scope of crowdsourcing in developing countries in Sect. 6.

2 Motivating Example – Public Bus Notification System

Buses are indispensible part of the public transport system. Bus transport services help in reducing private car usage and fuel consumption. Public transport bus notification services are generally based on predictable operation of transit buses along a route arriving at the specified bus stops according to a published public transport timetable. Many public bus services are run to a precise timetable giving specific departure and arrival times. These are often difficult to maintain in the event of traffic congestion, breakdowns, on/off bus incidents, road blockages or bad weather. Predictable effects such as morning and evening rush hour traffic are often accounted for in timetables using past experience of the effects. This prevents the drafting of a clock face time schedule where bus arrival is predictable at any time throughout the day. Providing precise arrival times of buses will advance user experience and attract more users to use buses. Nowadays, most of the public bus transport agencies provide their bus schedules on web but the bus schedule information are not real-time.

Increasingly, technology is being used to improve the information provided to bus users [10]. Vehicle tracking technologies are used to assist with scheduling and to achieve real time integration with passenger information systems. These information systems display service information at stops, inside buses, and to waiting passengers through personal mobile devices or text messaging, but this type of systems are costly. There are a few crowdsourcing applications such as Moovit [11] and Tiramisu [12], which collect bus passengers' real-time movements and share them with other users to track arriving bus. However, the existing crowdsourced applications does not work very well in developing countries because of various challenges listed in next section.

3 Challenges

The United Nations is using digital media and mobile phone technology to enable people from across the world to take part in setting the next generation of anti-poverty goals. Crowdsourcing applications are changing the way we work, hire, research, make and market. Many developing nations are trying to take advantage of crowdsourcing for information notification to make cost effective systems like real-time transit system, disaster notification system and other services which are to be available to the masses. Despite the fact that crowdsourcing is a well-known technique for information collection and providing real-time notifications, developing countries are still not able to completely benefit from it compared to developed nations. There are many limitations to using crowdsourcing for information gathering and providing real-time notification in developing countries.

3.1 The Lack of a Reliable Mobile Internet Infrastructure

Developed countries have decent Internet infrastructure, which can be used in combination with GPS technology to provide real-time notifications that are commonly used by many crowdsourcing applications [13]. For instance, in bus tracking systems, mobile apps can provide riders' locations when they are in transit by using their mobile data and GPS. Crowdsourcing systems can utilize this information to provide bus arrival times. On the other hand, however, developing countries do not always have the reliable Internet infrastructure which restricts the usage of the traditional crowdsourcing solutions.

3.2 The Lack of Contribution Motivation

Users participation drives the success for any online crowdsourcing applications. Participation becomes more difficult in developing countries where people have less knowledge of crowdsourcing. It is a big challenge to understand what motivates people to participate in online crowdsourcing platform. Even though it is easy to introduce a platform but how to attract a large number of crowds especially in this competitive market requires close study. For bus tracking system we can undertake people who use bus tracking systems to find the bus arrival time. However, in order to provide real-time bus arrival information, we need riders inside bus to use app to provide real-time notification and updates.

3.3   The Lack of Active Crowd Source

When working with real-time mobile crowdsourced based applications, it is highly probable to see gaps in real-time notification because of lack of crowd source (participating people). It is a challenge to retain existing users (crowd) if system does not provide real-time information. To give an example in bus tracking system, there could be possible time or day when there are none or only a few people travelling in bus and proving tracking information and system would not sufficient data to provide bus arrival times.

4   Solution

We have implemented smart mobile crowdsourcing platform for developing countries which includes reliable communication infrastructure, reward system for motivating users and intelligent system which can predict information and interested rewards.

4.1   Reliable Communication Infrastructure

We know with Internet we can easily develop any crowdsourcing application because of the ability to communicate information (like device location and small data) from crowd (people) to system and vice-versa. It would take decades to replicate a developed country's equivalent infrastructure in developing countries. Based on our reviews and our experience of using Short Message Service (SMS), we know SMS is still widely used and remains one of the most reliable channels of communication so far [14]. We have designed and implemented a SMS protocol to send data over mobile networks. SMS can work in combination with other methods of communication in areas of developing countries that have fractional infrastructure for Internet. Figure 2 is the high-level flow of a crowd-sourced bus tracking system where users with mobile apps are requesting and providing tracking information.

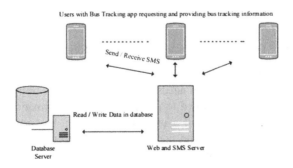

Fig. 2. Bus tracking system in developing countries based on SMS protocol

In our implementation we identify each rider by their phone number used in mobile phone. To get the bus arrival information, registered rider uses our mobile app to select the bus number, mobile app internally checks for internet connectivity and if internet is not available then it forms a SMS which is combination of request type, user location and bus number and send it to server. Server process the request based on request type and send the response back in the form of SMS. Bus arrival response SMS include response type, bus number, bus arrival time, distance of user from nearest bus stop and some advertisement text. When user's mobile phone receives the SMS, mobile application recognizes and reads it then displays message in user interface based on response type. If rider's mobile device has internet connection then mobile application can directly make web API call to communicate with server instead of communicating via SMS.

4.2 Reward System for Contribution Motivation

Users' participation in mobile crowdsourcing platforms is vital as the success of platforms largely depends on the presence of their members. Researchers found that the most frequently mentioned motives of users participating in crowdsourcing are: money, altruism, fun, reputation/attention, and learning. Many scholars of crowdsourcing suggest that there are both intrinsic and extrinsic motivations that cause people to contribute to crowdsourced tasks and that these factors influence different types of contributors [7]. We implemented a reward-based system in the crowdsourcing application to motivate people in developing countries. This reward should be mapped to financial incentives like bonus, coupons, deals with lower cost, free product, free service etc. The reward value should be divided based on type of activity. For example, reward value should be in the descending order of: user referral, users helping to provide notification, users using applications for some need etc.

To make reward system more interesting, we have managed to get commercial companies to participate in offering the rewards. For example, when users request for bus arrival time we would process the request, but along with bus arrival response we would add advertisement data to display on user's screen.

4.3 Prediction of the Information and Interested Rewards

The lack of crowdsource or participating people is one of the biggest factors for failure of a crowdsourcing application that needs real-time information from participants. We cannot completely rely on the crowd to provide real-time information in to minimize any response gaps. For example, if a user is requesting a bus arrival time and the system does not have any real-time update, then the system would not be able to provide a response. This limitation would discourage the crowd from using the application. This would again decrease real-time notifications. Machine learning algorithms work best to predict information based on historical data. Here we implemented information prediction based on the historical data using Naïve Bayes [26] algorithm. For an example user requesting Bus No "A1" arrival time, if system does not find real-time information of Bus No "A1" then machine learning algorithm would predict the Bus No "A1" arrival based on historical Bus No "A1" historical arrival data. It is really important to pick the correct machine learning algorithm that fits this particular need. The accuracy of algorithm should be known before it is put in live application. Figure 3 shows the basic workflow of the request and response to understand the bus tracking example. In the figure we have request and response on right side which is sent to the tracking request web API which has logic inside to handle the response.

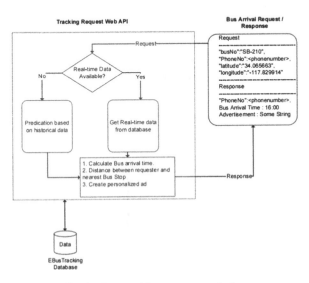

Fig. 3. Bus tracking request work flow

The system also predict user's interested reward category for the reward based system. For example, in the bus tracking application, we can categorize riders based on their ride actions such as bus stops, riding days, riding times, information from user profiles captured during registration. If rider is requesting bus arrival times for a bus stop which is near a college, and request is within college working hours, and the rider's age is 18, then it is highly possible the rider is student. Figure 4 shows the machine learning basic workflow to predict the user's category for bus tracking system.

To find the optimal training approach for the user's category prediction, a group of sample data containing 220 instances is processed using Weka. In total, 35 different classifiers are tested under the same qualification of 10-fold cross-validation, and the percentage of correctly classified instances is recorded as the accuracy of the classifier. The results are summarized in Fig. 5.

The black bars represent the classifiers that yield relatively high accuracies, while the light gray bars represent the classifiers that yield relatively low accuracies. From the diagram, LogiBoost, Random Committee and Random Forest correctly classify the largest amount of instances with an accuracy of 80%, while nine other classifiers, such as Naïve Bayes, KStar, Multi Scheme and several Meta Classifiers, have the lowest

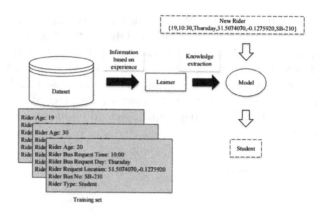

Fig. 4. Predict user's category for interested reward

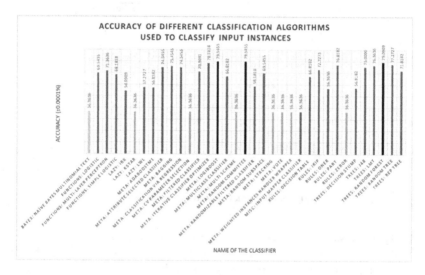

Fig. 5. The performance of different machine learning algorithms

accuracy of 36.3636%. Noticeably, the average accuracies of the Functions-based classifiers and the Trees-based classifiers are relatively higher than other types. The main reason is that Random Forest is a highly suitable and accurate algorithm to predict the user's profile as it has an internal system to get an unbiased estimate of test error and a useful tool called proximities [19]. In random forest, as many classification trees as needed based on the number of data are grown which each gives a classification. After each tree "votes" for a class, the forest chooses the classification having the most votes. During this process, out-of-bag error for each point is recorded and averaged over the forest. Meanwhile, proximities are computed for each pair of cases and normalized by dividing by the number of trees. Due to these features, random forest is able to compute prototypes of high accuracy, 79.0909% in this case, to model the relation between the variables and the classification efficiently.

On the other hand, Naïve Bayes Multinomial Text is not the classifier that best fits this set of data due to the severe assumptions it makes that would adversely affect the quality of its results. As designed, it can model the distribution of words in a document as a multinomial with high efficiency [20]. However, its inherent systemic errors lead to the low accuracy of the classification model it gives. One systemic error is the production of biased decision boundary weights due to skewed data. As there are more training examples for the class of student than other classes, the data is skewed to one direction and therefore the accuracy of Naïve Bayes decreases. Moreover, the production of different magnitude classification weights based on the independence assumption is another problem that impedes the accurate modeling.

However, there are classification models simply inapplicable to this particular training set such as SMO, Additive Regression, and Simple Linear Regression. The primary underlying reason is the mismatch between how the classifier works and the nature of the training data.

5 Related Work

There are quite a few initiatives by organizations and researchers to enable crowdsourcing as a platform to solve developing countries' needs.

NextDrop [15], a social enterprise which is streamlining urban water collection in India. The enterprise collects and shares water delivery information with city residents and water utilities. In this way, efficiency and transparency are improved upon. NextDrop started as a pilot in Hubli-Dharwad. These twin cities in the Indian state of Karnataka, which have seen rapid urban growth in recent years. There have been decent growth like shopping malls and American fast food chains, but access to water is still a challenge. It is therefore unsurprising that NextDrop's Smart Water Supply Message Service has thousands of household subscribers paying a monthly nominal fee to receive advance water alerts. NextDrop's messaging system uses SMS to notify subscribers about when they will get water, when there is an interruption, when pipe damage is likely to affect them, and when someone in the community has water updates to share. However, they do not use machine learning to predict the information.

Medic Mobile [16] uses mobile technology to create connected, coordinated health systems. Medic Mobile is specifically designed for health worker and medical system to offer a free, scalable software toolkit that combines messaging, data collection, and analytics. It is intended for hard-to-reach areas and supports may languages. It works with or without internet connectivity, locally or in the cloud. The tools run on basic phones, smartphones, tablets, and computers. In six months, their pilot program in Malawi had saved hospital staff 1,200 h of follow-up time and over US$3,000 in motorbike fuel. More than 100 patients started tuberculosis treatment after their symptoms were noticed by community health workers and reported by text message. Medic Mobile uses crowd-sources translation, categorization and geo-tagging which helps to create quick reports for first responders in emergency situations [17]. Compared with our solution, Medic Mobile did not have intelligent reward based system and real-time information prediction.

6   Conclusion and Future Work

In this paper, we have designed and developed an intelligent method to implement crowdsourcing applications in developing countries. A series of solutions have been proposed to address the challenges we identified. Firstly, we chose the use of SMS as a reliable communication channel that is not dependent on the Internet for mobile crowdsourcing communication. Secondly, we proposed a motivating reward system which predicts the rewards that interest our crowd. Thirdly, we performed predications based on historical data by using relevant machine learning algorithms to fill the gaps in real-time information. The goal of this study helps researchers, nonprofit organizations and commercial companies to plan better crowdsourcing applications in developing countries.

The research presented in this thesis seems to have raised more questions that it has answered. There are several lines of research arising from this work which should be pursued. Firstly, we know the SMS is a reliable communication channel but there is need for dynamic thresholding to minimize the SMS traffic when it is used for crowdsourcing. The second line of research is designing one crowdsourcing application to solve multiple needs in developing countries. This would provide us more information for better information prediction (e.g. multiple user actions would provide better prediction to classify users).

References

1. What is a notification system. http://www.wisegeek.com/what-is-a-notification-system.htm
2. Brabham, D.: Crowdsourcing as a model for problem solving: an introduction and cases. Convergence **14**(1), 75–90 (2008)
3. Why crowdsourcing is the next cloud computing. http://blog.innocentive.com/2013/10/14/why-crowdsourcing-is-the-next-cloud-computing
4. Wikipedia. https://en.wikipedia.org
5. Waze. https://www.waze.com
6. After Waze, What else can mobile crowdsourcing do - Liz Gannes. http://allthingsd.com/20130719/after-waze-what-else-can-mobile-crowdsourcing-do
7. Hossain, M.: Users' motivation to participate in online crowdsourcing platforms. In: 2012 International Conference on Innovation Management and Technology Research (ICIMTR), Malacca, pp. 310–315 (2012)
8. Google. https://www.google.com
9. Facebook. https://www.facebook.com
10. Camacho, T., Foth, M., Rakotonirainy, A.: Pervasive technology and public transport: Opportunities beyond telematics. IEEE Pervasive Comput. **12**(1), 18–25 (2013)
11. Moovit. http://moovitapp.com
12. Tiramisu. http://www.tiramisutransit.com
13. Howell, J., Frolik, J.: An internet-based, inverse-GPS system for monitoring and tracking mobile aquatic sensors. In: Proceedings of IEEE Sensors 2002, vol. 2, pp. 1734–1739 (2002). doi:10.1109/ICSENS.2002.1037386
14. Why are people still using SMS in 2015. http://thenextweb.com/future-of-communications/2015/02/16/people-still-using-sms-2015/#gref?

15. NextDrop. https://nextdrop.co
16. Medic mobile. http://medicmobile.org
17. Medic mobile. http://skoll.org/organization/medic-mobile
18. Main participants and abstract data flow. https://www.horizon.ac.uk/wp-content/uploads/2015/01/Crowd-sourcing-images-700x539.jpg
19. Additive model, 2 November 2009. doi:10.1007/0-387-33960-4_5
20. Breiman, L., Cutler, A.: Random forests Leo Breiman and Adele Cutler. http://www.stat.berkeley.edu/~breiman/RandomForests/cc_home.htm#workings

Support Vector Machine Based Range-Free Localization Algorithm in Wireless Sensor Network

Tao Tang[1(✉)], Haicheng Liu[1], Haiyan Song[1], and Bao Peng[2]

[1] College of Electrical and Information Engineering,
Heilongjiang Institute of Technology, Harbin 150070, Heilongjiang, China
Tt8854@126.com
[2] School of Electronic and Communication,
Shenzhen Institute of Information Technology, Shenzhen 518172, China

Abstract. Localization method is critical issues in Wireless Sensor Network (WSN) system. The existing node localization algorithms, especially range-based algorithms, did not consider the distances measured error and this may result in severe location errors that degrade the WSN performance. In this paper, a new algorithm called Support Vector Machine based Range-free localization (RFSVM) algorithm in WSN is proposed. This algorithm introduced a new matrix called transmit matrix which maps the relationship between the hops and distance. And use the SVM model to estimate the position of unknown nodes. This algorithm does not need any addition hardware, and the experiments shows that it can lead to the localization accuracy character good enough.

Keywords: Wireless Sensor Networks · Localization · Range-free · Support Vector Machine

1 Introduction

One of the critical issues in Wireless Sensor Network (WSN) research is to determine the physical positions of nodes [1]. This is because: information from the sensors is useful only if node location information is also available; additionally, some routing protocols use position to determine viable routes [2, 3]. The process of determining the position of WSN nodes is known as localization. The objective of localization is to determine the virtual or physical coordinates of each node in the network [4]. Along with the employing in military activities such as reconnaissance, surveillance [5], and target acquisition [6], environmental activities [7], or civil engineering such as structural health measurement [8], the localization system becomes a significant base in WSN.

A simple way to ensure this is to equip every node in the WSN with GPS. Due to economical issues, only a small subset of the network can affordably be equipped with GPS; such nodes are called anchor nodes, or beacons. An automatic localization process is required for the rest of the nodes (unknown nodes) in the network, using the anchor nodes as reference points. This process is commonly known as location

© ICST Institute for Computer Sciences, Social Informatics and Telecommunications Engineering 2017
X.-L. Huang (Ed.): MLICOM 2016, LNICST 183, pp. 150–158, 2017.
DOI: 10.1007/978-3-319-52730-7_15

discovery (LD) [9]. The existing node localization algorithms can be divided into Range-based and Range-free algorithms.

Range-based employed distance measurements between sensor nodes through a received signal strength indicator (RSSI), time of arrival (TOA), or time difference of arrival (TDOA), the angle can be measured by the angle of arrival (AOA) [10] and et al. However, the distances measured among the nodes in a WSN usually contain errors. These errors range from slight errors to large ones, and are difficult to characterize by a standard model such as Gaussian. These distance measurement errors may result in severe location errors that degrade the WSN performance, and therefore should be taken into account during the LD. Range-free algorithms do not need absolute range information. Typical range-free algorithms including centroid, DV-Hop [11], amorphous [12], APIT [13], etc. The authors in [14] proposed a coarse grained range-free algorithm to lower the uncertainty of nodes positions using radio connectivity constraints. In [15], the authors used geometry method to determine the sensor node location based on the cross point of the two chords in a circle. The range-free algorithms are more economical, cost-effective, and feasible for the large-scale WSN. However, the Range-free always have accuracy problem.

In this paper, a new algorithm called Support Vector Machine based Range-free localization (RFSVM) algorithm in WSN is proposed. This algorithm first introduced a new matrix called transmit matrix which maps the relationship between the hops and distance. Then use the transmit matrix to calculate the training vector data and test vector. Finally, establish the SVM model and estimate the position of unknown nodes.

2 Network Model and Assumptions

Suppose that we have $M + N$ nodes, $\{n_1, n_2, \cdots, n_{M-1}, n_M, n_{M+1}, \cdots, n_{M+N}\}$, random deployed in a two-dimension space $[0, D] \times [0, D]$. Among them, M nodes $\{n_1, n_2, \cdots, n_M\}$ are anchor nodes, the locations of which are supposed to be known. The anchors are randomly static deployed in WSN with density ρ_L. Other nodes $\{n_{M+1}, n_{M+2}, \cdots, n_{M+N}\}$ are regarded as unknown nodes, and they are randomly deployed with a density $\rho_S, \rho_S \gg \rho_L$. We assume the communication radius of each node is R.

3 The Design of RFSVM Algorithm

The localization process of RFSVM algorithm can be divided into three steps, which are transmit matrix obtained, SVM [16] model build and localization. The WSN is initialized and transmit matrix is got in the first step. Then, the localization system get the SVM model through the training vector date which is calculated by the transmit matrix. Finally, the unknown nodes justify which area is it in using SVM method. The whole process of RFSVM algorithm is shown in Fig. 1.

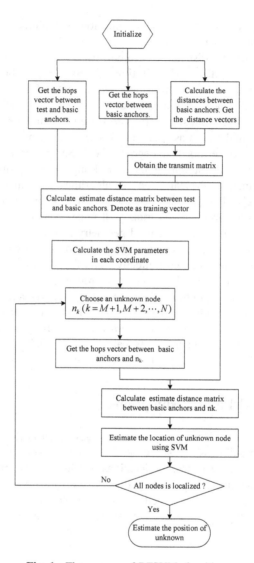

Fig. 1. The process of RFSVM algorithm

3.1   Transmit Matrix Obtained

First of all, the anchors are separated into two parts: basic anchor $n_i(i = 1, 2, \cdots, K)$ and test anchor $n_j(j = K + 1, K + 2, \cdots, M)$. Let **H** be the hops vector between the ith basic anchor node and the other basic anchor nodes, therefore

$$\mathbf{H} = \begin{bmatrix} \mathbf{h}_1 \\ \mathbf{h}_2 \\ \vdots \\ \mathbf{h}_K \end{bmatrix} = \begin{bmatrix} h_{1,1} & h_{2,1} & \cdots & h_{K,1} \\ h_{1,2} & h_{2,2} & \cdots & h_{K,2} \\ \vdots & \vdots & \ddots & \vdots \\ h_{1,K} & h_{2,K} & \cdots & h_{K,K} \end{bmatrix} \in Z^{K \times K}$$

The distance vector between them can be calculated as

$$\mathbf{D} = \begin{bmatrix} d_{1,1} & d_{2,1} & \cdots & d_{K,1} \\ d_{1,2} & d_{2,2} & \cdots & d_{K,2} \\ \vdots & \vdots & \ddots & \vdots \\ d_{1,K} & d_{2,K} & \cdots & d_{K,K} \end{bmatrix} \in Z^{K \times K}$$

Then, let us think about a linear matrix called transmit matrix $\mathbf{T}$ which optimally maps the hops vector $\mathbf{H}$ to the geographical distance $\mathbf{D}$,

$$\mathbf{T} = \begin{bmatrix} \mathbf{t}_1 \\ \mathbf{t}_2 \\ \vdots \\ \mathbf{t}_k \end{bmatrix} = \begin{bmatrix} t_{1,1} & t_{2,1} & \cdots & t_{K,1} \\ t_{1,2} & t_{2,2} & \cdots & t_{K,2} \\ \vdots & \vdots & \ddots & \vdots \\ t_{1,K} & t_{2,K} & \cdots & t_{K,K} \end{bmatrix} \in Z^{K \times K}$$

where $t_{i,j}$ represents the effect of proximity to the jth basic anchor node on the geographic distance to the ith basic anchor node. We derive the transition by minimizing the following square error for $i = 1, 2, \cdots, K$

$$e_i = \sum_{k=1}^{K} (d_{i,k} - \mathbf{t}_i \mathbf{h}_k)^2 = \left\| \mathbf{d}_i^{\mathrm{T}} - \mathbf{t}_i \mathbf{H} \right\| \tag{1}$$

Finally, our goal is to minimize (1). Therefore, the transmit matrix $\mathbf{T}$ could be obtained through this equals.

3.2   SVM Model Constitution

In this step, each dimension of WSN coordinate value is separated into M pieces for M classes. So, in the two-dimension space $[0, D] \times [0, D]$, we have M classes $\{cx_1, cx_2, \cdots, cx_M\}$ in x dimension and M classes $\{cy_1, cy_2, \cdots, cy_M\}$ in y dimension. Then, each anchor n_j $(j = K+1, K+2, \cdots, M)$ computed the distance to all other anchor nodes n_i $(i = 1, 2, \cdots, K)$ using the transmit matrix $\mathbf{T}$. Supposed that the hops between n_j and n_i is h_{ij}, therefore

$$\mathbf{H} = \begin{bmatrix} h_{K+1,1} & h_{K+2,1} & \cdots & h_{M,1} \\ h_{K+1,2} & h_{K+2,2} & \cdots & h_{M,2} \\ \vdots & \vdots & \ddots & \vdots \\ h_{K+1,K} & h_{K+2,K} & \cdots & h_{M,K} \end{bmatrix} \in Z^{K \times (M-K)}$$

And the estimate distance matrix between n_j and n_i is

$$\mathbf{D}' = \mathbf{T} \times \mathbf{H} = \begin{bmatrix} d'_{K+1,1} & d'_{K+2,1} & \cdots & d'_{M,1} \\ d'_{K+1,2} & d'_{K+2,2} & \cdots & d'_{M,2} \\ \vdots & \vdots & \ddots & \vdots \\ d'_{K+1,K} & d'_{K+2,K} & \cdots & d'_{M,K} \end{bmatrix} = \begin{bmatrix} \mathbf{D}'_{K+1} & \mathbf{D}'_{K+2} & \cdots & \mathbf{D}'_M \end{bmatrix} \in Z^{K \times (M-K)}$$

(3)

Where, $\mathbf{D}'_j = \begin{bmatrix} d'_{j1} & d'_{j2} & \cdots & d'_{jk} \end{bmatrix}^{\mathrm{T}}$ $(j = K+1, K+2, \cdots, M)$. Thus, a distance vector $\begin{bmatrix} d'_{j1}, d'_{j2}, \cdots d'_{jK} \end{bmatrix}$ is constructed in each anchor n_j, d'_{ji} distance between test anchor n_j and basic anchor n_i.

Because of the position acknowledgement in test anchor n_j, the location classes of it known as (cx_j, cy_j). Then, transform the anchor nodes n_j distance vector $\begin{bmatrix} d'_{j1}, d'_{j2}, \cdots, d'_{jK} \end{bmatrix}$ into the training data which are $\left\{ \begin{bmatrix} d'_{j1}, d'_{j2}, \cdots d'_{jK} \end{bmatrix}, cx_j \right\}$ in x dimension and $\left\{ \begin{bmatrix} d'_{j1}, d'_{j2}, \cdots d'_{jK} \end{bmatrix}, cx_j \right\}$ in y dimension. Therefore, two SVMs will be built with a Gauss kernel function as formula (4)

$$K(x, x_i) = \exp\{-\frac{|x - x_i|^2}{\sigma^2}\}$$

(4)

In each SVM, arbitrary test data x can be classified by

$$f(x) = sgn\{\sum_{i=1}^{l} \alpha_i^* y_i K(x_i, x) + b^*\}$$

(5)

Where x_i is support vector, $y_i \in \{-1, 1\}$ is class label corresponded to x_i, $K(x, x_i)$ is kernel function, α_i^* is Lagrange multiplier corresponded to x_i, b^* is classification threshold value and sgn is symbol function.

Through calculating the training data, support vector x_i, Lagrange multiplier α_i^*; and classification threshold value b^* are achieved for two SVMs respectively.

3.3   Localization

Like step 2, first we estimate the distance between the unknown nodes $n_k(k = M+1, M+2, \cdots, N)$ and the basic anchors $n_i(i = 1, 2, \cdots, K)$. Supposed that the hops between n_k and n_i is h_{ik}, therefore

$$\mathbf{H} = \begin{bmatrix} \mathbf{H}_{M+1} & \mathbf{H}_{M+2} & \cdots & \mathbf{H}_{M+N} \end{bmatrix} \in Z^{K \times N} \tag{6}$$

And the estimate distance matrix between $n_k S$ and $n_i S$ is

$$\mathbf{D}' = \mathbf{T} \times \mathbf{H} = \begin{bmatrix} d'_{M+1,1} & d'_{M+2,1} & \cdots & d'_{M+N,1} \\ d'_{M+1,2} & d'_{M+2,2} & \cdots & d'_{M+N,2} \\ \vdots & \vdots & \ddots & \vdots \\ d'_{M+1,K} & d'_{M+2,K} & \cdots & d'_{M+N,K} \end{bmatrix} = \begin{bmatrix} \mathbf{D}'_{M+1} & \mathbf{D}'_{M+2} & \cdots & \mathbf{D}'_{M+N} \end{bmatrix} \in Z^{K \times N} \tag{7}$$

Where, $\mathbf{D}'_k = \begin{bmatrix} d'_{k1} & d'_{k2} & \cdots & d'_{kK} \end{bmatrix}^{\mathrm{T}} (k = M+1, M+2, \cdots, M+N)$.

Thus, the test vector date of n_k can be denoted as $\left\{ \begin{bmatrix} d'_{k1}, d'_{k2}, \cdots d'_{kK} \end{bmatrix} \right\}$. By considering the distance vectors in unknown nodes as testing data x, the classes of two coordinate values are attained according to Eq. (5). If the classes of two coordinate values of node n_k are (cx_i, cy_j), which are decided by SVMs, the node n_k is believed to locate in the cubic unit $[(i-1)D/M, iD/M] \times [(j-1)D/M, jD/M]$. The cubic unit centroid $\left[(i - \frac{1}{2})D/M, (j - \frac{1}{2})D/M \right]$ is used as the estimated position of the node n_k. The maximum error of localization for node n_k is $\sqrt{2}D/2M$, when the SVMs perform correct classification.

The process of SVM model constitution and localization step is shown in Fig. 2.

4 Emulation and Analysis

In this paper, MATLAB simulation environment is used to simulate and analyze the two-dimension localization algorithm. Supposing all nodes is distributed randomly in a two-dimension space, which is 50 m x 50 m in size. The two-dimension space is divided into 25 small cubic units, which are 10 m x 10 m in size. Thus, the distance of every dimension is D = 50 m, and the number of classes for every dimension is M = 5. In all of the experiments, the total number of nodes is set to N = 250 (50 anchor and 200 unknown node). The communication radius is assumed to be R = 25 m. Due to the random distribution of sensor nodes, the randomness of the localization result derived from the experiment can not be ignored. The experiment is repeated for 50 times on the same network parameter settings and use statistical methods to obtain accurate experimental results. All nodes will be laid in two-dimension region uniformly and randomly in each experiment. Therefore, the error of all unknown nodes location is shown in Fig. 3.

As shown in Fig. 3, the estimate errors of unknown nodes are less than 5 m when the communication range is 25 m. In the other words, the accuracy of the RFSVM algorithm is more than 80%. And if we set more anchors in the networks, we will get more accuracy of the nodes localizations.

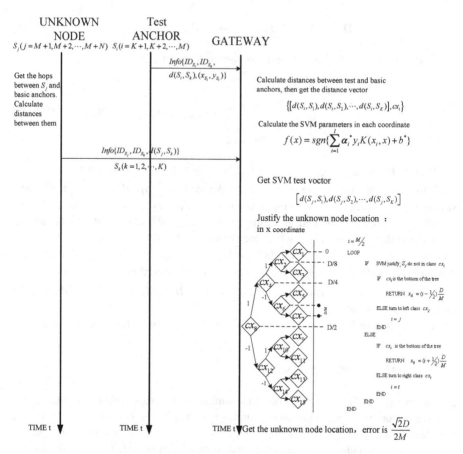

Fig. 2. The process of SVM model constitution and localization

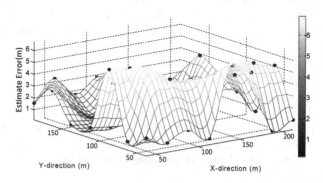

Fig. 3. Unknown nodes location error

5 Conclusions

In this paper, a new algorithm called Support Vector Machine based Range-free localization (RFSVM) algorithm in WSN is proposed. This algorithm introduced a new matrix called transmit matrix which maps the relationship between the hops and distance. Then use the transmit matrix and SVM method to get the unknown nodes position. The whole algorithm does not need any addition hardware such as RSSI, antennas et al. The experiment shows that the accuracy of the RFSVM algorithm is more than 80%. Along with the anchor number increasing, the location of nodes will be more accuracy.

Acknowledgments. This paper is by Research Fund for the Doctoral Program of HLJIT under Grant No. 2014BJ10; University Nursing Program for Young Scholars with Creative Talents in Heilongjiang Province UNPYSCT-201501.

Also supported by The Basic Research Plan in Shenzhen City under Grant No. JCYJ20130401100512995.

References

1. Culler, D., Estrin, D., Srivastava, M.: Overview of sensor networks. IEEE Comput. **37**(8), 41–49 (2004)
2. Li, J., Jannotti, J., De Couto, D., Karger, D., Morris, R.: A scalable location service for geographic ad-hoc routing. In: Proceedings of the 6th ACM International Conference on Mobile Computing and Networking (MobiCom 2000), pp. 120–130 (2000)
3. Amouris, K., Papavassiliou, S., Li, M.: A position-based multi-zone routing protocol for wide area mobile ad-hoc networks. In: Proceedings of the Vehicular Technology Conference, Houston, TX (2010)
4. Bäck, T.: Evolutionary Algorithms in Theory and Practice: Evolution Strategies, Evolutionary Programming, Genetic Algorithms. Oxford University Press, New York (1996)
5. Lédeczi, Á., Nádas, A., Völgyesi, P., Balogh, G., Kusy, B., Sallai, J., Pap, G., Dóra, S., Molnár, K., Maróti, M., Simon, G.: Countersniper system for urban warfare. ACM Trans. Sen. Netw. **1**(2), 153–177 (2015)
6. Nemeroff, J., Garcia, L., Hampel, D., DiPierro, S.: Application of sensor network communications. In: Military Communications Conference, 2001, MILCOM 2001, Communications for Network-Centric Operations: Creating the Information Force, vol. 1, pp. 336–341. IEEE (2011)
7. Mladineo, N., Knezic, S.: Optimisation of forest fire sensor network using GIS technology. In: Proceedings of the 22nd International Conference on Information Technology Interfaces, ITI 13–16 June 2000, pp. 391–396 (2000)
8. Xu, N., Rangwala, S., Chintalapudi, K.K., Ganesan, D., Broad, A., Govindan, R., Estrin, D.: A wireless sensor network for structural monitoring. In: SenSys 2004: Proceedings of the 2nd International Conference on Embedded Networked Sensor Systems, pp. 13–24. ACM, New York (2004)
9. Molina, G., Alba, E.: Location discovery in Wireless Sensor Networks using metaheuristics. Appl. Soft Comput. **11**, 1223–1240 (2011)

10. Mao, G., Fidan, B., Anderson, B.D.O.: Wireless sensor network localization techniques. Comput. Netw. **51**(10), 2529–2553 (2007)
11. Bulusu, N., Heidemann, J., Estrin, D.: GPS-less low cost outdoor localization for very small devices. IEEE Personal Commun. Mag. **7**(5), 28–34 (2010)
12. Nagpal, R.: Organizing a global coordinate system from local information on an amporphous computer. Technical Report AI Memo 1666, MIT Artificial Intelligence Laboratory (1999)
13. He, T., Huang, C.D., Blum, B.M.: Range-free localization schemes for large scale sensor networks. In: Proceedings of the 9th Annual International Conference on Mobile Computing and Networking, pp. 81–95. ACM, San Diego (2013)
14. Galstyan, A., Krishnamachari, B., Lerman, K., Pattem, S.: Distributed online localization in sensor networks using a moving target. In: Proceedings of the 3rd International Symposium on Information Processing in Sensor Networks. ACM, 61–70, Barkeley (2004)
15. Su, K.F., Ou, C.H., Jiau, H.C.: Localization with mobile anchor points in wireless sensor networks. IEEE Trans. Veh. Technol. **54**(3), 1187–1197 (2015)
16. Tran, D.A., Nguyen, T.: Support vector classification strategies for localization in sensor networks. In: Proceedings of the Communications and Electronics, Hanoi (2006)

Reducing Calibration Effort for Indoor WLAN Localization Using Hybrid Fingerprint Database

Mu Zhou, Yunxia Tang$^{(\boxtimes)}$, Zengshan Tian, and Feng Qiu

Chongqing Key Lab of Mobile Communications Technology,
Chongqing University of Posts and Telecommunications, Chongqing 400065, China
{zhoumu,tianzs}@cqupt.edu.cn, 13629735505@139.com,
qiufeng245@outlook.com

Abstract. Due to the implementation ease and cost-efficiency, the indoor Wireless Local Area Network (WLAN) fingerprint based localization approach is preferred compared with the conventional trilateration localization approaches. In this paper, we propose a new semi-supervised learning algorithm based on manifold alignment with cubic spline interpolation to reduce the offline calibration effort for indoor WLAN localization using hybrid fingerprint database. The proposed approach significantly reduces the number of labeled training samples collected at each survey location by constructing the hybrid database via interpolation and semi-supervised manifold learning. We carry out extensive experiments in a ground-truth indoor environment to examine the localization accuracy of the proposed approach. The experimental results demonstrate that our approach can effectively reduce the calibration effort, as well as achieve high localization accuracy.

Keywords: WLAN · Location fingerprint · Interpolation · Semi-supervised learning · Manifold alignment

1 Introduction

With the development of light-weighted mobile devices, Location-based Services (LBSs) have gained considerable attention over the last decade due to the potential in the technology and the significant challenges facing this area of research [1]. The popular Global Positioning System (GPS) has been recognized as a success for outdoor localization, but it is generally not applicable for the indoor environment. The conventional indoor localization systems based on the infrared ray [2], ultrasound [3], video [4], and Radio Frequency (RF) techniques [5–10] have been widely studied. The RF technique has the advantage of ubiquitous coverage by using the inexpensive Wireless Local Area Network (WLAN). Due to the considerations of cost overhead and localization accuracy, the fingerprint database based indoor WLAN localization has been widely studied. Two phases are involved in the fingerprint database based indoor WLAN localization, namely offline phase and online phase. In off-line phase, we calibrate a series of Reference

© ICST Institute for Computer Sciences, Social Informatics and Telecommunications Engineering 2017
X.-L. Huang (Ed.): MLICOM 2016, LNICST 183, pp. 159–168, 2017.
DOI: 10.1007/978-3-319-52730-7_16

Points (RPs) in target area, and then collect the Received Signal Strength (RSS) measurements from hearable Access Points (APs) at each RP to construct the fingerprint database, namely radio-map. In on-line phase, we match the newly collected RSS measurements against the radio map to estimate target location.

Since the point-by-point RP calibration is time-consuming, we aim to reduce the offline calibration effort, as well as maintain the high-enough localization accuracy. The proposed approach reduces both the number of labeled samples collected at survey points and number of survey points. However, reducing the number of the labeled samples and survey points may result in the inaccurate radio-map and deteriorate the localization accuracy. To solve this problem, the proposed approach relies on the cubic spline interpolation algorithm to obtain the predicted location fingerprints, and employs the manifold alignment (MA) algorithm [10] to label the locations at which we collect the sequences of RSS measurements according to the users motion traces. Since the users motion traces can be recorded easily without the labeling process, our approach is able to reduce the labor cost, as well as improve the accuracy of the reconstructed fingerprint database.

The remainder of the paper is organized as follows. Section reviews some related work. In Section, we introduce the proposed approach in detail. Section conducts the performance evaluation under different parameters and shows the experimental results. Finally, the conclusion is provided in Section.

2   Related Work

There are bathes of studies focusing on the indoor WLAN localization, such as the RADAR [5] and Horus [6]. Although the approaches in [5] and [6] can achieve high localization accuracy, a large number of RSS measurements are required to be collected and manually labeled at survey points. Since the labeled RSS measurements collection is time consuming and labor intensive, the existing literatures mainly focused on using the unlabeled data to reduce the time overhead involved in offline phase. In [7], the authors addressed a label propagation algorithm based semi-supervised learning approach to construct a hybrid database of labeled and unlabeled data using the concept that the similar data are corresponding to the similar labels. In [8], the authors proposed a hybrid generative and discriminative semi-supervised learning algorithm by predicting a large amount of unlabeled data to replenish the sparse labeled database, and meanwhile the online test data are selected as the offline unlabeled data and the labels of unlabeled data are learned from the labeled data. The authors in [9] exploited a new approach in which a manifold-based model is built from a batch of labeled and unlabeled data in offline phase, and then the weighted K Nearest Neighbor (KNN) algorithm is used to estimate the target locations in online phase. However, since the aforementioned approaches significantly depend on the RSS measurements, the performance could be seriously degraded when the RSS changes abruptly among the neighboring RPs. The main contribution of this paper is that we build a more accurate and reliable radio map by using

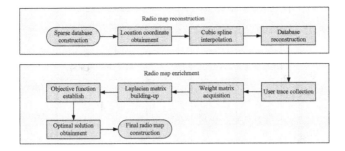

Fig. 1. Flow chart of the proposed approach.

the cubic spline interpolation and manifold learning algorithms to supplement the sparse fingerprint database in indoor WLAN localization. In addition, we propose to use the timestamp information during the process of intra-manifold graph construction in manifold learning, which can avoid the sharp deterioration of localization accuracy when the RSS changes abruptly.

3   Algorithm Description

3.1   Algorithm Overview

In this paper, we propose to use the cubic spline interpolation and manifold learning algorithms to construct the hybrid fingerprint database. In concrete terms, we rely on the cubic spline interpolation algorithm to enrich the sparse fingerprint database. After that, we apply the manifold learning algorithm to label the unlabeled trace locations based on the known RPs and the corresponding RSS sequences. The flow chart of the proposed approach is shown in Fig. 1.

3.2   Radio Map Reconstruction

To study the performance of the proposed approach, we carry out the experiments in a real indoor WLAN environment under different interpolation algorithms. Figure 2 shows the mean of errors with different radios of RPs used for the Radial Basis Functions (RBF), Linear, and Cubic Spline Interpolation (CUBIC) interpolation algorithms respectively. From this result, we can find that the CUBIC interpolation algorithm achieves the highest localization accuracy.

3.3   Radio Map Enrichment

Since the labeled and unlabeled data are collected in the same environment, the RSS measurements share the similar properties of the low-dimensional manifold [10]. Based on this, we label the unlabeled data by aligning their corresponding manifolds in the physical location space. In concrete terms, we build two intra-manifold graphs with respect to the labeled and unlabeled data respectively, as well as one inter-manifold graph between them. Then, we construct the weighted

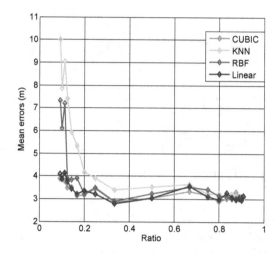

Fig. 2. Mean of errors under different interpolation algorithms.

graph matrices to describe the relations of RSS measurements, timestamps, and physical locations respectively, i.e., W_r, W_t, W_{loc}.

To build the graph matrix W_r, we simply connect each RSS vector with its K nearest neighbors, and then assign a value to each pair of vectors by using the heat Kernel [10], such that

$$W_r\,(i,j) = \begin{cases} e^{-\frac{\|x_i - x_j\|^2}{\theta_r}}, \text{if } i \text{ and } j \text{ are connected} \\ 0, \text{otherwise} \end{cases} \quad (1)$$

where θ_r is the heat kernel of radio space.

The graph matrix W_t is built for the unlabeled traces by using the timestamps. Based on the assumption that the samples collected within short time duration are corresponding to the physically adjacent locations, we construct this matrix according to the time difference between every two samples. For the k-th trace, the W_t is defined as

$$W_t^{u_k}\,(i,j) = \begin{cases} e^{-\frac{|t_i - t_j|^2}{\theta_t}}, \; |t_i - t_j| \le T_{thr} \\ 0, \text{otherwise} \end{cases} \quad (2)$$

where θ_t and T_{thr} are the heat kernel and threshold of time space respectively.

We build W_{loc} only for the labeled data with known location information. We define that the two samples are connected when they are physically closest. Thus, W_{loc} is defined as

$$W_{loc}^l\,(i,j) = \begin{cases} e^{-\frac{dist(l_i,l_j)^2}{\theta_{dist}}}, \text{if } i \text{ and } j \text{ are connected} \\ 0, \text{otherwise} \end{cases} \quad (3)$$

where θ_{dist} is the heat kernel of physical location space.

After that, we construct two intra-manifold graphs for the labeled and unlabeled data by using a relative weight $\alpha \in [0,1]$, as shown in (4) and (5) respectively.

$$W^l = \alpha W_r^l + (1 - \alpha) W_{loc}^l \tag{4}$$

$$W^{u_k} = \alpha W_r^{u_k} + (1 - \alpha) W_t^{u_k} \tag{5}$$

Since the labeled and unlabeled data have the common feature space with RSS measurements in the inter-manifold graph, we link the labeled and unlabeled data by using the properties of RSS measurements.

Then, we can obtain the graph Laplacian matrices for the labeled data, unlabeled data, and the corresponding inter-dataset respectively as follows,

$$L^l_{\,N_l \times N_l} = D^l - W^l \tag{6}$$

$$L^{u_k}_{\,T_k \times T_k} = D^{u_k} - W^{u_k} \tag{7}$$

$$L^{lu_k}_{\,(N_l+T_k) \times (N_l+T_k)} = \begin{bmatrix} D^{lu_k} & -W^{lu_k} \\ -W^{lu_k\,T} & D^{u_k l} \end{bmatrix} \tag{8}$$

where D^l, D^{u_k}, D^{lu_k}, and $D^{u_k l}$ are the diagonal matrixes with the diagonal elements where D^l is a diagonal matrix with diagonal elements $D^l(i,i) = \sum_{j=1}^{N_l} W^l(i,j)\,(i=1,...,N_l)$, D^{u_k} is a diagonal matrix with diagonal elements $D^{u_k}(i,i) = \sum_{j=1}^{T_k} W^{u_k}(i,j)\,(i=1,...,T_k)$, D^{lu_k} is a diagonal matrix with diagonal elements $D^{lu_k}(i,i) = \sum_{j=1}^{T_k} W^{lu_k}(i,j)\,(i=1,...,N_l)$, and $D^{u_k l}$ is a diagonal matrix with diagonal elements $D^{u_k l}(i,i) = \sum_{j=1}^{N_l} W^{lu_k}(j,i)(i=1,...,T_k)$.

We continue to combine the intra-manifold graphs with inter-manifold graph by using a relative weight μ for manifold alignment. The composite graph Laplacian is described as

$$L_k = \begin{bmatrix} L^l & 0 \\ 0 & L^{u_k} \end{bmatrix} + \mu L^{lu_k} \tag{9}$$

We choose the two-dimensional physical space as the common low-dimensional latent space. By denoting the coordinate matrix for the new space of the labeled data and k-th trace as $q_k \in R^{(N_l+T_k)\times 2}$, we formulate the optional manifold alignment problem as

$$\hat{q}_k^{(h)} = \arg\min_{q_k^{(h)}} \left(q_k^{(h)} - Y_k^{(h)} \right)^{\mathrm{T}} J_q \left(q_k^{(h)} - Y_k^{(h)} \right) + \gamma q_k^{(h)\mathrm{T}} L_k q_k^{(h)}, h = 1,2 \tag{10}$$

where

$$J_q = \begin{bmatrix} I_{N_l \times N_l} & 0 \\ 0 & 0 \end{bmatrix} \tag{11}$$

$A^{(h)}$ is the h-th column of matrix A, and $I_{N_l \times N_l}$ is the $N_l \times N_l$ identity matrix. In $Y_k \in R^{(N_l+T_k)\times 2}$, the previous N_l rows are the coordinates of the labeled data, i.e., $Y^l = [l_1, l_2, ..., l_{N_l}]^T$, while the latter T_k rows are the arbitrary values. Based

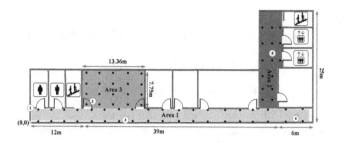

Fig. 3. Physical layout of target environment.

on the objective function in (10), the first term stands for the fitting error of the labeled data, while the second term enforces the smoothness of the manifold. The parameter γ controls the relative strength of location constraint.

Then, we calculate the objective function $q_k^{(h)}$ as

$$q_k^{(h)} = (J_q + \gamma L_k)^{-1} J_q Y_k^{(h)}, h = 1, 2 \tag{12}$$

Based on (12), we can assign location coordinates to the unlabeled data in the low-dimensional space. Finally, after the process of labeling the unlabeled data, we use the KNN to estimate the target locations.

4   Performance Evalution

4.1   Experimental Setup

To investigate the performance of the proposed approach, we conduct the experiments in a real indoor WLAN environment with the size of 57 m by 25 m on the fifth floor of an office building, as shown in Fig. 3. The target area is covered by five APs. A Samsung S7568 mobile phone is selected as the receiver installed with our developed Wi-Fi localization software. The data are stored as the TXT files. We calibrate 73 RPs with the same interval of 3 m in three subareas, namely Area 1, 2, and 3. In addition, we record 30 traces without location information for the testing.

4.2   Parameters in Manifold Learning

In our approach, the parameters in manifold learning are significantly important and require to be carefully studied. Figure 4 shows the impact of different parameters in manifold learning on localization errors. In Fig. 4, we can find that as the values of θ_r, θ_t, α, and γ increase, the localization error decreases first and then slightly increases or approximately maintains the same. For the parameter μ, the localization error reaches the lowest when the value μ is greater than 3. Our approach achieves the best performance when the parameters $\theta_r = 50$, $\theta_t = 0.1$, $\alpha = 0.1$, $\gamma = 0.1$, and $\mu = 3$, and the number of neighbors used in manifold learning equals 4, i.e., k = 4. Since the variation of value θ_{dist} has no impact on localization error, we set $\theta_{dist} = 1$ in the results that follow.

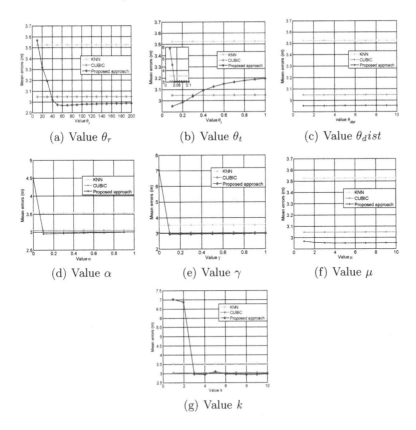

(a) Value θ_r (b) Value θ_t (c) Value $\theta_d ist$

(d) Value α (e) Value γ (f) Value μ

(g) Value k

Fig. 4. Mean of errors under different parameters in manifold learning.

4.3 Localization Algorithms

We compare the Cumulative Distribution Functions (CDFs) of errors by the proposed approach and the conventional KNN, MA, and CUBIC approaches in Fig. 5. From this figure, we can find that our approach performs best in localization accuracy, and meanwhile it reduces the mean of errors by 16.4 % compared with the result without our approach.

4.4 Number of RPs

Figure 6 compares the mean of errors under different ratios of the number of RPs by the proposed, KNN, MA, and CUBIC approaches. From this figure, we can find that in the small ratios condition, the increase of the number of RPs significantly reduces the localization error, whereas when the ratio is over 0.5, the variation of the number of RPs generally has slight impact on localization error.

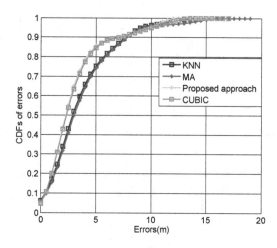

Fig. 5. CDFs of errors under different localization algorithms.

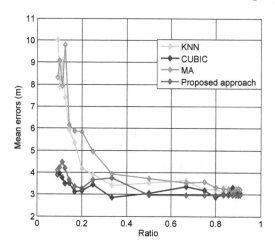

Fig. 6. CDFs of errors under different localization algorithms.

4.5   Number of Unlabeled Traces

Figure 7 compares the mean of errors under different number of unlabeled traces by the proposed, KNN, and CUBIC approaches. From this figure, we can find that the increase of the number of traces reduces the localization error by the proposed approach, and meanwhile when the number of traces is over 12, the number of trace has slight impact on localization error.

4.6   Impact of Timestamps

Figure 8 compares the CDFs of errors with and without timestamps by the proposed approach. From this figure, we can find that the localization performance

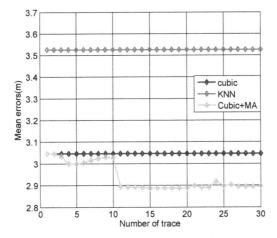

Fig. 7. Mean of errors under different number of traces.

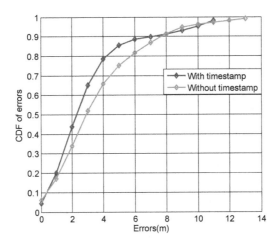

Fig. 8. CDFs of errors with and without timestamps.

is enhanced when the timestamp is considered. This result can be interpreted by the fact that the timestamp of traces is able to strengthen the correlation between the successively collected RSS measurements.

5   Conclusion

To reduce the labor effort involved in indoor WLAN fingerprint based localization, we propose a new integrated cubic spline interpolation approach with manifold learning from the low-overhead unlabeled traces of users in target environment. The experiments conducted in a real indoor WLAN environment demonstrate that our approach can not only reduce the density of RPs used for radio map construction, but also improve the accuracy of indoor WLAN fingerprint based localization. In future, the application of this approach to a larger-scale or multi-floor indoor WLAN environment forms an interesting work.

Acknowledgment. This work was supported in part by the Program for Changjiang Scholars and Innovative Research Team in University (IRT1299), National Natural Science Foundation of China (61301126), Special Fund of Chongqing Key Laboratory (CSTC), and Fundamental and Frontier Research Project of Chongqing (cstc2013jcyjA40041, cstc2015jcyjBX0065).

References

1. Axel, K.: Location-Based Services: Fundamentals and Operation, pp. 185–245. Wiley, New York (2005)
2. Want, R., Hopper, A., Falcao, V., Gibbons, J.: The active badge location system. ACM Trans. Inf. Syst. **10**(1), 91–102 (1992)
3. Hazas, M., Ward, A.: A novel broadband ultrasonic location system. In: Borriello, G., Holmquist, L.E. (eds.) UbiComp 2002. LNCS, vol. 2498, pp. 264–280. Springer, Heidelberg (2002). doi:10.1007/3-540-45809-3_21
4. Darrell, T., Gordon, G., Harville, M., Woodfill, J.: Integrated person tracking using stereo, color, and pattern detection. Int. J. Comput. Vis. **37**(2), 175–185 (2000)
5. Bahl, P., Padmanabhan, V.N.: RADAR: an in-building RF-based user location and tracking system. In: IEEE INFOCOM, pp. 775–784 (2002)
6. Youssef, M., Agrawala, A.: The Horus WLAN location determination system. Wirel. Netw. **13**(3), 357–374 (2005)
7. Liu, S., Luo, H., Zou, S.: A low-cost and accurate indoor localization algorithm using label propagation based semi-supervised learning. In: International Conference on Mobile Ad-hoc and Sensor Networks, pp. 108–111 (2009)
8. Ouyang, R., Wong, A., Lea, C., Chiang, M.: Indoor location estimation with reduced calibration exploiting unlabeled data via hybrid generative/discriminative learning. IEEE Trans. Mob. Comput. **11**(11), 1613–1626 (2011)
9. Pan, J.J., Pan, S.J., Yin, J., Ni, L.M., Yang, Q.: Tracking mobile users in wireless networks via semi-supervised colocalization. IEEE Trans. Pattern Anal. Mach. Intell. **34**(3), 587–600 (2012)
10. Mikhail, B., Partha, N.: Laplacian eigenmaps for dimensionality reduction and data representation. Neural Comput. **15**(6), 1373–1396 (2003)

Accuracy Enhancement with Integrated Database Construction for Indoor WLAN Localization

Qiao Zhang$^{(\boxtimes)}$, Mu Zhou, and Zengshan Tian

Chongqing Key Lab of Mobile Communications Technology,
Chongqing University of Posts and Telecommunications,
Chongqing 400065, China
18716322725@139.com, {zhoumu,tianzs}@cqupt.edu.cn

Abstract. In this paper, we rely on the neighborhood relations of the physically adjacent Reference Points (RPs) to construct a physical neighborhood database with the purpose of enhancing the accuracy of the Receive Signal Strength (RSS) fingerprint based localization algorithms in Wireless Local Area Network (WLAN) environment. First of all, based on the Most Adjacent Points (MAPs) and their corresponding Physically Adjacent Points (PAPs), we construct the Feature Groups (FGs), and then calculate the New Reference Point (NRP) with respect to each FG. Second, the RSS at each NRP is estimated by using the least square method based surface interpolation algorithm. Finally, we apply the K Nearest Neighbor (KNN), Weighted KNN (WKNN), and Bayesian inference algorithms to locate the target. The experimental results show that the proposed integrated database construction helps a lot in improving the localization accuracy of the widely-used KNN, WKNN, and Bayesian inference algorithms.

Keywords: WLAN localization · Location fingerprinting · Physical neighborhood · Reference Points · Received Signal Strength

1 Introduction

With the significant development of Wireless Local Area Network (WLAN) technique and the wide deployment of WLAN Access Points (APs) in public environments, it is particularly valuable and cost-efficient to rely on the WLAN infrastructures and the off-the-shelf smartphones to conduct the people's location tracking [1]. By employing the WLAN Received Signal Strength (RSS), the WLAN fingerprint based localization techniques have been carefully studied in recent decade due to the advantages of the free ISM band and high enough accuracy performance [2–4]. And as far as we know, most of the existing RSS fingerprint based localization techniques do not pay much attention to the physical adjacency relations of Reference Points (RPs), while in fact, these relations can help a lot in improving the localization accuracy [5, 6]. On this basis, we construct an integrated physical neighborhood and location fingerprinting database in off-line phase. Then, in on-line phase, we first select the k RPs with the RSSs having the smallest distances from the newly recorded RSS by the target as the k Most Adjacent Points (MAPs). Second, based on the physical neighborhood

© ICST Institute for Computer Sciences, Social Informatics and Telecommunications Engineering 2017
X.-L. Huang (Ed.): MLICOM 2016, LNICST 183, pp. 169–177, 2017.
DOI: 10.1007/978-3-319-52730-7_17

database, we search the n Physically Adjacent Points (PAPs) corresponding to the k MAPs. Third, we use every c MAPs and PAPs (also named as Feature Points (FPs)) to construct a Feature Group (FG). Obviously, the number of FGs equals to C_{k+n}^c. Fourth, in each FG, we calculate a New Reference Point (NRP), as well as estimate the RSS at the NRP by using the least square method based surface interpolation algorithm. Finally, based on the C_{k+n}^c NRPs, we apply the RSS fingerprint based localization algorithm (e.g., K Nearest Neighbor (KNN), Weighted KNN (WKNN), and Bayesian inference) to locate the target.

The rest of this paper is structrued as follows. In Sect. 2, we show the steps of the integrated physical neighborhood and location fingerprinting database construction. In Sect. 3, the performance of the integrated database for indoor WLAN localization is examined. The experimental results with the WLAN RSSs recorded in an actual indoor WLAN environment are provided in Sect. 4. Finally, Sect. 5 concludes the paper and provides some future directions.

2   Integrated Database Construction

The physical neighborhood database is constructed based on the physical layout of the target environment. In concrete terms, for the physical layout of an actual indoor WLAN environment [9–17], we represent each office room or each segment of straight corridors as a representative node. Every two adjacent representative nodes are connected by an edge. Then we obtain a physical graph describing the physical layout of the target environment. To construct the physical neighborhood database, we first label each RP with a unique Reference Point Identifier (RPID). Second, based on the geographic relations of RPs described in physical graph, we construct a set of r adjacent PAPs with respect to each RP. Finally, we construct the physical neighborhood database consisting of the sets of PAPs for all the RPs. And the construction of location fingerprinting database consists of two main steps as follow. First of all, we record a sequence of RSS measurements at each RP, notated as $\{S_1 = (\text{rss}_{11}, \text{rss}_{12, \dots}, \text{rss}_{1w})$, $S_2 = (\text{rss}_{21}, \text{rss}_{22}, \dots, \text{rss}_{2w}), \dots\}$, where w is the number of APs and rss_{ij} is the RSS value from the j-th AP in S_i. Second, we calculate the mean and standard deviation of RSS at each RP to form a RSS fingerprint. Finally, the location fingerprinting database is constructed to describe the relationship between the RSS fingerprints and the locations of RPs.

3   Accuracy Enhancement for Indoor WLAN Localization

To enhance the localization accuracy of the KNN, WKNN, and Bayesian inference algorithms, we first select the k MAPs with respect to each newly recorded RSS. Second, using the physical neighborhood database, we search the n PAPs corresponding to the k MAPs, and then calculate the coordinates of the C_{k+n}^c NRPs. Finally, the localization algorithm (e.g., KNN, WKNN, and Bayesian inference) is applied to estimate the locations of the target.

In the results that follow, we mainly focus on the three typical combinational localization algorithms: (i) KNN based WKNN (i.e., KNN is applied to calculate the NRPs, and then WKNN is used to estimate the locates of the target), named as KbW; (ii) WKNN based Bayesian inference (i.e., WKNN is applied to calculate the NRPs, and then Bayesian inference is used to estimate the locates of the target), named as WbB; and (iii) Bayesian inference based KNN (i.e., Bayesian inference is applied to calculate the NRPs, and then KNN is used to estimate the locates of the target), named as BbK.

3.1   Combinational Localization Algorithms

Steps of KbW. After the k MAPs are selected, we search the n PAPs corresponding to the k MAPs by using the physical neighborhood database. After that, we construct C_{k+n}^c FGs, and the NRP in each FG, P_{ref}, is calculated by

$$P_{ref} = \frac{1}{c} \sum\nolimits_{i=1}^{c} (x_{zi}, y_{zi}) \tag{1}$$

where (x_{zi}, y_{zi}) is the 2-dimensional (2-D) coordinates of the i-th FP in the z-th FG. To estimate the RSS at P_{ref}, we assume that in each FG, the relationship between the coordinates of FPs and their corresponding RSSs satisfies

$$s_{zi-j} = ax_{zi} + by_{zi} + d + \delta_{zi} \tag{2}$$

where s_{zi-j} is the RSS from the j-th AP at the i-th FP in the z-th FG. δ_{zi} is the RSS distance between s_{zi-j} and the estimated RSS at the i-th FP. By using the least square method based surface interpolation algorithm, the coefficients a, b, and d in (2) are calculate by

$$\begin{cases} a = \frac{c(sy)c(xy) - c(sx)d(y)}{(c(xy))^2 - d(x)d(y)}, \\ b = \frac{c(sx)c(xy) - c(sy)d(x)}{(c(xy))^2 - d(x)d(y)}, \\ d = \bar{s} - a\bar{x} - b\bar{y}. \end{cases} \tag{3}$$

where

$$\begin{cases} c(sy) = \sum_{i=1}^{c} \left(s_{zi-j} - \bar{s}\right)\left(y_{zi} - \bar{y}\right), \\ c(xy) = \sum_{i=1}^{c} \left(x_{zi} - \bar{x}\right)\left(y_{zi} - \bar{y}\right), \\ c(sx) = \sum_{i=1}^{c} \left(s_{zi-j} - \bar{s}\right)\left(x_{zi} - \bar{x}\right), \\ d(x) = \sum_{i=1}^{c} \left(x_{zi} - \bar{x}\right)^2, \\ d(y) = \sum_{i=1}^{c} \left(y_{zi} - \bar{y}\right)^2, \\ \bar{s} = \frac{1}{c}\sum_{i=1}^{c} s_{zi-j}, \bar{x} = \frac{1}{c}\sum_{i=1}^{c} x_{zi}, \bar{y} = \frac{1}{c}\sum_{i=1}^{c} y_{zi}. \end{cases} \tag{4}$$

After the coefficients in (2) are calculated, we estimate the RSS at P_{ref} based on the fitted surface function [8], $z = ax + by + d$. Finally, WKNN is applied to estimate the locations of the target P_{user}, as described in (5).

$$P_{user} = \frac{\sum_{i=1}^{q} (1/d_i)(x_i, y_i)}{\sum_{i=1}^{q} (1/d_i)} \tag{5}$$

where d_i is the distance between the estimated RSS at the i-th selected NRP, (x_i, y_i), and the newly recorded RSS. In KbW, the selected NRPs are the NRPs with the RSSs having the smallest distances from the newly recorded RSS.

Steps of WbB. In WbB, we apply WKNN to calculate the NRP in each FG, as shown in (6).

$$P_{ref} = \frac{\sum_{i=1}^{c} (1/d_{zi})(x_{zi}, y_{zi})}{\sum_{i=1}^{c} (1/d_{zi})} \tag{6}$$

where d_{zi} is the distance between the estimated RSS at the i-th FP in the z-th FG, (x_{zi}, y_{zi}), and the newly recorded RSS. The estimation of RSS at each NRP in WbB follows the same steps involved in KbW. After that, we use Bayesian inference to calculate the posterior probability of each NRP with respect to each newly recorded RSSs. We take the NRP L_f as an example. By using the Bayesian inference, the posterior probability of L_f with respect to s, $P(L_f|s)$, is equivalent to the product of the prior probabilities, as shown in (7).

$$P(s|L_f) = P(s_1|L_f)P(s_2|L_f)...P(s_w|L_f) \tag{7}$$

where s_j is the newly recorded RSS from the j-th AP. By assuming that the RSS distribution at each NRP obeys a Gaussian distribution, we have

$$P(s_j|L_f) = \frac{1}{\sqrt{2\pi}\delta} \exp[\frac{-(s_j - \mu)^2}{2\delta^2}] \tag{8}$$

where μ and δ are the mean and standard deviation of the RSS distribution from the j-th AP at L_f respectively. Then, we rely on the q selected NRPs with the largest posterior probabilities with respect to the newly recorded RSS to estimate the locations of the target, as shown in (9).

$$P_{user} = \frac{\sum_{i=1}^{q} pro_i(x_i, y_i)}{\sum_{i=1}^{q} pro_i} \tag{9}$$

where pro_i and (x_i, y_i) are the posterior probability and the 2-D coordinates of the i-th selected NRP respectively.

Steps of BbK. In BbK, we calculate the NRP P_{ref} in each FG by

$$P_{ref} = \frac{\sum_{i=1}^{c} \text{pro}_{zi}(x_{zi}, y_{zi})}{\sum_{i=1}^{c} \text{pro}_{zi}} \tag{10}$$

where pro_{zi} and (x_{zi}, y_{zi}) are the posterior probability and the 2-D coordinates of the i-th FP in the z-th FG. The steps of the estimation of RSS at each NRP in BbK are the same to the ones in KbW and WbB. After the estimated RSS at each NRP is obtained, we can estimate the locations of the target by

$$P_{user} = \frac{1}{q} \sum_{i=1}^{q} (x_i, y_i) \tag{11}$$

where the 2-D coordinates of the q selected NRPs are denoted as $(x_i, y_i)(i = 1, , q)$. In BbK, the selected NRPs are the NRPs with the RSSs having the smallest distances from the newly recorded RSS.

3.2 Modified Bayesian Inference

In WbB, the RSS distribution at each location is assumed to obey a Gaussian distribution, while in fact, the Gaussian distribution of RSS cannot always be approximately obeyed especially in the Non-Line-of-Sight (NLOS) scenario. To solve this problem, we use (12) to calculate the similarity between the RSS distributions at each RP and the distributions of the newly recorded RSSs.

$$S_i = \frac{1}{\sum_{j=1}^{w} \left(\sum_{x=\text{RSS}_{\text{lower}}}^{\text{RSS}_{\text{upper}}} P_{on-j}(x) \ln \frac{P_{on-j}(x)}{Q_{ij}(x)} \right)} \tag{12}$$

where $Q_{ij}(x)$ and $P_{on-j}(x)$ are the RSS distribution from AP j at the i-th RP and the distribution of the newly recorded RSSs from AP j respectively. The RSS value x is in the range of $[\text{RSS}_{\text{lower}}, \text{RSS}_{\text{upper}}]$.

4 Experimental Results

4.1 Accuracy Discussion

We conduct the experiments in an actual indoor WLAN environment with the dimensions of 66 m × 22 m. The target environment is covered by 9 Cisco WRT54G APs which are placed on the same floor in a building [7], as shown in Fig. 1. The 182 RPs are uniformly calibrated with the same interval of 1 m and the 81 test points (TPs) are randomly selected in five straight corridors for the testing.

Figure 2 compare the Cumulative Distribution Functions (CDFs) of errors between the proposed Combinational Localization Algorithms (i.e., KbW, WbB, and BbK) and the conventional WKNN, Bayesian inference, and KNN, named as C-W, C-B, and C-K

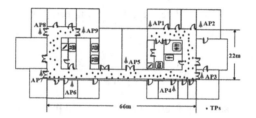

Fig. 1. Experimental layout.

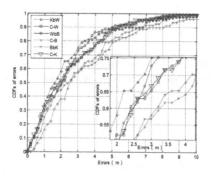

Fig. 2. CDFs of errors by KbW, WbB, BbK, C-W, C-B, and C-K.

respectively. As can be seen from Fig. 2, the proposed algorithms generally perform better than the conventional WKNN, Bayesian inference, and KNN in localization accuracy. We take KbW as an example. By using KbW, the probabilities of errors within 3 m and 2.5 m are about 10% and 5% more than the ones achieved by C-W respectively.

4.2   Parameter Discussion

To examine the performance of the proposed integrated database construction for indoor WLAN localization, we use the control variable approach to investigate the relationship between the localization errors and the four parameters as follows: (i) number of MAPs, k; (ii) number of the adjacent PAPs for each RP, r; (iii) number of FPs in each FG, c; and (iv) number of NRPs, q.

The optimal parameters which are corresponding to the smallest mean of errors for all the combinational localization algorithms and the conventional WKNN, Bayesian inference, and KNN are shown in Table 1.

From Table 1, we can observe that: (i) by using the integrated database, most of the combinational localization algorithms achieves lower mean of errors compared to C-W, C-B, and C-K; and (ii) the lowest mean of errors, 2.2946 m, is obtained by KbK. On this basis, the integrated physical neighborhood and location fingerprinting database is

Table 1. Parameters vs. Errors

Algorithms	Optimal parameters	Mean of errors (m)
KbW	$k = 9, r = 3, c = 5, q = 17$	2.2962
WbW	$k = 5, r = 5, c = 4, q = 17$	2.3416
BbW	$k = 17, r = 3, c = 2, q = 5$	3.1147
C-W	$k = 9$	2.5213
KbB	$k = 9, r = 3, c = 5, q = 1$	2.5333
WbB	$k = 5, r = 5, c = 4, q = 17$	2.4483
BbB	$k = 17, r = 3, c = 2, q = 5$	3.3674
C-B	$k = 7$	3.4197
KbK	$k = 9, r = 3, c = 5, q = 17$	2.2946
WbK	$k = 5, r = 5, c = 4, q = 17$	2.3418
BbK	$k = 17, r = 3, c = 2, q = 5$	3.1453
C-K	$k = 9$	2.5346

proved to be able to enhance the accuracy of the conventional indoor WLAN RSS fingerprint based localization algorithms.

Finally, to verify the efficiency of the modified Bayesian inference for indoor WLAN RSS fingerprint based localization, we compare the CDFs of errors by C-B, WbB, and the WbB using the modified Bayesian inference. In Fig. 3, we find that: (i) the C-B performs poorest in localization accuracy with the probabilities of errors within 3 m lower than 60%; (ii) compared to the C-B, the higher localization accuracy is achieved by using the proposed WbB with the probabilities of errors within 3 m more than 70%; and (iii) there is a further improvement in localization accuracy when the WbB using the modified Bayesian inference is adopted.

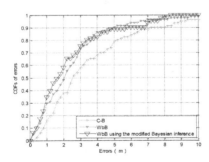

Fig. 3. CDFs of errors by C-B, WbB, and the WbB using the modified Bayesian inference.

5   Conclusion

A novel approach to improve the accuracy of the indoor WLAN RSS fingerprint based localization by using the integrated physical neighborhood and location fingerprinting database is proposed in this paper. We not only construct location fingerprinting database, but also utilize the physical adjacency relations of RPs to construct the physical neighborhood database. The extensive experiments demonstrate that the integration of the physical neighborhood database and location fingerprinting database can help a lot in improving the accuracy of the widely-used KNN, WKNN, and Bayesian inference algorithms. For the future work, the more accurate and efficient estimation of the RSS at each NRP forms an interesting topic.

Acknowledgment. This work was supported in part by the Program for Changjiang Scholars and Innovative Research Team in University (IRT1299), National Natural Science Foundation of China (61301126), Special Fund of Chongqing Key Laboratory (CSTC), and Fundamental and Frontier Research Project of Chongqing (cstc2013jcyjA40041 and cstc2013jcyjA40032) and Postgraduate Scientific Research and Innovation Project of Chongqing (CYS16157).

References

1. Zhou, M., Wong, A.K., Tian, Z., Zhang, V.Y., Yu, X., Luo, X.: Adaptive mobility mapping for people tracking using unlabelled Wi-Fi shotgun reads. IEEE Commun. Lett. **17**(1), 87–90 (2013)
2. Bahl, P., Padmanabhan, V.N.: RADAR: an in-building RF-based user locationand tracking system. In: IEEE INFOCOM, vol. 2, pp. 775–784 (2000)
3. Zhou, M., Wong, A.K., Tian, Z., Luo, X., Xu, K., Shi, R.: Personal mobility mapconstruction for crowd-sourced Wi-Fi based indoor mapping. IEEE Commun. Lett. **18**(8), 1427–1430 (2014)
4. Sun, Y., Xu, Y., Ma, L., Deng, Z.: KNN-FCM hybrid algorithm for indoor location on WLAN. In: The 2nd International Conference on Power Electronics and Intelligent Transportation System, pp. 251–254 (2009)
5. Tang, D., Liu, D., Xu, Z., Jang, P.: Research on indoor localization technology based on nearest neighbor points database. Transducer Microsyst. Technol. **32**(9), 69–71 (2013)
6. Alasti, H., Xu, K., Dang, Z.: Efficient experimental path loss exponent measurement for uniformly attenuated indoor radio channels. In: The 10th IEEE Southeast Conference, pp. 255–260 (2009)
7. Ma, L., Xu, Y., Wu, D.: A novel two-step WLAN indoor positioning method. J. Comput. Inf. Syst. **6**(14), 4627–4636 (2010)
8. Tian, L., Liu, Z.: Least-squares method piecewise linear fitting. Comput. Sci. **39**(6A), 482–484 (2012)
9. Zhou, M., Zhang, Q., Tian, Z., Qiu, F., Wu, Q.: Integrated location fingerprinting and physical neighborhood for WLAN probabilistic localization. In: International Conference on Computing, Communication and Networking Technologies, pp. 1–5 (2014)
10. Zhou, M., Qiu, F., Kunjie, X., Tian, Z., Haibo, W.: Error bound analysis of indoor Wi-Fi location fingerprint based positioning for intelligent access point optimization via fisher information. Comput. Commun. **86**, 57–74 (2016)
11. Zhou, M., Zhang, Q., Tian, Z., Xu, K., Qiu, F., Wu, H.: IMLours: indoor mapping and localization using time-stamped WLAN received signal strength. In: IEEE Wireless Communications and Networking Conference, pp. 1817–1822 (2015)
12. Jiang, Q., Li, K., Zhou, M., Tian, Z., Xiang, M.: Competitive agglomeration based KNN in indoor WLAN localization environment. In: 10th International Conference on Communications and Networking in China, pp. 338–342 (2015)

13. Zhou, M., Zhang, Q., Tian, Z., Qiu, F., Wu, Q.: Correlated received signal strength correction for radio-map based indoor Wi-Fi localization. In: International Conference on Computing, Communication and Networking Technologies, pp. 1–6 (2014)
14. Zhou, M., Qiu, F., Tian, Z., Haibo, W., Zhang, Q., He, W.: An information-based approach to precision analysis of indoor WLAN localization using location fingerprint. Entropy **17** (12), 8031–8055 (2015)
15. Tian, Z., Liu, X., Zhou, M., Xu, K.: Mobility tracking by fingerprint-based KNN/PF approach in cellular networks. In: IEEE Wireless Communications and Networking Conference, pp. 4570–4575 (2013)
16. Zhou, M., Tian, Z., Kunjie, X., Xiang, Yu., Haibo, W.: Theoretical entropy assessment of fingerprint-based Wi-Fi localization accuracy. Expert Syst. Appl. **40**(15), 6136–6149 (2013)
17. Zhou, M., Zhang, Q., Kunjie, X., Tian, Z., Wang, Y., He, W.: PRIMAL: page rank-based indoor mapping and localization using gene-sequenced unlabeled WLAN received signal strength. Sensors **15**(10), 24791–24817 (2015)

Intelligent Spectrum (or Resource Block) Allocation Schemes

Sensing-Throughput Tradeoff in Spectrum Handoff-Based Cognitive Radio

Xin Liu[1,2(✉)], Weidang Lu[3], and Feng Li[3]

[1] School of Information and Communication Engineering,
Dalian University of Technology, Dalian 116024, People's Republic of China
liuxinstar1984@dlut.edu.cn
[2] College of Astronautics, Nanjing University of Aeronautics and Astronautics,
Nanjing 210016, China
[3] College of Information Engineering, Zhejiang University of Technology,
Hangzhou 310014, People's Republic of China

Abstract. In cognitive radio (CR), there is a tradeoff between spectrum sensing time and throughput of secondary user (SU). In order to improve the SU's throughput, a sensing-throughput tradeoff in spectrum-handoff based CR is proposed, which allows the SU to search and transfer to a new idle channel to continue communication, when the PU is present. An optimization problem is proposed to maximize the SU's throughput in the proposed scheme through jointly optimizing the sensing time, the searching time and the number of available channels subject to the detection probability to the PU. The simulation results show that there exist the optimal solutions to the proposed scheme and the proposed scheme outperforms the conventional scheme notably.

Keywords: Cognitive radio · Energy sensing · Throughput · Sampling frequency · Detection probability

1 Introduction

In order to improve the current spectrum utilization, CR allows the secondary user (SU) to dynamically access the idle spectrum that isn't temporarily used by the PU, providing that the SU will not disturb the normal communication of the PU [1–3]. To control interference to PU, the SU detects whether the PU exists in the channel depending on the spectrum sensing technology; if the absence of the PU is detected, the SU can transmit data in the channel, and otherwise it must stop communicating [4, 5].

Energy sensing, as an effective spectrum sensing technology, is frequently used in CR, which estimates the presence of the signal source through comparing the energy statistics of the received signal to a presettled threshold [6–8]. The SU often uses the PU's spectrum by the listen-before-transmit strategy. At the beginning of every frame, the SU firstly detects the absence of the PU in the sensing slot, and then forwards data in the transmission slot [9]. An optimization scheme of energy sensing-throughput tradeoff is proposed in [10–13], whose optimal solutions are theoretically proven to be existent; however, the authors assume that when the presence of the PU is detected, the SU must stop transmitting and wait until the new detection result is achieved in the following frame, therefore yielding the great loss of the SU's throughput.

© ICST Institute for Computer Sciences, Social Informatics and Telecommunications Engineering 2017
X.-L. Huang (Ed.): MLICOM 2016, LNICST 183, pp. 181–188, 2017.
DOI: 10.1007/978-3-319-52730-7_18

In the proposed scheme, the SU is allowed to transfer to another idle channel to continue communication through spectrum searching, when the presence of the PU is detected. We have considered the spectrum handoff and proposed a joint optimization scheme of sensing time and searching time [14]. Through the joint optimization, the total sensing delay including local spectrum sensing and idle channel searching can be decreased and the achievable throughput of the SU in a specific frame can be maximized.

2   Sensing-Throughput Tradeoff Scheme

2.1   Energy Sensing

In energy sensing, according to the two different states: the absence of the PU (denoted by H_0) and the presence of the PU (denoted by H_1), the received sampling signal $y(m)$ is given by

$$y(m) = \begin{cases} h(m)s(m) + n(m), H_1 \\ n(m), H_0 \end{cases}, m = 1, 2, \ldots, M \tag{1}$$

where $s(m)$ is the PU's signal with the power of p_s, $n(m)$ is the Gaussian noise with the power of σ_n^2, $h(m)$ is the channel gain between the SU and the PU, and M is the number of the sampling nodes. By supposing that the sampling frequency is f_s and the spectrum sensing time is τ, M is given by

$$M = \tau f_s. \tag{2}$$

Energy sensing firstly calculates the energy statistics of the SU's received signal as follows

$$Z(y) = \frac{1}{M} \sum_{m=1}^{M} \|y(m)\|^2. \tag{3}$$

By comparing $Z(y)$ to a presettled threshold λ, the presence of the PU is decided by $Z(y) \geq \lambda$, while the absence of the PU is determined by $Z(y) < \lambda$. Hence, the false alarm probability P_f and the detection probability P_d are the functions related with the sensing time τ as follows

$$P_f(\tau) = Q\left(\left(\frac{\lambda}{\sigma_n^2} - 1\right)\sqrt{\tau f_s}\right); \quad P_d(\tau) = Q\left(\left(\frac{\lambda}{\sigma_n^2(1 + \gamma_p)} - 1\right)\sqrt{\tau f_s}\right) \tag{4}$$

where the SNR of the communication link between the PU and the SU is $\gamma_p = h^2 p_s / \sigma_n^2$ and the function $Q(x) = \frac{1}{\sqrt{2\pi}} \int_x^{+\infty} \exp(-\frac{y^2}{2}) \, dy$. According to (4), P_f can be denoted by P_d as follows

$$P_f(\tau) = Q\left(Q^{-1}(P_d)(1 + \gamma_p) + \gamma_p \sqrt{\tau f_s}\right). \tag{5}$$

2.2   System Model

By supposing that the length of the SU's transmission frame is T, the conventional sensing-throughput tradeoff scheme is shown in Fig. 1. In the conventional scheme, every frame of the SU includes one sensing slot and one transmission slot. In the sensing slot, the SU firstly senses the PU, and if the absence of the PU is detected, the SU then forwards data in the transmission slot, otherwise it stops transmitting any data in this frame and waits to redetect the absence of the PU in the following frame [11]. In the conventional scheme, when the presence of the PU is detected, the SU cannot continue communication, which yields the great throughput loss and even interrupts the normal communication of the PU.

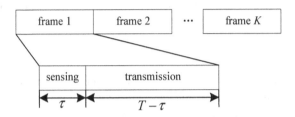

Fig. 1. Conventional sensing-throughput tradeoff scheme

The proposed spectrum handoff-based sensing-throughput tradeoff scheme is shown in Fig. 2, wherein one searching slot is added following the sensing slot, if the presence of the PU is detected in the sensing slot. By supposing that the frequency band available for the SU includes L channels, the SU will detect the left $L-1$ channels one by one for choosing a new idle channel in the searching slot, and then transfer to this idle channel to continue communication in the following transmission slot. In this figure, we assume that the time for searching one channel is ξ.

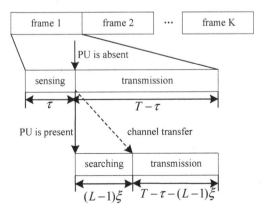

Fig. 2. Spectrum handoff-based sensing-throughput tradeoff scheme

We suppose that the communication rates of the SU at the absence and presence of the PU are C_0 and C_1, respectively, and $P(H_1)$ and $P(H_0)$ are the present and absent probabilities of the PU, respectively. Then we have

$$P(H_1) + P(H_0) = 1. \tag{6}$$

In the conventional scheme of Fig. 1, the SU may forward data in the transmission slot with the time of $T - \tau$, when the absence of the PU is detected. Hence, the SU's throughput in the unit bandwidth is given by

$$R_0(\tau) = (T - \tau) \ (P(H_0)C_0(1 - P_f(\tau)) + P(H_1)C_1(1 - P_d(\tau))) \tag{7}$$

When the SU detects the presence of the PU, it stops communicating in the transmission slot with the probability of

$$P_u(\tau) = P(H_0)P_f(\tau) + P(H_1)P_d(\tau). \tag{8}$$

In the proposed scheme of Fig. 2, when the SU detects the presence of the PU, it may choose and transfer to a new idle channel to continue communication, providing that there exists one idle channel among the left L–1 channels. The existent probability of one idle channel is given as follows

$$P_v(L) = 1 - P(H_1)^{L-1}. \tag{9}$$

After transferring to a new idle channel, the SU will forward data in the transmission slot of this channel with the time of $T - \tau - (L - 1)\xi$. As in (9), the SU's throughput in the unit bandwidth of the new channel is given by

$$R_1(\tau, \xi, L) = (T - \tau - (L - 1)\xi)P_u(\tau)P_v(L) \times \\ (P(H_0)C_0(1 - P_f(\xi)) + P(H_1)C_1(1 - P_d(\xi))) \tag{10}$$

Hence, the aggregate throughput of the SU in the proposed scheme is given by

$$R(\tau, \xi, L) = R_0(\tau) + R_1(\tau, \xi, L). \tag{11}$$

2.3   Scheme Optimization

With the fixed frame length, the transmission time available for the SU decreases with the increasing of the spectrum sensing time, yielding the great throughput loss; on the other hand, since $Q(x)$ is a monotone decreasing function, according to (5), with the fixed detection probability, the false alarm probability decreases with the increasing of the spectrum sensing time, yielding the great increasing of the spectrum access opportunity of the SU. Hence, there exists a tradeoff between the spectrum sensing time and the SU's throughput, and the achieved throughput can be improved through reasonably optimizing the sensing time.

Our goal to optimize the proposed scheme is to maximize the SU's throughput R through jointly optimizing the sensing time τ, the searching time ξ and the number of available channels L, providing that the detection probability to the PU is guaranteed. The optimization problem is described as follows

$$
\begin{aligned}
&\max_{\tau,\xi,L} R(\tau,\xi,L) \\
&\text{s.t. } P_d(\tau) \geq \bar{P}_d \\
&\quad\quad P_d(\xi) \geq \bar{P}_d \\
&\quad\quad \tau + (L-1)\xi \leq T \\
&\quad\quad \tau \geq 0,\ \xi \geq 0 \\
&\quad\quad L \geq 1
\end{aligned}
\tag{12}
$$

Since L is an integer, it is not computationally expensive to search L one by one. Hence, by fixing L, (12) is a double-variable optimization problem of τ and ξ, which can be solved by adopting alternating direction optimization, i.e., we formulate two sub-optimization problems about one of the two parameters, respectively, by fixing the other parameter with an initial value, and then obtain the joint optimal solutions by repeating to optimize these two sub-optimization problems iteratively. We initialize $\tau = \tau_0$ where $\tau_0 \in [0, T]$ and satisfies $P_d(\tau_0) \geq \bar{P}_d$, and therefore $R_0(\tau_0)$ is a constant that can be ignored in the optimization process. We can calculate the approximate solution by using the half searching algorithm as shown in Table 1. The time complexity of the half searching algorithm is related with the computation accuracy δ, which is denoted by $O(log_2\frac{1}{\delta})$.

Table 1. Half searching algorithm

---

(1) initialize $\xi_{\min} = 0, \xi_{\max} = T'$ and $\delta = 10^{-3}$;

(2) let $\xi = (\xi_{\min} + \xi_{\max})/2$;

(3) if $\nabla R_1(\xi) \geq 0$, let $\xi_{\min} = \xi$; otherwise let $\xi_{\max} = \xi$;

(4) repeat to implement (2) and (3) until $|\xi_{\max} - \xi_{\min}| < \delta$;

(5) let $\xi^{\#} = (\xi_{\min} + \xi_{\max})/2$.

Then fixing $\xi = \xi^{\#}$ and substituting it into (11), we have

$$
\begin{aligned}
R(\tau,\xi^{\#}) = &(T-\tau)(P(H_0)C_0(1-P_f(\tau)) + P(H_1)C_1(1-P_d(\tau))) + \\
&P_v(L)G(\xi^{\#})(T-\tau-(L-1)\xi^{\#})(P(H_0)P_f(\tau) + P(H_1)P_d(\tau))
\end{aligned}
\tag{13}
$$

where $G(\xi^{\#}) = P(H_0)C_0(1-P_f(\xi^{\#})) + P(H_1)C_1(1-P_d(\xi^{\#}))$. Obviously $G(\xi^{\#}) < C_0$, and thus $1 - G(\xi^{\#})/C_0 > 0$. There is also an optimal $\tau^{\#} \in [0, T'']$ that makes

$\bar{R}(\tau^{\#})$ achieve the maximum, and $\tau^{\#}$ can also be obtained through the half searching algorithm. By supposing that k is the number of the iterations, the joint optimization algorithm of τ and ξ is shown in Table 2.

Table 2. Joint optimization algorithm

---

(1)   initialize $k=1$, $\tau^{(k)} = \tau_0$, $\xi^{(k)} = 0$ and $\delta = 10^{-3}$;

(2)   with the given $\tau^{(k)}$, calculate the optimal $\xi^{\#}$ by using the half searching algorithm;

(3)   let $\xi^{(k+1)} = \xi^{\#}$, calculate the optimal $\tau^{\#}$ by using the half searching algorithm;

(4)   let $\tau^{(k+1)} = \tau^{\#}$ and $k=k+1$;

(5)   repeat to implement (2)~(4) until $\left|\tau^{(k)} - \tau^{(k-1)}\right| < \delta$ and $\left|\xi^{(k)} - \xi^{(k-1)}\right| < \delta$;

(6)   output the optimal solutions $\tau^* = \tau^{(k)}$ and $\xi^* = \xi^{(k)}$.

The iterative complexity of the joint optimization algorithm is $O(1/\delta^2)$, and the half searching algorithm is implemented in each iteration. Hence, the aggregate time complexity is given by $O\left(\frac{1}{\delta^2}\log_2\frac{1}{\delta}\right)$.

3   Simulation Results

In the simulations, we suppose that the frame length $T = 10$ s, the PU's state probabilities $P(H_0) = 0.2$ and $P(H_1) = 0.8$, and the SNR between the SU's transmitter and receiver $\gamma_s = 5$ dB.

Figure 3 shows the SU's throughput R changing with the spectrum sensing time τ and the channel searching time ξ, with the fixed number of channels $L = 10$. In this figure we also set the sampling frequency $f_s = 1$ kHz, the detection probability $P_d = 0.99$, and the SNR between the SU and the PU $\gamma_p = -10$ dB. It is seen that R is a convexity, which indicates that there deed exist the optimal τ and ξ that make R reach the maximum. When $\tau = 1.8$ s and $\xi = 0.56$ s, the maximum $R = 4.23$ bit·Hz^{-1}. Figure 4 shows R versus $f_s = (200, 400, 600, 800, 1\ k)$ Hz with $\gamma_p = -15 \sim -5$ dB. It is seen that R improves with the increasing of f_s, because the spectrum sensing performance of detecting PU also improves. However, compared to $f_s = 800$ Hz, the improvement on the throughput of $f_s = 1$ kHz is not obvious, which indicates that overmany sampling nodes will not improve the sensing performance notably, but increase the difficulty of designing hardware. Hence, it is very important to choose an appropriate sampling frequency according to our demands.

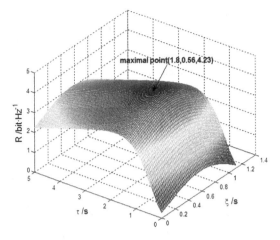

Fig. 3. Throughput changing with sensing time and searching time

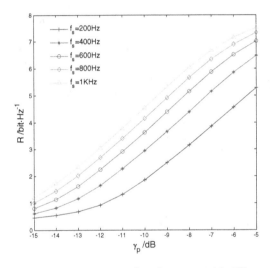

Fig. 4. Throughput versus sampling frequency with different SNR

4   Conclusions

In the proposed sensing-throughput tradeoff scheme, the SU is allowed to transfer to a new idle channel to continue communication through spectrum searching, when the presence of the PU is detected. The SU's throughput is maximized through jointly optimizing the sensing time, the searching time and the number of available channels. Through analyzing the simulation results, we get the following outlines: there deed exist the optimal solutions to the proposed scheme; the proposed scheme outperforms the conventional scheme obviously; an appropriate sampling frequency should be chosen according to our demands.

Acknowledgments. This work was supported by the National Natural Science Foundations of China under Grant No. 61601221; the Natural Science Foundation of Jiangsu Province under Grant No. BK20140828; the Chinese Postdoctoral Science Foundation under Grant No. 2015M580425; the Fundamental Research Funds for the Central Universities under Grant No. DUT16RC(3)045; the Scientific Research Foundation for the Returned Overseas Chinese Scholars of State Education Ministry.

References

1. Mitola, J., Maguire, G.Q.: Cognitive radio: making software radios more personal. IEEE Person. Commun. **6**(4), 13–18 (1999)
2. Haykin, S.: Cognitive radio: brain-empowered wireless communications. IEEE J. Selected Areas Commun. **23**(2), 201–220 (2005)
3. Liu, X., Jia, M., Gu, X., et al.: Optimal spectrum sensing and transmission power allocation in energy-efficiency multichannel cognitive radio with energy harvesting. Intl. J. Commun. Syst. (in press). doi:10.1002/dac.3044
4. Ghasemi, A., Sousa, E.S.: Spectrum sensing in cognitive radio networks: requirements, challenges and design tradeoffs. IEEE Communications Mag. **46**(4), 32–39 (2008)
5. Busson, A., Jabbari, B., Babaei, A., et al.: Interference and throughput in spectrum sensing cognitive radio networks using point processes. J. Commun. Netw. **16**(1), 67–80 (2014)
6. Farag, H.M., Mohamed, E.M.: Improved cognitive radio energy detection algorithm based upon noise uncertainty estimation. In: Proceedings of National Radio Science Conference (NRSC), 28–30 April 2014, Cairo, Egypt, pp. 107–115 (2014)
7. Liu, X., Jia, M., Gu, X., et al.: Optimal periodic cooperative spectrum sensing based on weight fusion in cognitive radio networks. Sensors **13**(4), 5251–5272 (2013)
8. Han, W., Li, J., Li, Z., et al.: Spatial false alarm in cognitive radio network. IEEE Trans. Sig. Process. **61**(6), 1375–1388 (2013)
9. Marinho, J., Monteiro, E.: Cooperative sensing-before-transmit in ad-hoc multi-hop cognitive radio scenarios. In: Proceedings of Wired Wireless Internet Communications (WWIC), 6–8 June 2012, Santorini island, Greece, pp. 186–197 (2012)
10. Edward, C.Y.P., Liang, Y., Guan, Y.L.: Optimization of cooperative sensing in cognitive radio networks: a sensing-throughput tradeoff view. IEEE Trans. Vehicular Technol. **58**(9), 5294–5299 (2009)
11. Liang, Y., Zeng, Y., Edward, C.Y.P., et al.: Sensing-throughput tradeoff for cognitive radio networks. IEEE Trans. Wireless Commun. **7**(4), 1326–1336 (2008)
12. Pei, Y., Hoang, A.T., Liang, Y.: Sensing-throughput tradeoff in cognitive radio networks: how frequently should spectrum sensing be carried out. In: Proceedings of Personal, Indoor and Mobile Radio Communications (PIMRC), 3–7 September 2007, Athens, Greece, pp. 1–5 (2007)
13. Liu, X., Na, Z., Jia, M., et al.: Multislot simultaneous spectrum sensing and energy harvesting in cognitive radio. Energies **9**(7), 1–13 (2016)
14. Liu, X., Li, F., Lu, W.: A novel spectrum handoff-based sensing-throughput tradeoff scheme in cognitive radio. China Commun. (2016, in press)

Self-similar Traffic Prediction Scheme Based on Wavelet Transform for Satellite Internet Services

Yu Han, Dezhi Li, Qing Guo$^{(\boxtimes)}$, Zhenyong Wang, and Deyang Kong

Harbin Institute of Technology, Harbin, China
{13B905008,lidezhi,qguo,ZYWang,kongdyang}@hit.edu.cn

Abstract. With service types and requirements of broadband satellite internet continuously increasing, improving QoS (Quality of service) of satellite internet has attracted extensive attention. To reduce the impact of self-similarity caused by various of service traffic sources converge on satellite communication system, we propose a novel model from the perspective of self-similar traffic prediction. Combining wavelet transform and ARIMA (Autoregressive Integrated Moving Average) model to predict self-similar traffic of satellite internet is proposed. The optimal model to the problem is presented. The number selection of prediction samples and the impact of prediction steps on the accuracy of the prediction system are discussed, and the parameters are addressed. Simulation results show ARIMA model can achieve a better prediction effect with a combination of wavelet transform than that of the traditional autoregressive model not utilizing wavelet technology.

Keywords: Satellite internet · Self-similar traffic prediction · Wavelet transform · ARIMA model

1 Introduction

Nowadays the prosperity of Internet promotes the rapid development of Next Generation Network which consists of air-space-ground integrated network. [1] As a kind of auxiliary network of terrestrial networks, broadband satellite network which has advantages in global coverage, can release the pressure of terrestrial networks. However, it also face a series of challenging problems to be solved. [2,3] The requirements of different service, such as bandwidth and delay, are obviously different. Convergency of various of service traffic sources results self-similarity of the service traffic on satellite communication system [4], having a clear difference from the traditional Poisson characteristics. The self-similar process, which can reflect the long range dependence of the traffic, is one of simple models with long range dependence, reflecting the performance of traffic at any time.

   With service types and quantity growing, the self-similar degree will continuously increase. As a result, the self-similarity of traffic has the adverse effect on satellite communication system, aggravating congestion of the network nodes, as

© ICST Institute for Computer Sciences, Social Informatics and Telecommunications Engineering 2017
X.-L. Huang (Ed.): MLICOM 2016, LNICST 183, pp. 189–197, 2017.
DOI: 10.1007/978-3-319-52730-7_19

well as vibration and vibration delay. A large number of studies show that the main reason of the situations, such as high rate of buffer overflow, lengthened transmission delay and the continuous increasingly periodic congestion, is the self-similarity of network traffic. The traffic self-similarity directly affects structure design, control mechanism, analysis methods and management measures of the next generation network communication system. Besides, packet loss rate rises along with the increase of the self-similarity degree, while the bandwidth utilization decreases [5–8]. Therefore, it is necessary to predict the network traffic reasonably and accurately, in order to provide the prediction queue information to the scheduling star and reliable data for the cross-layer scheduling of satellite Internet. Accurate traffic prediction results can ease the network congestion and optimize the resource allocation to improve the performance of satellite network.

Currently, the research on prediction for internet self-similar traffic has made a lot of achievements both domestically and externally, such as the wavelet prediction model in [9], the neural network prediction model and various time series prediction models in [10], as well as the multi-fractal wavelet model [11], and the finite impulse response neural network prediction model of the multi fractal [12], etc. However, the prediction accuracy of the proposed prediction models, on whose condition variables there is no detailed discussion, is not very satisfactory. It is extremely difficult to ensure the accuracy of real-time prediction for network traffic in the case of high bit error rate on such satellite networks.

In this paper, the method combining wavelet transform and ARIMA model is proposed to predict self-similar traffic of the satellite network. Wavelet decomposition makes the non-stationary, periodic and self-similar network traffic stationary and the degree of self-similarity reduced to achieve traffic prediction reasonably and accurately. With a large number of simulation experiments, the parameters of prediction model are analyzed in details and selected, so as to obtain the optimal prediction model of network traffic improving the accuracy of the prediction. Therefore, it is required to perform a reasonable and accurate prediction for satellite network self-similar traffic, aiming to enhance the QoS of satellite internet.

2 ARIMA Model

ARIMA model is employed to process non-stationary time series. And white noise is processed to obtain historical independence white noise, so as to improve the prediction accuracy. Compared with the traditional prediction models such as AR, MA, ARMA, ARIMA model established based on the Markov process can accurately capture several features on the network. Processing the collected traffic data with the prediction model established to predict the traffic can get the high prediction accuracy.

The basic idea of ARIMA is to use a specific math model to approximate time series. The model can predict the future value through past values of the time series and the moment values as long as recognized.

ARIMA can be described as,

$$\Phi(B)\nabla^d X_t = \Theta(B)\varepsilon_t \tag{1}$$

where $\Phi(B)$ and $\theta(B)$ represent the p and q polynomial respectively. The expressions are shown as,

$$\Phi(B) = 1 - \Phi_1(B) - \ldots - \Phi_p(B)^p \tag{2}$$

$$\Theta(B) = 1 + \Theta_1(B) + \ldots + \Theta_q(B)^q \tag{3}$$

Besides, $Bx_t = x_{t-1}$, where B is the delay factor, ∇^d represents the differential factor of d order, and the relation between them is $\nabla = 1 - B$, where ∇ is the differential operator. The binomial expansion is,

$$\nabla^d = (1 - B)^d = \sum_{k=0}^{\infty} \begin{bmatrix} d \\ k \end{bmatrix} (-B)^k \tag{4}$$

where $\begin{bmatrix} d \\ k \end{bmatrix} = \frac{d(d+1)\ldots(d+k-1)}{k!}$. In general, d is zero or one in ARIMA(p, d, q) model. Then non-stationary time series will be changed into stationary time series [13].

3   ARIMA Prediction Model Based on Wavelet Transform

Wavelet transform can exhibit both global and local characteristics of traffic, and reduce the self-similarity degree of signal decomposed on the different frequency ranges. High frequency part of the decomposed signal is short range dependence, and therefore prediction can be made directly using ARMA model. Nevertheless, the low frequency part having long range dependence is not able to be predicted accurately only utilizing ARIMA model. Thus, the signal is decomposed using wavelet transform level by level, then the levels are predicted respectively and recombined together to achieve network traffic prediction. Wavelet transform improves network traffic prediction accuracy.

Self-similar traffic is decomposed by wavelet transform of different decomposition level, and wavelet function takes sym $N = 2, 3, 4$. By calculating the decomposition of the Hurst, as shown in Table 1, we can see that the correlation of the details of the wavelet decomposition is short, and only the approximate part has long-range dependence. Therefore, this paper selects three level wavelet decomposition to process the self-similar network traffic, so as to achieve the time series of network traffic stationary and weaken the correlation, and guarantee the efficiency of system.

In ARIMA(p, d, q) model, a first-order difference can achieve the expected effect that high degree of traffic self-similarity is changed into low degree of the one, and the data reaches a plateau. Therefore, we take $d = 1$. The autocorrelation function and partial correlation function of the details and the difference

Table 1. Level 2–4 wavelet decomposition with hurst parameter

Level 2	Level 3	Level 4
$H_{A2} = 0.7296$	$H_{A3} = 0.7209$	$H_{A4} = 0.7132$
$H_{D1} = 0.2700$	$H_{D1} = 0.3009$	$H_{D1} = 0.4078$
$H_{D2} = 0.2822$	$H_{D2} = 0.3239$	$H_{D2} = 0.1851$
	$H_{D3} = 0.3711$	$H_{D3} = 0.3873$

approximation part, which are decomposed by three level wavelet decomposition, both show smear characteristic, therefore ARMA model should be chosen. To determine the order p, q of the model the AIC (Akaike information criterion) criterion will be used for order selection. The order of each part of the wavelet decomposition model is shown in Table 2.

Table 2. Order of each part of the wacelet decomposition model

	D1	D2	D3	D4
order	ARMA(2,1)	ARMA(3,1)	ARMA(2,3)	ARMA(1,3)

After determining the order of the prediction model, $\text{ARIMA}(p, d, q)$ model is established. Then the parameter estimation method of ARMA model will be presented. Autocovariance function of $\text{ARMA}(p, q)$ series satisfies extending Yule-Walker equations,

$$
\begin{bmatrix} \gamma_{q+1} \\ \gamma_{q+2} \\ \cdots \\ \gamma_{p+q} \end{bmatrix}
=
\begin{bmatrix} \gamma_q & \gamma_{q-1} & \cdots & \gamma_{q-p+1} \\ \gamma_{q+1} & \gamma_q & \cdots & \gamma_{q-p+2} \\ & \cdots & & \\ \gamma_{q+p-1} & \gamma_{q+p-2} & \cdots & \gamma_q \end{bmatrix}
\begin{bmatrix} a_1 \\ a_2 \\ \cdots \\ a_p \end{bmatrix}
\tag{5}
$$

This is the method to estimate parameters, in which obtaining moment estimation of a

$$
\begin{bmatrix} \hat{a}_1 \\ \hat{a}_2 \\ \cdots \\ \hat{a}_p \end{bmatrix}
=
\begin{bmatrix} \hat{\gamma}_q & \hat{\gamma}_{q-1} & \cdots & \hat{\gamma}_{q-p+1} \\ \hat{\gamma}_{q+1} & \hat{\gamma}_q & \cdots & \hat{\gamma}_{q-p+2} \\ & \cdots & & \\ \hat{\gamma}_{q+p-1} & \hat{\gamma}_{q+p-2} & \cdots & \hat{\gamma}_q \end{bmatrix}^{-1}
\begin{bmatrix} \hat{\gamma}_{q+1} \\ \hat{\gamma}_{q+2} \\ \cdots \\ \hat{\gamma}_{q+p} \end{bmatrix}
\tag{6}
$$

The $p \times p$ matrix $\Gamma_{p,q}$ is invertible in (5). When $N \to \infty$ in $\text{ARMA}(p, q)$ model, the $\Gamma_{p,q}$ in (6) is invertible. Based on the above conditions, moment estimation (6) are consistent, namely $\lim_{N \to \infty} \hat{a}_j = a_j$, $1 \le j \le p$. Next we will estimate partial parameters in $\text{MA}(q)$, namely $\hat{a}_1, \hat{a}_2, ..., \hat{a}_p$. Then

$$
z_t = x_t - \sum_{j=1}^{p} \hat{a}_j x_{t-j}, t = p+1, p+2, ..., N
\tag{7}
$$

(7) is an approximate measurement data to MA(q). The auto covariance function is given in (8) with $\widehat{a}_0 = -1$,

$$\widehat{\gamma}_z(k) = \sum_{j=0}^{p} \sum_{l=0}^{p} \widehat{a}_j \widehat{a}_l \widehat{\gamma}_{k+j-l} \tag{8}$$

Finally, the stationary time series is predicted. In order to predict X_{n+k} with $X_n = (X_1, X_2, \ldots, X_n)^T$, first we use ARMA sequence to achieve optimal linear prediction based on $Y_{d+1}, Y_{d+2}, \ldots, Y_{d+k}$,

$$\widetilde{Y}_{n+j} = L(Y_{n+j}|Y_{d+1}, Y_{d+2}, \ldots, Y_n), \; j = 1, 2, \ldots, k \tag{9}$$

Based on the formula,

$$(1 - B)^d \widehat{X}_t = \widetilde{Y}_t \;, \; t = n+1, n+2, \ldots, n+k \tag{10}$$

It comes to the recursive formula

$$\widehat{X}_{n+k} = \widetilde{Y}_{n+k} - \sum_{j=1}^{d} C_d^j (-1)^j \widehat{X}_{n+k-j} \;, \; k \geq 1 \tag{11}$$

where $\widehat{X}_{n-j} = X_{n-j} \;, j \geq 0$.

4   Simulation and Analysis

Since traffic data of satellite communication system can not be obtained, this paper will use the data collected in the ground backbone network from http:// mawi.wide.ad.jp/mawi/ditl/ditl2009/. The data is collected continuously for 96 h in every 15 min, which is 380 data in total. After calculation, the average rate of the ground backbone network can reach 98 Mbps, slightly less than the traffic receiving rate of 100 Mbps on satellite switches of broadband satellite network. So it can simulate the traffic in broadband satellite network with the set of data to establish prediction model. The traffic with a day (24 h) period as time unit is non-stationary and has a very obvious sudden. By the simulation, the Hurst parameters $H \approx 0.7308$, according to $H \in (0.5, 1)$ can observe that the network traffic shows an obvious self-similarity.

This paper will use this traffic data. The impact of prediction step k and the number n selection of predicted samples to the traffic data prediction accuracy of prediction system is discussed.

First, the impact of the number n selection of predicted samples on prediction accuracy is analyzed. The smaller the relative root mean square error (RRMSE) of prediction results is, the higher the prediction accuracy is.

The original traffic data is decomposed with the three level wavelet transform, and the prediction step in prediction system is set to 1. By simulations, the RRMSE between the prediction value from different number selection of predicted samples and the true value is obtained, as shown in Fig. 1.

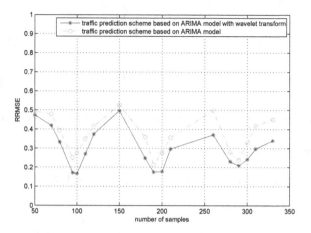

Fig. 1. RRMSE of different number of predicted samples.

Network traffic with self-similarity is subdivided into different frequency range by utilizing wavelet decomposition in order to observe the self-similarity of detail and approximate parts. The decomposition signal with the short range dependence is to be predicted in the next step, while doing difference to the other one with the long range dependence to produce the signal which is short or approximately short range dependence. The accuracy of analysis is improved because of the stationary and multi-resolution processing to original signal.

As observed in Fig. 1, the better prediction results lie in the range of the interval $[95, 100]$, $[190, 200]$, $[290, 300]$. The number of samples in the range of $[95, 100]$ have the best effect, and the RRMSE reaches the minimum value. The

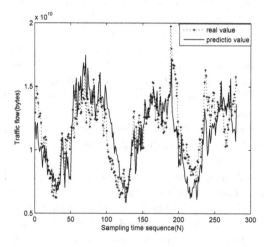

Fig. 2. Comparison between the real and prediction traffic (number of samples is 100).

three interval in which prediction accuracy is high is approximately a multiple of 96, mainly because traffic data used in the model is collected every 15 min, then record 96 times within 24 h, therefore in the original model data 96 is a data cycle. The number selection of predicted samples is related to periodicity of network traffic data itself, and choosing the number of predicted samples which is in accordance with periodicity can obtain high accuracy prediction results.

Figure 2 is the comparison between the real traffic and the prediction traffic when the number of samples is set to 100. As demonstrated in Fig. 2, the fitting effect choosing this number of samples is fine, and the predictive value and the real value maintain a high degree of consistency in the burst point, while it also reflects the periodicity of the original traffic data.

Next, the impact of different prediction steps k on prediction accuracy is analyzed. The number of predicted samples is set to 100.

As shown in Table 3, it can be seen that the smaller prediction steps are, the better prediction effect is. However, difference accross RRMSE of one-step prediction, two-steps and five-steps prediction is not great. If using one-step prediction which has the best effect, the ratio between the number of predicted samples and the predicted number is 100: 1, but the efficiency of prediction model is very low.

Table 3. RRMSE of different predicted steps

	One-step	Two-steps	Five-steps	Ten-steps
RRMSE	0.1663	0.1682	0.1713	0.2330

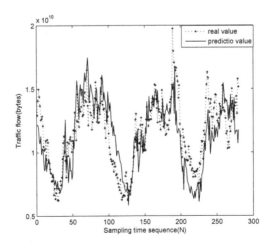

Fig. 3. Comparison between the real and prediction traffic in two-steps prediction

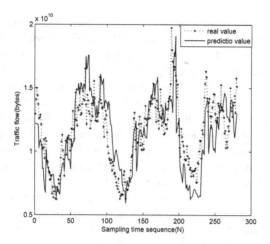

Fig. 4. Comparison between the real and prediction traffic in five-steps prediction

In Figs. 3 and 4, in fitting between the prediction value and the real value, there is not much difference between two-steps prediction and five-steps prediction. In particular, the five-steps prediction is better than the two-steps one in the burst traffic data prediction. Compared with the ratio between the number of two-steps predicted samples and the predicted number is 50: 1, the ratio is 20: 1 in five-steps prediction. Thus, the prediction step is assigned to 5, and time series in autoregressive prediction model can get the best prediction value.

5   Conclusion

This paper presents the self-similarity of traffic has the adverse effect on satellite communication system, and traffic prediction model based on ARIMA model with wavelet transform is established to reduce the impact of self-similarity. The impact of prediction steps and number selection of prediction samples on prediction accuracy of the model is analyzed, while obtaining the optimal traffic prediction model improving the accuracy of the prediction. Simulation results show that RRMSE between the prediction and real value is 0.1663, using 100 as prediction samples which is periodicity of network traffic data itself approximately and predicting 5 values per step. With a combination of wavelet technology, ARIMA model can achieve a better prediction effect, aiming to enhance the QoS of satellite internet.

References

1. Sallai, G.: Chapters of future internet research. In: 2013 IEEE 4th International Conference. IEEE (2013)
2. Fan, L., Cruickshank, H., Sun, Z.: IP Networking of Next-Generation Satellite Systems. Springer, Heidelberg (2007). pp. 1–14

3. Ngo, T.A., Tummala, M., McEachen, J.C.: Optimal wireless aerial sensor node positioning for randomly deployed planar collaborative beamforming. In: 2014 47th Hawaii International Conference on System Sciences (HICSS), pp. 5122–5128, January 2014

4. Urke, A.R., Braten, L.E., Ovsthus, K.: TCP challenges in hybrid military satellite networks; measurements and comparison. In: Military Communications Conference, pp. 1–6. IEEE (2012)

5. Xuan, Y., Tao, S.: Research on self-similarity network group. In: 2011 IEEE International Conference on Computer Science and Automation Engineering (CSAE), vol. 3, pp. 52–56. IEEE (2011)

6. Leland, W.E., Taqqu, M.S., Willinger, W., Wilson, D.V.: On the self-similar nature of ethernet traffic. IEEE/ACM Trans. Netw. **2**(1), 1–15 (1994)

7. Zhen-Yu, N., Zi-He, G., Qing, G.: Performance analysis of self-similar traffic in LEO satellite network. In: 2007 International Conference on Machine Learning and Cybernetics, pp. 2649–2652 (2007)

8. Song, S., Thompson, J.S., Chung, P.-J., Grant, P.M.: Ber analysis for distributed beamforming with phase errors. IEEE Trans. Veh. Technol. **59**(8), 4169–4174 (2010)

9. Vankka, J.: Performance of satellite gateway over geostationary satellite links. In: Military Communications Conference, MILCOM 2013, pp. 289–292. IEEE (2013)

10. Neng, Z., Jianfeng, G., Changgiao, X.: Traffic prediction model for cognitive networks. In: 2011 International Conference on Advanced Intelligence and Awareness Internet (AIAI 2011), pp. 76–80 (2011)

11. Yongtao, W., Jinkuan, W., Cuirong, W., et al.: Network traffic prediction by traffic decomposition. In: 2012 Fifth International Conference on Intelligent Networks and Intelligent Systems (ICINIS), pp. 158–161 (2012)

12. Chen, D., Feng, H., Lin, Q., et al.: Multi-scale internet traffic prediction using wavelet neural network combined model. In: First International Conference on Communications and Networking in China, ChinaCom 2006, pp. 1–5 (2006)

13. Feng, H.A., Shlt, Y.T.: Study on network traffic prediction techniques. In: Wireless Communications, Networking and Mobile Computing Proceedings, vol. 2, pp. 995–998 (2005)

Machine Learning Algorithm
and Cognitive Radio Networks

Space-Based Information Integrated Network Technology and Performance Analysis Based on Cognitive Radio

Shuai Liu[✉], Hu-mei Wang, Shi-tao Wang, and Ming-ming Bian

Beijing Institute of Spacecraft System Engineering, Beijing 100094, China
lsshr@163.com

Abstract. In the demonstration and construction process of the space-based remote sensing, communication, navigation and information integration network, there are several problems such as: the network system architecture is complex, the information fusion efficiency is low and the anti-jamming ability is poor and so on. Based on space-based cognitive radio technology, the paper proposed the method to optimize and enhance the space-based information network performance by adding cognitive satellite nodes in the network. Mathematical modeling and performance analysis results show that the space-based cognitive integrated information network can optimize network configuration, improve the task affect speed, use the data resources efficiently. Meanwhile, the method can enhance the anti-jamming performance of the entire space-based information network greatly.

Keywords: Space-based · Integrated information network · Performance analysis · Cognitive radio

1 Introduction

Recently, the number of terrestrial internet and mobile users of China become a world leader with the rapid development of information network construction. However, the development of the space-based information network, the internet and the mobile communication network is very uneven, showing "weak space strong ground" characteristics. In the space-based, Although China has initially built Beidou navigation and position system, earth observation and remote sensing system, data communication relay system in the space, while each system can form internal space-based network. However, since the beginning of the construction of major systems is relatively independent, the satellite types are relatively simple, there is no inter-satellite network, the space-based information network can not give out the comprehensive performance efficiently.

With the rapid development of network technology, especially the emergence of inter-satellite links (ISL), space-based information system gradually to form network [1]. During this time we worked a lot of top-level design of space-based information network and general structure research, divided system functional specifications, and defined the system composition and function [2]. Typical space-based information

© ICST Institute for Computer Sciences, Social Informatics and Telecommunications Engineering 2017
X.-L. Huang (Ed.): MLICOM 2016, LNICST 183, pp. 201–208, 2017.
DOI: 10.1007/978-3-319-52730-7_20

network systems in foreign include the Iridium system, Advanced EHF satellite communication systems, global position system (GPS) and so on. We focused on information integration of the space-based systems of communication, navigation and remote sensing in China.

Although we have a good foundation in the space-based information network, and have made many useful ideas and concepts of space-based information fusion of three networks, it is necessary to solve single information network system itself and also to solve the new problems posed by the integration of three information networks. Currently, the space-based information network argumentation problems mainly as follows: the complex network architecture, the long task response time in network, low allocation efficiency of resources within the network, the poor environmental adaptability, weak anti-interference ability and so on [3]. Therefore, new means and methods to demonstrate and optimize the space-based information network system are urgently needed.

Based on cognitive radio technology, the paper proposed a new method to build cognitive network nodes in the network [4], which can optimize the network architecture, improve the system response time, optimize the resources allocation, and enhance the anti-jamming capability of the entire network greatly [5].

2 Space-Based Cognitive Radio Technology

Cognitive radio is an intelligent wireless communication system which can sense and learn the spectral characteristics of the surrounding. According to adjust the specific transmission parameters such as transmit power, carrier frequency, modulation, etc. in real time, the internal states can adapt to the external input radio incentives, which can achieve high reliability and efficiency use of spectrum resources whenever and wherever in communication system.

Based on the traditional cognitive radio, space-based cognitive radio technology can use their own learning and reasoning ability, to deploy the available network resources adaptively, fusion the network-wide data effectively, response rapidly and complete the tasks efficiently by cognitiving tasks intelligently and learning surrounding wireless environment quickly. Meanwhile, it can greatly improve the anti-jamming performance in the space-based information network [6].

At the same time, the space-based remote sensing, communication, navigation and integrated information network systems are very complex, we need to analysis the system performance adding the cognitive nodes before and after. System performance analysis process contains a number of uncertainties, such as fuzzy comprehensive evaluation, artificial neural networks and so on [7]. In this paper, we carry out theoretical mathematical model and analysis the space-based cognitive-information network performance [8].

3 Space-Based Cognitive-Integrated Network Technology and Performance Analysis

With the strong demand and rapid develop of space-based integrated information network, it's necessary to build the integrated network quickly. At this stage, the space-based integration information network system mainly uses the "local distribution station, network in the space" mode [9]. With the support of terrestrial network systems, we focus on building space-based backbone network, space-based access network, space-based remote sensing network and space-based space-time reference network. The network can propose information acquisition, transmission and temporal reference services by the inter-satellite link connection. The space-based integrated information network architecture shown as Fig. 1 [10].

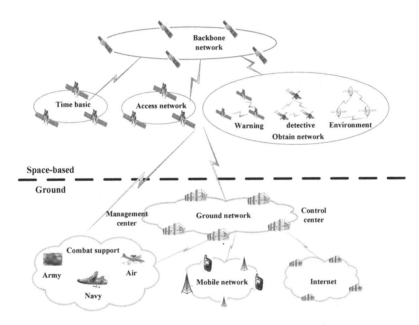

Fig. 1. Space-based integrated information network architecture

Of course, it's necessary to build the space-based integrated information network depending on the existing space-based equipment and facilities. Therefore, the space-based information obtain network based on remote sensing network, the space-based space-time reference network based on navigation network, the space-based backbone network and access network based on communication network to construct and demonstrate.

3.1   Space-Based Cognitive-Integrated Network Technology

Based on cognitive radio technology, we can optimize the integrated information network architecture to increase the flexibility and efficiency of the whole network system. The space-based cognitive-integrated information network architecture shown as below (Fig. 2):

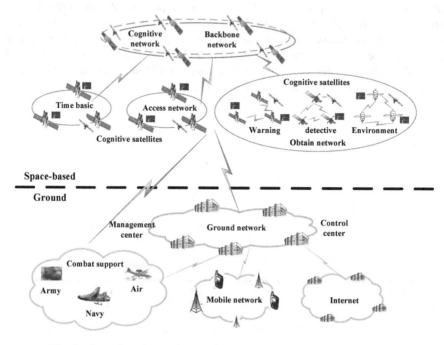

Fig. 2. Space-based cognitive-integrated information network architecture

As we can seen, although the space-based integrated information network is powerful, it's architecture is quite complex. The network is unable to maximize the throughout performance apparently if only building ISLs among several independent information networks. The space-based cognitive-integrated network has the following main functions: the network can respond to external tasks quickly and efficiently in order to provide services to users on demand; based on satellite data and external tasks, the network can adjust the internal data by allocating resources dynamically, which can complete the tasks efficiently in a higher data fusion rate; according to sense the external environment and interference in real time, the network can avoid interference or use the undisturbed resources to complete the tasks by planning in advanced [11].

3.2   Space-Based Cognitive-Integrated Network Performance Analysis

The space-based integrated information network is built of remote sensing, communication and navigation network, so the network performance analysis indicators include data fusion capability, task execution efficiency and robustness of the network and so on except the remote sensing, communication and navigation indicators.

Take the average response time for example, we compared the system performance between the space-based integrated information network and cognitive-integrated information network. The method is divided into the following steps [12]:

(1) Space-based integrated information and cognitive-integrated information data from STK simulation;
(2) Choose 10 agencies/experts to process and evaluate the index values. Set the dimensionless credibility of 10 agencies are a1 = 0.95, a2 = 0.90, a3 = 0.70, a4 = 0.95, a5 = 0.85, a6 = 0.80, a7 = 0.80, a8 = 0.75, a9 = 0.85, a10 = 0.90. The judgment values of 10 agencies are p1 = 0.1, p2 = 0.1, p3 = 0.09, p4 = 0.08, p5 = 0.1, p6 = 0.11, p7 = 0.1, p8 = 0.12, p9 = 0.08, p10 = 0.09.
(3) We analyze the performance of space-based integrated information network firstly, Table 1 is the normalization allocation table of 10 agencies in space-based remote sensing network.

Table 1. Space-based integrated information network normalization allocation

	Excellent	Good	Average	Bad	Poor	Uncertain
y1	0.19	0.19	0.285	0.19	0.095	0.05
y2	0.27	0.18	0.18	0.09	0.18	0.1
y3	0.21	0.28	0.105	0.07	0.035	0.3
y4	0.095	0.19	0.38	0.19	0.095	0.05
y5	0.17	0.255	0.17	0.17	0.085	0.15
y6	0.08	0.32	0.16	0.16	0.08	0.2
y7	0.08	0.16	0.24	0.16	0.16	0.2
y8	0.3	0.225	0.075	0.075	0.075	0.25
y9	0.34	0.17	0.17	0.085	0.085	0.15
y10	0.18	0.18	0.315	0.135	0.09	0.1

(4) Select the performance analysis methods, we can obtain the excellent/good/average/bad/poor/uncertain results after 9 synthesis according to mathematical models and probability distribution function, the results are: (0.2466, 0.3354, 0.2762, 0.0833, 0.0525, 0.006).
(5) 10 agencies evaluate the importance of the impact indicators independently, Table 2 shows the importance and the important coefficient of the average remote sensing response time.
(6) Repeating this process, we can obtain the integrated system performance evaluation results are (0.2327, 0.3611, 0.2543, 0.1027, 0.0452, 0.004), the excellent degree is 23.27%, good degree is 36.11%, average degree is 25.43%, bad degree is 10.27%, poor degree is 4.52%.

Similarly, we can obtain the integrated system performance evaluation results of space-based cognitive-integrated information network are (0.2740, 0.3857, 0.2507, 0.0655, 0.0206, 0.0035).

Table 2. Importance factor of average response time

	Most important (0.30)	Middle value (0.24)	Very important (0.18)	Middle value (0.15)	Important (0.09)	No important (0.04)
y1	✓					
y2	✓					
y3	✓					
y4		✓				
y5		✓				
y6	✓					
y7			✓			
y8		✓				
y9		✓				
y10	✓					

According to space-based integrated information system performance analysis method, we can obtain the integrated network and cognitive-integrated network analysis as follow (Fig. 3):

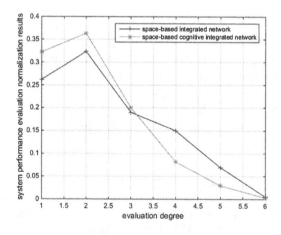

Fig. 3. System performance evaluation normalization results

From the figure we can see that the excellent and good results of space-based cognitive-integrated information network are better than space-based integrated network, bad and poor results have a greater degree of decline, the overall system performance is developed greatly.

In addition, the anti-jamming performance of space-based cognitive-integrated information network is shown as follow [13]. We can see that the error rates of cognitive-integrated information network have a obvious decline compared with integrated information network (Fig. 4).

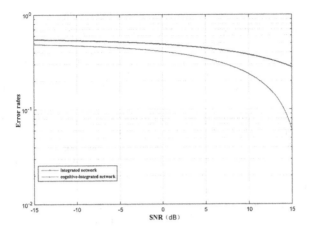

Fig. 4. Error rates results

4 Conclusion

In this paper, we proposed a new method to develop the system performance of integrated information network based on cognitive radio technology. According to constructing cognitive backbone network and adding cognitive satellite nodes in the network architecture, we can response the tasks rapidly, complete the data fusion effectively and develop the whole network system performance finally. In addition, adding the cognitive radio nodes can improve the anti-jamming performance greatly in the space-based information network.

References

1. Chini, P., Giambene, G., Kota, S.: A survey on mobile satellite systems. Int. J. Satell. Commun. Netw. **28**(1), 29–57 (2010)
2. Alagoz, F., Korcak, O., Jamalipour, A.: Exploring the routing strategies in next-generation satellite networks. IEEE Wirel. Commun. **14**(3), 79–88 (2007)
3. Hu, B., Li, F., Zhou, H.S.: Robustness of complex networks under attack and repair. Chin. Phys. Lett. **26**(12), 12–18 (2009). Beijing
4. Hoytya, M., Kyrolainen, J., Hulkkonen, A., et al.: Application of cognitive radio techniques to satellite communication. In: IEEE International Symposium on Dynamic Spectrum Access Networks, Bellevue, Washington, USA, pp. 540–551 (2012)
5. Biglieri, E.: An overview of cognitive radio for satellite communications. In: 2012 IEEE First AESS Europe an Conference on Satellite Telecommunications (ESTEL), Rome, Italy, p. 13 (2012)
6. Shree, K.S., Symeon, C.: Cognitive radio techniques for satellite communication systems. IEEE Trans. Commun. **21**(6), 781–787 (2013)
7. Nishiyama, H., Tada, Y., Kato, N., et al.: Toward optimized traffic distribution for efficient network capacity utilization in two-layered satellite networks. IEEE Trans. Vehicular Technol. **62**(3), 1303–1313 (2013)

8. Urquizo Medina, A.N., Qiang, G.: QoS routing for LEO satellite networks. In: Zu, Q., Hu, B., Elçi, A. (eds.) ICPCA/SWS 2012. LNCS, vol. 7719, pp. 482–494. Springer, Heidelberg (2013). doi:10.1007/978-3-642-37015-1_43

9. Uchida, N., Takahata, K., Shibata, Y., et al.: Never die network extended with cognitive wireless network for disaster information system. In: 2011 International Conference on Complex, Intelligent and Software Intensive Systems (CISIS), pp. 24–31. IEEE, New York (2011)

10. Ciftci, S., Torlak, M.: A comparison of energy detectability models for cognitive radios in fading environments. Wireless Pers. Commun. **68**(3), 553–574 (2013)

11. An, X., Zhao, Y., Yang, L., Zhang, W.: Simulation of effectiveness evaluation for satellite systems based on fuzzy theory. J. Syst. Simul. (S1004-731X) **18**(8), 2334–2337 (2006)

12. Sithamparanathan, K., Nardis, L.D., Benedetto, M.G.D., et al.: Cognitive satellite terrestrial radios. In: IEEE Globecom, pp. 1–6 (2010)

13. Sharma, S.K., Chatzinotas, S., Ottersten, B.: Satellite cognitive communications: interference modeling and techniques selection. In: Advanced Satellite Multimedia Systems Conference (ASMS) and 12th Signal Processing for Space Communications Workshop (SPSC), Munich, German, pp. 111–118 (2012)

Koch Fractal-Based LED Lamp Appearance Design Method

Xin Cao, Xufen Xie$^{(\boxtimes)}$, Weihao Xiao, Nianyu Zou, and Xiaoyang He

Research Institute of Photonics, Dalian Polytechnic University,
Dalian 116034, China
xiexf@dlpu.edu.cn

Abstract. Along with developing of LED lamps, more technical appearance design methods become possible. A lamp appearance design method based on fractal technology is researched in this paper. Firstly, the famous Koch fractal is analyzed at different generators. Then, the Koch fractal pattern is generated by the relevant parameters. Finally, the lamps appearance was designed by the Koch fractal characteristics. The design and simulation results show that the fractal model can be applied to the design of lamps appearance.

Keywords: Koch fractal · Lamp appearance · LED

1 Introduction

LED lighting has become a trend in 21st century. Especially in bedroom lighting, LED lamps will replace traditional incandescent and fluorescent lamps. LED technical features have made the lighting design of contents and modes change substantially. The new light-resource of LED promotes innovations in lighting design and development. To a certain extent, LED technique also changes our concept of lighting. We are gradually liberated from traditional, linear light limitation. Lamp designs of the language and concept can be free to play and reshape with greater and flexible space in the pursuit of creative expression of natural beauty [1, 2].

The fractal theory is a very active branch in Nonlinear Science, which mainly study irregular and rough geometric shapes in nature and non-linear systems. Existing research shows that natural scene and contours possess non-stationary, self-similarity and multiple scales characteristics. These characteristics are described by the stochastic fractal [3–10]. Therefore, Natural scenery and some graphics can be produced by fractal technique [11–13]. It can bring out a person feeling the characteristics of natural beauty.

Koch curve is a very important fractal curve in geometry, with clearly mathematical description and simple programming. In this paper, the deformed Koch curve composes of regular polygons by edges. Because of regular polygon ring closed and symmetrical, the curve changes with some parameters selecting the patterns that can make design possess natural beauty and form new light-outlook. Meanwhile, due to the geometric shapes are constructed by model it makes lamps shape with natural beauty [14, 15].

© ICST Institute for Computer Sciences, Social Informatics and Telecommunications Engineering 2017
X.-L. Huang (Ed.): MLICOM 2016, LNICST 183, pp. 209–216, 2017.
DOI: 10.1007/978-3-319-52730-7_21

2 The Generation Method of Koch Fractal Curve for Lamp Appearance

Koch curve is typical of fractal curve, its structure process is through to repeatedly replacing each line with similar graphics of generator, thus graphics of each part are same with shape itself called as self-similarity, which is also an most important feature of fractal pattern, its structural process also decides the way of making the curve in computer that should be recursive method, that is function himself calls the process of himself. Transformation rules of Koch curves R is produced by fractal generator, the basic characteristics are decided by fractal generator completely. We can generate a variety of fractal graphics by some generator, the process is shown as in Fig. 1.

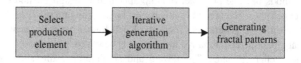

Fig. 1. Forming method of Koch fractal patterns

2.1 Koch Fractal Generators

(1) Triangle generators

Given a straight F0, dividing the line into three fractions equally and replacing the middle fraction with other two edges of the equilateral triangle constituted of this line, F1 can be get. According to above way, we modify each piece of fig F1 in this manner, until infinitum, the ultimate Fn of the curve is finally get. It is called as the Koch curve. This process is shown in Fig. 2.

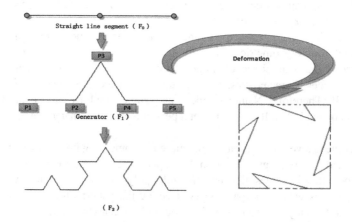

Fig. 2. Triangle generator

(2) Square generators

With the same as the construction method of the triangle Koch curve generator, the line is divided into three fractions equally. The middle fraction is replaced by the square. We can get the F1. F1 is formed substantially by triangle generator whose triangular parts are replaced by square, it is noted that the height of square should be one-third of the original line, finally we can get first order of curve constituted with five equal lines (shown as in Fig. 3a). Then, according to above way, we modify each piece of fig F1, until infinitum, we finally get the ultimate curve. It is called as the Koch curve with square generators. In this manner, if it takes a square as the original graph and divides each edge of the square into three parts, then similarly, the middle section is used to form another square generator. We can get another fractal pattern. It is shown as Fig. 3(b).

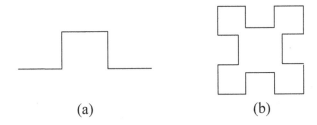

(a)   (b)

Fig. 3. Square generator

(3) Tree generators

Tree generators start from a line and marking two endpoints, as well as 1/3, 2/3, then starting with 1/3 points and 2/3 points to extend out several branches, length of the branch is 1/3 of the initial linear, We finally get branch generators as shown in Fig. 4. The one-third point and the two-thirds points of the line is respectively shown as point 2 and 6.

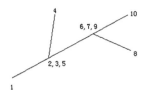

Fig. 4. Fractal generator

2.2 Koch Fractal Pattern Generating Algorithm

Koch fractal curve construction begins with a linear segment. The construction progress is shown as Fig. 5:

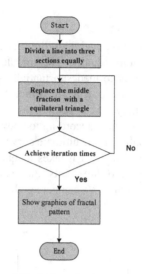

Fig. 5. Process of Koch fractal pattern generating algorithm

(1) Segment

Considering the process of a straight line segment (2 points) producing the first shape (5 points). In Fig. 2, assuming that P1 and P2 is the two start points of the original line segment, then we need to assert three points P1, P2 and P3. And obviously, P2 is in one-third line, P4 locates in the two-thirds line segment.

(2) Rotation

The location of the P3 point can be considered that P4 rotates counter-clockwise by P2 as the axis. Rotation can be achieved by an orthogonal matrix. It can be described as Eq. (1):

$$A = \begin{pmatrix} \cos\alpha & -\sin\alpha \\ \sin\alpha & \cos\alpha \end{pmatrix}. \tag{1}$$

According to original data, the algorithm produces 5 joint points in the Fig. 2. An array of nodes forms a 5×2 matrix, the coordinate of the first behavior p1, the coordinate of the second, and so on, until the coordinates of p5. Matrix elements in the first column are the 5 nodes' x coordinate, the elements in the second column are the 5 nodes' y coordinate.

(3) Iterations

Further consideration is the regularity of node number in the process of forming Koch curve. Supposing that the k iteration produce n_k nodes, and the at time k+1 iteration, n_{k+1} nodes generates. Namely the recurrence relation between n_k and n_{k+1} is

$$n_{k+1} = 4n_k - 3. \tag{2}$$

2.3   Fractal Pattern of Koch

As mentioned above, the fractal function is programmed. Koch curve is generated by MATLAB. In order to present the change of points in the recursive process intuitively and dynamically, we use vector x, y. Finally it can be shown intuitively by the graph. We can see that the key step of pattern generation is algorithm design.

Triangle generator generating fractal patterns are shown as Fig. 6. In order to generate the final Koch snowflake pattern by the original lines or graphics, we should start from the definition of the Koch curve, giving a equilateral triangle that the length of side is and then adding a side a/3 equilateral triangle in each center of side, determining to rotate Orthogonal matrix in algorithm analysis, considering the regularity of the number of nodes' change in the process of forming Koch curve, and infinite iterations, finally it can form the following snowflakes graphics.

The train of thought for generating fractal patterns by a rectangular generator or branch generator is basically the same, only need to adjust rotating orthogonal matrix and the corresponding part of the function. Patterns generated by square generators are shown as Fig. 7. Patterns generated by tree fractal generators are shown as Fig. 8.

(1) The triangle generator

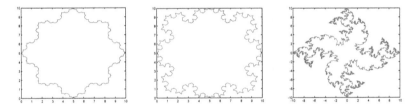

Fig. 6. Patterns generated by triangle generators

(2) The square generator

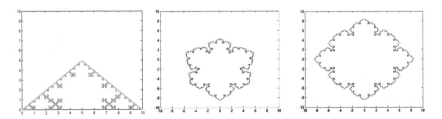

Fig. 7. Patterns generated by square generators

(3) The tree fractal generator

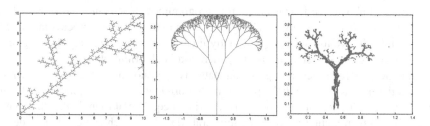

Fig. 8. Patterns generated by tree fractal generators

3 Method of Lamp Outlook-Design Based on Koch Fractal Patterns

The outlook-design of lamps is based on the Rhinoceros software modeling and apply KeyShot software in simulated lighting image in this paper. Rhinoceros software is suit at modeling of product appearances. Therefore, it is always being used in industrial design is an interactive ray tracing and rendering programs with the full domain, without complicated setting, it can produce 3D rendered image (Fig. 9).

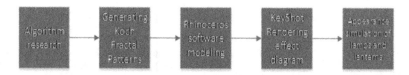

Fig. 9. Modeling and Simulation of lighting model based on Koch fractal

Imitating the natural shape is one of the simplest and most direct methods of lighting design. It not only affects people's aesthetic consciousness, but also promotes a natural, eco-design, which retains the spiritual and natural breath of lamps. The generation of fractal graphics relies on iterative function, which can express infinite subtle structure, if the computer accuracy is not restricted, it can infinitely magnify the boundary of fractal graph and a region in the graph to show a new structure element. From the overall visual effect, we can see that pattern generated by the fractal graph is with more abstract, art, regularity than the traditionally manual drawing pattern, which make up the traditional decorative pattern. Eventually we will apply fractal graphics to the structure of lamps' shape, this is a process to transform the flat pattern into three-dimensional pattern, in this process, and different three-dimensional methods will produce different modes of lamps, which give us as a wider space to play in the mode design. The LED light-source has a small size and many other features. Thus, the lamp shape can break the limits of traditional lamps to design a variety of lamps and the reflective and casted type of lamps can be perfectly applied to lamps that use Koch curve as a prototype.

Reflective mode is the light provided through reflector lighting for residents. Using secondary optical design, we make the light spot from light-source form selected curve shape of Koch curve. And then put the light-source into lamps so that all of the light can be casted onto the reflector. Then, reflectors cast the light onto the ground. Light is provided for residents or decorative use.

Casted pattern is directly using the light-source through the outer shade to cast onto the ground to provide lighting for the residents. It can be achieved in two methods one way is to filter out of the Koch curve made as lamp covers, then the light-source through to the shell light interior. This way generally uses in the production of ceiling lamps. The other way is through the use of LED light source characteristics of small volume of LED light-source to arrange the Koch as selected and finally show the curve outline. This method should generally be used for ceiling lamps.

4   Design Results of LED Lamps Combined with Koch Fractal Pattern

As mentioned above, Koch curve patterns is an input on the Rhinoceros software, multiple linear functions is used to draw the simulation. Then, we can get the outlook of lamps, through to the operation of stretch. Finally, lamps are attached to materials in KeyShot the simulation is shown as Figs. 10, 11.

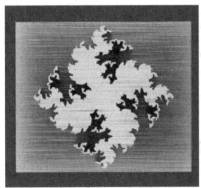 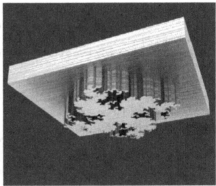

Fig. 10. Simulated lamp A

Fig. 11. Simulated lamp B

5 Conclusions

The fractal theory describes the feature of the natural scenery. It can construct the fractal pattern, while Koch fractal is a classic fractal pattern. This paper applies the characteristics of Koch fractal to design appearance of lighting. Design results show that the Koch fractal patterns can be used to design appearance, and fractal theory is suitable for designing appearance.

References

1. Cheng, Z., Han, C.: Household LED design methods and development prospect of. Sci. Technol. Econ. Market **11**, 81–82 (2007)
2. Fu, Z.: Discussion on development trend of modern lighting design. Stage Design **7**, 164–165 (2013)
3. Mandelbrot, B.B.: The Fractal Geometry of Nature, pp. 79–95. Freeman, San Francisco (1982)
4. Yue, S., Wan, D., Lu, J.: Application research on fractal graphics in packaging design. Packag. Eng. **10**, 67–69 (2011)
5. Ruderman, D.L.: Origins of scaling in natural images. Vis. Res. **37**, 3385–3398 (1996)
6. Al-Hamdan, M., Cruise, J., Rickman, D., et al.: Effects of spatial and spectral resolutions on fractal dimensions in forested landscapes. Remote Sens. **2**, 611–640 (2010)
7. Al-Hamdan, M.Z., Cruise, J.F., Rickman, D.L., et al.: Characterization of forested landscapes from remotely sensed data using fractals and spatial autocorrelation. Adv. Civ. Eng. **2012**, 1–15 (2012)
8. Ghosh, J.K., Somvanshi, A.: Fractal-based dimensionality reduction of hyperspectral images. J. Indian Soc. Remote Sens. **36**, 235–241 (2008)
9. Falconer, K.J.: Fractal Geometry Mathematical Foundations and Applications, pp. 122–125. Wiley, New York (1990)
10. Pentland, A.P.: Fractal-Based Description of Natural Scenes. Pattern Anal. Mach. Intell. **6**, 661–674 (1984)
11. Ueda, Y.: The Road to Chaos, pp. 55–74. Aerial Press, Santa Cruz (1992)
12. Reichl, L.E.: The Transition to Chaos, pp. 28–67. Springer, New York (1992)
13. Schuster, H.G.: Deterministic Chaos, 2nd edn, pp. 78–92. Phisik-Verlag, Weinheim (1988)
14. Xu, Y., Zhao, X.: Koch snowflake curve of production and its important conclusions. J. Changchun Teach. Univ. **2**, 6–8 (2003)
15. Qiu, W., An, N., Qi, X.: Koch fractal image generation based on MATLAB algorithm. Comput. Digit. Eng. **8**, 100–101 (2010)

Bandwidth and Power Allocation for Wireless Cognitive Network with Eavesdropper

Kecai Gu[1(✉)], Weidang Lu[1], Guomin Zhou[2], Hong Peng[1], Zhijiang Xu[1], and Xin Liu[3]

[1] School of Information Engineering,
Zhejiang University of Technology, Hangzhou, China
celus@zjut.edu.cn
[2] Zhejiang Police College, Hangzhou, China
[3] School of Information and Communication Engineering,
Dalian University of Technology, Dalian 116024, China

Abstract. In this paper, we consider secure communications for a five-node cognitive wireless network system including one primary user (PU) pair and one secondary user (SU) pair in presence of one eavesdropper. The secrecy transmission process departs into two equal time phases. To ensure transmission process safety, the primary source and receiver are allowed to deliver artificial noise to interfere the eavesdropper. To obtain higher spectrum efficiency, we propose an anti-interference spectrum access strategy with cooperative trusted DF relaying over flat fading channel, in which secondary user forward primary information and deliver its own information with different part of licensed spectrum. We study how to optimize the bandwidth and power allocation ratio to maximize the secondary user rate while guaranteeing the primary system to achieve its target secrecy rate. The expression of the optimal bandwidth allocation ratio is derived. Simulation results demonstrate that proposed strategy can achieve win-win result.

Keywords: Cognitive radio · Physical layer security · Artificial noise · Achievable secrecy rate · Power allocation

1 Introduction

Spectrum utilization has received a lot of attention during the past decade due to the rarity of radio spectrum and the fixed spectrum allocation strategy which divide the spectrum into two parts: licensed spectrum and unlicensed spectrum [1]. Traditional fixed spectrum allocation strategy authorizes the specific communication system use the specific spectrum but doesn't allow others to use it even when licensed user doesn't use the spectrum sometimes. It leads to low utilization and waste of spectrum resource in time and space. Cognitive radio [2] (CR) is a promising technology to improve the wireless spectrum utilization by supporting the unlicensed systems access to the same spectrum resource already licensed to the primary systems while not degrading the performance of primary system. However, there are two main problems in the existing underlay spectrum access strategy in CR network. One is that there will be interference

© ICST Institute for Computer Sciences, Social Informatics and Telecommunications Engineering 2017
X.-L. Huang (Ed.): MLICOM 2016, LNICST 183, pp. 217–227, 2017.
DOI: 10.1007/978-3-319-52730-7_22

between the primary system and secondary system when the cognitive user forward the primary user information and deliver own information simultaneously. The other is the secondary user is allowed to access to the licensed spectrum if and only if the channel of primary system is good enough. Cooperative diversity has been proposed as a spatial diversity technique to solve the above problems [3]. Because it can degrade the influence of path loss in wireless links. Thus we exploit the cooperative diversity technology to overcome the existing shortcomings and improve the utilization of licensed spectrum.

Another issue in wireless communication environment is the security [4]. Wireless communication is not secure as wire communication due to the openness of the wireless medium. Some illegal receivers within the communication range may wiretap and decode the secrecy information, which easily lead to the information leakage. The security of traditional wireless communication depends on the upper layers of the protocol stack through the use of encryption algorithms [5, 6]. But there are still some challenges such as secret key management complexity, key transmission and distribution security issues in open wireless communication environment and so on. Significant works have been done on physical (PHY) layer security and various advanced signal processing and coding techniques have been proposed to improve the secrecy of the wireless communication in the presence of some eavesdroppers. Shannon firstly investigates information theoretic security in 1949 and Wyner introduce the conception of secrecy capacity [7]. The secrecy rate is defined as the difference between achievable rates of the main channel and the wiretap channel with the Gaussian code-book and the maximum of secrecy rate is defined as secrecy capacity. Positive secrecy rate only exist when the main channel is more advantage than wiretap channel. But now, we can achieve a positive secrecy rate even when the main channel is worth than wiretap channel with using the nodes cooperative technology. Generally, there are two main methods to improve the information security. One is cooperative note plays as a jammer to deteriorate the wiretap channel. The other is cooperative node plays as trusted relay to help the primary system improve the channel quality to the legitimate user. Cooperative jamming has been studied in paper [8, 9] to maximum the achievable secrecy rate While Cooperative beam-forming (CB) are studied in [10, 11]. Cooperative nodes can forward the confidential information in above both manners based on DF or AF ways.

In this paper, we exploit the artificial noise [12–15] to confuse the eavesdropper to ensure the security transmission. To improve the utilization of spectrum resource and eliminate the mutual interference between primary and secondary system, a kind of effective spectrum access strategy have been proposed, in which secondary system is allowed to transmit primary and its own information with different bandwidth on the condition that it gets access to the licensed spectrum. Our goal is to study how to optimize the bandwidth and power allocation ratio to maximize the secondary user rate while guaranteeing the primary system to achieve its target secrecy rate.

2 System Model and Problem Formulation

2.1 System Model

The system configuration of the proposed anti-wiretapping access strategy is show in Fig. 1. The whole system consists of primary system including one PT (Primary Transmitter) and one PR (Primary Receiver) and secondary system including one ST (Secondary Transmitter) and one SR (Secondary Receiver) in the presence of one eavesdropper. We assume that ST is trusted. The PR can transmit and receive simultaneously while others operate in a half-duplex mode. All notes are equipped with a single antenna. In this paper, we assume the channel are quasi-static Rayleigh channel, the channel coefficient $h_i > 0$ where $i = \{1, 2, 3, 4, 5, 6\}$ and $r_i = |h_i|^2$ represents the instantaneous channel gain, d_i imply distance between two nodes and v is the path-loss exponent (typically value from 2 to 6). All channels coefficient remain unchanged during both phases. For simplicity, we assume that the noise at all the nodes is complex addictive white Gaussian noise (AWGN) with zero means and variance. The total power of primary and secondary system is constrained by p and p_s respectively.

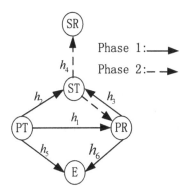

Fig. 1. System model

In order to transmit information safely, the transmission process departs into two equal time phases. We assume transmission time is 1, each phase accounted for 1/2. In phase1, PT delivers the information including artificial noise and the secrecy signal to PR and ST while the PR transmits the no-information-bearing artificial noise concurrently to PR. The eavesdropper is passive and only phase 1 could be tapped. In phase 2, ST forward the message to PR in DF fashion with part of licensed bandwidth and use the rest spectrum to transmit own information. There would be no interference between primary and secondary user with using different spectrum. Note that ST is permitted to operate in the licensed spectrum if and only if it can guarantee the secrecy Rate of the primary system.

2.2 Problem Formulation

The instantaneous secrecy rate of primary system is defined as

$$R_Q = (R_p - R_E)^+ \tag{1}$$

where R_p and R_E represents the instantaneous rate of the primary system and eaves-dropper. $(x)^+ = \max(x, 0)$.

Firstly, we consider that the PT only delivers the secrecy information to PR without the help of secondary user. And the received signal at E and PR are given as follows

$$y_{sd} = \sqrt{p}h_2 x + n_{sd} \tag{2}$$

$$y_{se} = \sqrt{p}h_5 x + n_{se} \tag{3}$$

where n_{sd} and n_{se} is noise and follows $CN(0, \sigma^2)$. x implies the secrecy signal. So the rate of the primary system R_D and eavesdropper R_E can be written as

$$R_D = W \log_2(1 + \frac{P_p \gamma_1}{\sigma^2}) \tag{4}$$

$$R_E = W \log_2(1 + \frac{P_p \gamma_5}{\sigma^2}) \tag{5}$$

when $R_D - R_E < R_T$, where R_T is the secrecy rate threshold of the primary system, primary user seeks help from around user. The secondary user judge whether or not it can access to the licensed spectrum through the two time slots.

In phase 1, the transmitted signal by PT(x_1) and PR(x_2) are respectively given by

$$x_1 = \sqrt{p\alpha}s + \sqrt{p(1-\alpha)}u_1 z \tag{6}$$

$$x_2 = \sqrt{p(1-\alpha)}u_2 z \tag{7}$$

where x_1 is a mixture of the information signal and the jamming signal and x_2 is purely artificial jamming signal designed to cancel out the interference at ST while further confuse the eavesdropper. α implies the power allocation ratio of between the information signal s and jamming signal z. both of them are unit-power. u_1 and u_2 are the weight coefficients and satisfy

$$|u_1|^2 + |u_2|^2 = 1 \tag{8}$$

The received signal at ST then given by

$$\begin{aligned} r_{ST} &= \sqrt{p\alpha}h_2 s + \sqrt{p(1-\alpha)}h_2 u_1 z + \sqrt{p(1-\alpha)}h_3 u_2 z + n_{sr} \\ &= \sqrt{p\alpha}h_2 s + \sqrt{p(1-\alpha)}(u_1 h_2 + u_2 h_3)z + n_{sr} \end{aligned} \tag{9}$$

The ST decodes the secrecy information and re-encodes it with the same code-words of source in phase 2. To avoid the interference of artificial noise at ST, we design

$$u_1 h_2 + u_2 h_3 = 0 \tag{10}$$

So the received signal can be rewritten as

$$r_{ST} = \sqrt{p\alpha} h_2 s + n_{sr} \tag{11}$$

The eavesdropper is passive and only wiretap the signal in phase 1 while keep silence in phase 2. It couldn't remove the jamming signal, so the received signal is given by

$$
\begin{aligned}
r_E &= \sqrt{p\alpha} h_5 s + \sqrt{p(1-\alpha)} h_5 u_1 z + \sqrt{p(1-\alpha)} h_6 u_2 z + n_{sr} \\
&= \sqrt{p\alpha} h_2 s + \sqrt{p(1-\alpha)}(u_1 h_5 + u_2 h_6) z + n_{sr}
\end{aligned}
\tag{12}
$$

Due to the mixture signal, PR receives the signal with artificial noise

$$r_d = \sqrt{p\alpha} h_1 s + \sqrt{p(1-\alpha)} h_1 u_1 z + n_{sd} \tag{13}$$

The rate of PR(R_d^1), ST(R_p^1) and E(R_E) are given respectively by

$$R_p^1 = \frac{1}{2} w \log_2(1 + \frac{p\alpha\gamma_2}{\sigma^2}), \tag{14}$$

$$
\begin{aligned}
R_E &= \frac{1}{2} w \log_2(\frac{P\alpha|h_5|^2}{\sigma^2 + p(1-\alpha)|u_1 h_5 + u_2 h_6|^2}) \\
&= \frac{1}{2} w \log_2(\frac{P\alpha\gamma_5}{\sigma^2 + p(1-\alpha)\gamma_m}),
\end{aligned}
\tag{15}
$$

$$R_d^1 = \frac{1}{2} w \log_2(1 + \frac{p\alpha\gamma_1}{\sigma^2 + p\alpha u_1^2 \gamma_1}). \tag{16}$$

where $\gamma_m = |u_1 h_5 + u_2 h_6|^2$, w represents the licensed bandwidth. The coefficient factor 1/2 is due to the fact that every transmission process needs two phases.

During phase 2, ST allocate a fraction of bandwidth and half of power to forward the secrecy message to PR, the rate $ST \rightarrow PR$ is

$$R_d^2 = \frac{1}{2} bw \log_2(1 + \frac{\frac{1}{2} P_s \gamma_3}{\sigma^2}) \tag{17}$$

where b represents the bandwidth allocation ratio between primary user and secondary user.

Then ST use the remaining bandwidth and the other half of power to transmit its own information, the rate $ST \rightarrow SR$ is given by

$$R_s = \frac{1}{2}(1-b)w\log_2(1+\frac{\frac{1}{2}P_s\gamma_4}{\sigma^2}) \tag{18}$$

The eavesdropper is not interested in ST and keeps silence in phase 2. If ST can decode successfully, PR apply the maximum ratio combination (MRC) to received message over two phases, then the primary system rate R_p^2 can be given as

$$
\begin{aligned}
R_p^2 &= \frac{1}{2}bw\log_2(1+\frac{P_s\gamma_3}{2\sigma_2}+\frac{p\gamma_1\alpha}{\sigma^2+P(1-\alpha)u_1^2\gamma_1}) \\
&+ \frac{1}{2}(1-b)w\log_2(1+\frac{p\gamma_1\alpha}{\sigma^2+P(1-\alpha)u_1^2\gamma_1})
\end{aligned}
\tag{19}
$$

So after the two transmission process, the primary system rate R_p can be written as $R_p = \min\{R_p^1, R_p^2\}$.

ST can forward the Primary user information only when ST can decode successfully. So operation symbol min means the performance of the primary link is limited to the worse the link of $PT \to PR$ and $PT \to ST$.

With the help of ST, if R_p can achieve the target secrecy rate, that is $R_p - R_E > R_T$, the primary system authorizes the secondary user to use the licensed spectrum. If not, the secondary user will do nothing.

3   Optimal Solution

In this section, we study how to optimize the allocation coefficient of the bandwidth b and power allocation ratio α to maximum the secondary user rate R_s while keep the primary system secrecy rate achieve the target secrecy rate threshold RT. First, we give the solution to the designed artificial noise parameters u_1, u_2

$$
s.t. \qquad \begin{cases} u_1h_2+u_2h_3 = 0 \\ |u_1|^2 + |u_2|^2 = 1 \end{cases}
\tag{20}
$$

We can easily solve the equation and get

$$
\begin{cases} u_1 = -\sqrt{\frac{|h_3|^2}{|h_2|^2+|h_3|^2}} \\ u_2 = \sqrt{\frac{|h_2|^2}{|h_2|^2+|h_3|^2}} \end{cases}
\tag{21}
$$

Or

$$\begin{cases} u_1 = \sqrt{\dfrac{|h_3|^2}{|h_2|^2 + |h_3|^2}} \\ u_2 = -\sqrt{\dfrac{|h_2|^2}{|h_2|^2 + |h_3|^2}} \end{cases} \tag{22}$$

In the following part, we derive the explicit expression of optimal b. The optimization problem can be translated into the follows

$$\max_{b,\alpha} R_s \tag{23}$$

It yields

$$\begin{cases} R_p - R_E \geq R_T \\ 0 < b < 1 \\ 0 < \alpha < 1 \end{cases} \tag{24}$$

For simplicity, this paper introduce some auxiliary variables R_2, R_3, R_4, R_d are given as following

$$\begin{cases} R_2 = w \log_2\left(1 + \frac{p\alpha\gamma_2}{\sigma^2}\right) \\ R_3 = w \log_2\left(1 + \frac{p_s\gamma_3}{2\sigma^2} + \frac{p\alpha\gamma_1}{p(1-\alpha)u_1^2\gamma_1 + \sigma^2}\right) \\ R_4 = w \log_2\left(1 + \frac{p_s\gamma_4}{\sigma^2}\right) \\ R_d = w \log_2\left(1 + \frac{p\alpha\gamma_1}{p(1-\alpha)u_1^2\gamma_1 + \sigma^2}\right) \end{cases} \tag{25}$$

Then we can rewrite R_p^1, R_p^2 and get $R_p^1 = \frac{1}{2}R_2$, $R_p^2 = \frac{1}{2}bR_3 + \frac{1}{2}(1-b)R_d$. According to the constraints, we can obtain

$$\begin{cases} \frac{1}{2}R_2 - R_E \geq R_T \\ \frac{1}{2}bR_3 + \frac{1}{2}(1-b)R_d - R_E \geq R_T \\ 0 < b < 1 \\ 0 < \alpha < 1 \end{cases} \tag{26}$$

From condition, we can derive the linear inequality about b given by

$$b \geq \frac{2(R_T + R_E) - R_d}{R_3 - R_d} \tag{27}$$

We can easily observe that R_s is monotonically decreasing function of b, so the optimal bandwidth allocation coefficient b can expressed

$$b^* = \frac{2(R_T + R_E) - R_d}{R_3 - R_d} \tag{28}$$

Subject to

$$\begin{cases} R_2 \geq 2(R_T + R_E) \\ R_D \leq 2(R_T + R_E) \\ R_3 \geq 2(R_T + R_E) \end{cases} \tag{29}$$

We can get the maximization rate of secondary user and is given by

$$R_s^* = \frac{[R_3 - 2(R_T + R_E)]R_4}{2(R_3 - R_d)} \tag{30}$$

Our goal is to maximize R_s, so we can obtain optimal α through minimize the power allocation ratio b.

4   Simulation Results

In this section, we investigate the performance of the proposed strategy numerically. The simulation setting is as follows. The five nodes are located in a 2-D square topology we set the PT, PR, ST in the same line. PT, PR are located in (0,0) and (1,0) respectively. So the distance between PT and PR is 1. ST moves from (0,0) to (1,0). The distance between ST and SR is constant $d_4 = 0.5$ and the eavesdropper is fixed in the place where $d_5 = 0.14$, $d_6 = 1$; In our simulation, we assume that the path-loss exponent v is −3, the licensed spectrum bandwidth is 1, and the noise variance is 1. The power of the primary system and secondary system is p = 8 dB, p_s = 10 dB respectively.

In the Fig. 2, we let x axis implies the location of ST, and y axis implies the optimal allocation ratio of bandwidth b. From the picture, when ST is close to PT, $b^* = 1$, $\alpha^* = 0$, $R_s = 0$. The reason is the primary system secrecy rate R_Q is small and ST can't help primary system achieve the target secrecy rate, so it can't get access to the licensed spectrum. With ST is far from PT, R_Q is getting large and exceed the target rate, so the secondary user allocates part of bandwidth to forward secrecy information to PR. However, R_Q is getting low again along with ST move further away from PT, R_2 will decrease and can't satisfy the condition $R_2 - 2R_E \geq 2R_T$, the secondary user can't acquire opportunity to access to the primary system, so b skip to 0. From the Fig. 2, the access range when $R_T = 1.2$ bps/HZ is large than $R_T = 1.5$ bps/HZ due to the secondary system is easier to help primary system get the lower target secrecy rate.

Figure 3 describes the secondary system rate under different target secrecy rate with the different location of ST, when $d_2 < 0.249$ under $R_T = 1.2$ bps/HZ, $R_s = 0$, this implies that R_Q is not large enough to support ST to help the primary user to achieve the target rate. With ST move far from PT and near to PR, R_Q is getting high. So the secondary user access to the licensed spectrum and R_s becomes positive. But when ST further gets close to PR, as $d_2 > 0.704$, R_s return to 0 due to R_Q decrease again. Therefore, secondary system can't provide support and get to access. Compared with $R_T = 1.2$ bps/HZ, secondary system get narrower access ranges when $R_T = 1.5$ bps/HZ. The higher R_T is, the more difficult for secondary system to help

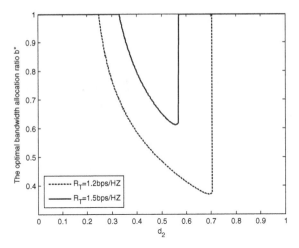

Fig. 2. The optimal bandwidth allocation ratio b^* vs. the distance between PT and ST d_2.

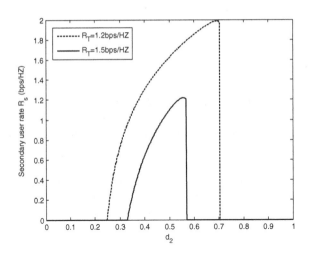

Fig. 3. Secondary user rate R_s vs. the distance between PT and ST d_2.

primary system to achieve the target secrecy rate. There are less left bandwidth for secondary system to transmit own information. So R_s is lower.

Figure 4 represents the optimal power allocation ratio between secrecy information and artificial noise for ST. The optimal α can be obtained through minimize the bandwidth allocation ratio. In other word, we get optimal a by maximizing the secondary user target rate R_s. In our communication scenario, the eavesdropper E is close to ST, to confuse eavesdropper and transmit confidential information, more power are allocated to transmit artificial noise. Therefore, we can see that α is relatively small and

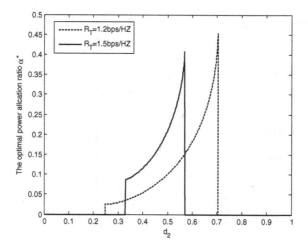

Fig. 4. The optimal power allocation ratio vs. the distance between PT and PR d_2

the maximum value of α is about 0.45 from the picture. Due to the same reason, α can get larger access ranges when $R_T = 1.2\,\text{bps/HZ}$.

5   Conclusion

In this paper, we have proposed an anti-interference strategy to solve the secure transmission problem in flat fading channel for the cognitive radio network in present of an eavesdropper. To improve the bandwidth utilization while ensuring information transmission safety, we allowed PT allocates part of available power for artificial noise to confuse the eavesdropper. Meanwhile, we derived the optimization bandwidth allocation ratio for ST and analyzed the optimal power allocation ratio of PT which can maximum the secondary transmission rate while the secrecy rate throughout constraint of primary user is satisfied. Further work includes the explicit derivation and analysis of power allocation problems.

Acknowledgments. This work was supported by China National Science Foundation under Grand No. 61402416 and 61303235, Natural Science Foundation of Zhejiang Province under Grant No. LQ14F010003 and LQ14F020005, NSFC-Zhejiang Joint Fund for·the Integration of Industrialization and Informatization under grant No. U1509219, Natural Science Foundation of Jiangsu Province under Grant No. BK20140828, the Fundamental Research Funds for the Central Universities under Grant No. DUT16RC(3)045 and the Scientific Foundation for the Returned Overseas Chinese Scholars of State Education Ministry.

References

1. Goldsmith, A., Jafar, S., Maric, I., Srinivasa, S.: Breaking spectrumgridlock with cognitive radios: an information theoretic perspective. Proc. IEEE **97**(5), 894–914 (2009)
2. Wang, C., Wang, H.M.: On the secrecy throughput maximization for MISO cognitive radio network in slow fading channels. IEEE Trans. Inf. Forensics Secur. **9**(11), 1814–1827 (2014)
3. Ikki, S.S., Ahmed, M.H.: Performance analysis of decode-and-forward incremental relaying cooperative-diversity networks over Rayleigh fading channels. In: 2009 IEEE 69th Vehicular Technology Conference, VTC Spring 2009, Barcelona, pp. 1–6 (2009)
4. Liu, R., Trappe, W. (eds.): Securing Wireless Communications at the Physical Layer. Springer, New York (2010)
5. Liang, Y., Poor, H.V., Shamai, S.: Information Theoretic Security. NowPublishers, Delft (2009)
6. Ramesh, A., Suruliandi, A.: Performance analysis of encryption algorithms for information security. In: 2013 International Conference on Circuits, Power and Computing Technologies (ICCPCT), Nagercoil, pp. 840–844 (2013)
7. Wyner, A.D.: The wire-tap channel. Bell Sys. Tech. J. **54**, 1355–1387 (1975)
8. Dong, L., Han, Z., Petropulu, A.P., Poor, H.V.: Improving wireless physical layer security via cooperating relays. IEEE Trans. Signal Process. **58**(3), 1875–1888 (2010)
9. Huang, J., Swindlehurst, A.L.: Cooperative jamming for secure communications in MIMO relay networks. IEEE Trans. Signal Process. **59**(10), 4871–4884 (2011)
10. AbolfathBeigi, M., Mohammad Razavizadeh, S.: Cooperative beamforming in cognitive radio networks. In: 2009 2nd IFIP Wireless Days (WD), Paris, pp. 1–5 (2009)
11. Yi, T., Guo, L., Niu, K., Cai, H., Lin, J., Ai, W.: Cooperative beam-forming in cognitive radio network with hybrid relay. In: 2012 19th International Conference on Telecommunications (ICT), Jounieh, pp. 1–5 (2012)
12. Kabeya, J., Takyu, O., Ohtsuki, T., Sasamori, F., Handa, S.: Performance evaluation of the physical layer security using artificial noise and relay station. In: 2015 Asia-Pacific Signal and Information Processing Association Annual Summit and Conference (APSIPA), Hong Kong, pp. 834–839 (2015)
13. Zhou, X., McKay, M.R.: Secure transmission with artificial noise over fading channels: achievable rate and optimal power allocation. IEEE Trans. Veh. Technol. **59**(8), 3831–3842 (2010)
14. Deng, H., Wang, H.M., Wang, W., Yin, Q.: Secrecy transmission with a helper: to relay or not to relay. In: 2014 IEEE International Conference on Communications Workshops (ICC), Sydney, NSW, pp. 825–830 (2014)
15. Deng, H., Wang, H.-M., Guo, W., Wang, W.: Secrecy transmission with a helper: to relay or to jam. IEEE Trans. Inf. Forensics Secur. **10**(2), 293–307 (2015)

A Machine Learning Based Forwarding Algorithm over Cognitive Radios in Wireless Mesh Networks

Jianjun Yang[1]([⊠]), Ju Shen[2], Ping Guo[3], Bryson Payne[1], and Tongquan Wei[4]

[1] University of North Georgia, Gainesville, GA, USA
jianjun.yang@ung.edu
[2] University of Dayton, Dayton, OH, USA
[3] University of Illinois at Springfield, Springfield, IL, USA
[4] East China Normal University, Shanghai, China

Abstract. Wireless Mesh Networks improve their capacities by equipping mesh nodes with multi-radios tuned to non-overlapping channels. Hence the data forwarding between two nodes has multiple selections of links and the bandwidth between the pair of nodes varies dynamically. Under this condition, a mesh node adopts machine learning mechanisms to choose the possible best next hop which has maximum bandwidth when it intends to forward data. In this paper, we present a machine learning based forwarding algorithm to let a forwarding node dynamically select the next hop with highest potential bandwidth capacity to resume communication based on learning algorithm. Key to this strategy is that a node only maintains three past status, and then it is able to learn and predict the potential bandwidth capacities of its links. Then, the node selects the next hop with potential maximal link bandwidth. Moreover, a geometrical based algorithm is developed to let the source node figure out the forwarding region in order to avoid flooding. Simulations demonstrate that our approach significantly speeds up the transmission and outperforms other peer algorithms.

Keywords: Mesh networks · Machine learning · Forwarding · Highest bandwidth capacity

1 Introduction

Mesh routers and client devices are self-organized and self-configured to form wireless mesh networks (WMNs) [1]. A device is called a node in WMNs. Each node is equipped with multiple radios to improve the whole capacities in WMNs [5]. The radios in WMNs are cognitive radios, by which the radio devices are capable of learning from their environment and adapting to the environment [2]. Cognitive radio is also called programmable radio because such radio has the ability of self-programming [3], learning and reasoning [2].

Machine learning has been studied for about 60 years. It evolved from simple artificial intelligence to a wide variety of applications in image processing,

© ICST Institute for Computer Sciences, Social Informatics and Telecommunications Engineering 2017
X.-L. Huang (Ed.): MLICOM 2016, LNICST 183, pp. 228–234, 2017.
DOI: 10.1007/978-3-319-52730-7_23

vision, networking, and pattern recognition. In this paper, we propose a learning algorithm for a forwarding node to find one of its links with possibly maximal bandwidth, and then choose next forwarding node and then forward the message to that node. Each node only saves the last three changed bandwidth status of its links. Then the forwarding node learns the three status and predict the potential bandwidth of its links. So the forwarding node is able to find the neighbor with highest link bandwidth as its next hop. We further devise an algorithm to let the source node figure out the forwarding region in order to avoid flooding.

The rest of the paper is organized as follows. Section 2 discusses the related research on this topic. Section 3 proposes our novel forwarding method that selects the best next hop. We evaluate the proposed schemes via simulations and describe the performance results in Sect. 4. Section 5 concludes the paper.

2   Related Work

Some approaches on machine learning, wireless forwarding and related work have been studied [6–11]. Wang et al. [12] proposed a machine learning mechanism to improve data transmission in sensor network. The predication of link quality was used to implement the approach. Additionally, they developed a protocol called MetricMap to maintain efficient routing in case the regular routing is not working.

Sawhney et al. [13] presented a machine learning algorithm to handle congestion controlling in wireless networks. Their approach learns many factors that have impact to congestion controlling, and then uses the parameters in a fuzzy logic to generate better result when congestion takes place. The efficiency is assessed with machine learning tools.

3   The Learning Based Forwarding Mechanism

3.1   The Forwarding Problem

In wireless mesh networks (WMNs), the communications are over links. Link bandwidth is critical for transmission speed. Since each node may be equipped with multiple network interfaces with different radios and the radios are switchable, the bandwidth over two neighbor nodes may vary from time to time. The radios in WMNs are cognitive radios and then the nodes are able to learn the changes of past bandwidths and can further predict and select the desired link with potential highest bandwidth.

Assuming a source node s intends to send data to a destination node d, many traditional routing algorithms set up the forwarding path by simply selecting the shortest route. For example, $s - c - g - h - d$ is the forwarding path in Fig. 1. However, it may not be the best path in WMNs. In WMNs, the bandwidth over two nodes changes frequently. The bandwidth of the link sc is possibly much lower than that of sa. Or the past bandwidth of sc is higher than sa but two

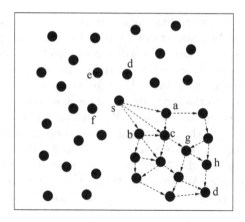

Fig. 1. Topology of a wireless mesh network

much traffic is over sc now so the available bandwidth of sc is going down while that of sa is going up.

Our goal is to let each forwarding node select the link for next hop with the highest potential bandwidth. In our approach, each node learns its links' past bandwidths and then predict their potential bandwidths. Then the forwarding node figures out its next hop with highest potential bandwidth.

3.2   Prediction for Future Bandwidth

Suppose node i saves the bandwidth changes of its links of the last three times t_0, t_1, and t_2. Then for any of its neighbor j, i predicts the potential bandwidth of link ij. By computational method [14], we define

$$\alpha_{i,j,k} = \sum_{m=0}^{k} \frac{B_{i,j,m}}{\prod_{n=0}^{k}(t_m - t_n)} \tag{1}$$

at time t_k, where $B_{i,j,m}$ is the bandwidth between node i and node j at time m. Then the bandwidth of link ij at future time p can be calculated and predicated as:

$$B_{i,j,p} = \alpha_{i,j,0} + \alpha_{i,j,1}(t_p - t_0) + \alpha_{i,j,2}(t_p - t_1)(t_p - t_0) \tag{2}$$

Algorithm 1 describes node i learns the bandwidth of link ij in the last three changes and then it predicts the bandwidth of next time p.

3.3   Forwarding Region

When a node s intends to send data to node d, it selects the neighbor node with highest potential link bandwidth as its next hop and then same metric continues to select the best next forwarding node. Apparently, s will not select any nodes

Algorithm 1. *Prediction for future bandwidth of link ij*

1: Learn and keep the bandwidth changes of link ij at the last three times 0,1, and 2.
2: Calculate $\alpha_{i,j,k}$ with equation (1)
3: Calculate the predicted bandwidth of link ij of time p with equation (2)

in the opposite direction from s to d. How is node s aware of the region where the next hop falls? In current WMNs, each device is equipped with GPS and hence it knows its location. We assume that the sender knows its own location and the location of the receiver. The assumption is very common in geographic routing [6]. Figure 2 shows the scenario. Suppose node s intends to send data to node d, it figures out the forwarding region as Algorithm 2.

Algorithm 2. *Figure out the region for next hop*

1: s connects d.
2: sd rotates 45 degrees anti-clockwise, the ray is the positive half of X axis.
3: sd rotates 45 degrees clockwise, the ray is the negative half of Y axis.
4: Oppositely extends the ray of X axis to generate the negative half of X axis.
5: Oppositely extends the ray of Y axis to generate the positive half of Y axis.
6: The plane is divided up to four quadrants. The 4th quadrant is where the forwarding will be conducted.

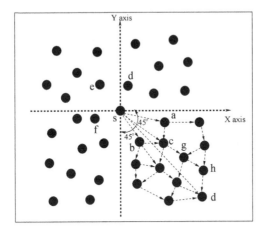

Fig. 2. Forwarding region

3.4   Forwarding Algorithm

Suppose each node in a Wireless Mesh Network regularly mains the last three changes of bandwidths of all its links that connect its neighbors. When node s intends to send data to node d, s first uses Algorithm 2 to figure out the region where the forwarding will be performed. Then s calls Algorithm 1 to find the node with potential highest bandwidth among all its neighbors as next hope. When the selected node relays the forwarding, it only considers its neighbors in the forwarding region as forwarding candidates, and it calls Algorithm 1 to forward the data to next hop with potential highest bandwidth. The forwarding resumes until the packets arrive destination node d.

Algorithm 3. *Forwarding algorithm*

1: s calls algorithm 2 to figure out the forwarding region.
2: s calls algorithm 1 to find the node n with potential highest bandwidth of link sn as next hope, where n is in the forwarding region.
3: if n is d, end the algorithm. Otherwise $s=n$, go to step 2.

---

4   Evaluation

We evaluated our mechanism in a simulated noiseless radio network environment by MATLAB. We create a topology that consists of a number of randomly distributed nodes. We compare our approach (ML Forwarding) with two other algorithms. One is congestion control and fuzzy logic with machine learning for wireless communications, say Fuzzy Logic. The other one is supervised learning approach for routing optimization in wireless networks, say Supervised Learning. The compared metrics are transmission delay (Milliseconds) and transmission speed (MBs/Millisecond). We performed a sequence of experiments in which the number of nodes varies from 100 to 300 in increments of 25 over an area of 100×100 m in the reference network. For each number of mobile users, we conduct our experiments 10 times and present the average value.

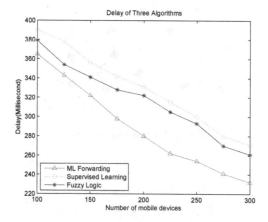

Fig. 3. Transmission delay of the three algorithms

Figure 3 shows that our approach results in the least delay. It is because our approach selects the link with potential maximum bandwidth of each hop. Figure 4 shows that with the same reason, our approach generates the maximal transmission speed among the three approaches.

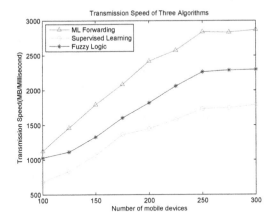

Fig. 4. Transmission speed of the three algorithms

5   Conclusion

A machine learning based forwarding algorithm in wireless mesh networks with cognitive radios is presented in this paper. In this algorithm, each mobile device keeps the last three times of bandwidth changes of its links that connect its neighbors. Then when a node intends to forward data, the node learns the historical changes of bandwidth and then predicts the possible future bandwidths of the links with neighbor nodes. Hence the forwarding node is able to select the next hop with highest bandwidth. We also designed a geometrical algorithm to let the source node figure out the forwarding region in order to avoid unnecessary flooding. Simulation results demonstrate that our approach outperforms peer approaches.

Acknowledgment. This work is supported in part by the Spanish government, Dirección General de Investigacin Cientfica y Técnica, a unit of the Ministerio de Economía y Competitividad, TIN2015-69542-C2-1-R (MINECO/FEDER), in collaboration with Universidad Rey Juan Carlos, Spain, under the project "Inteligencia Artificial y Métodos Matemáticos Avanzados para el Reconocimiento Automático de Actividades."

References

1. Akyildiz, I.F., Wang, X.: A Survey on wireless mesh networks. IEEE Commun. Mag. (2005)
2. Mitola, J.: Cognitive radio: an integrated agent architecture for software defined radio. Ph.D. dissertation, Royal Institute of Technology (KTH), Stockholm, Sweden (2000)
3. Costlow, T.: Cognitive radios will adapt to users. IEEE Intell. Sys. **18**(3), 7 (2003)
4. Ayodele, T.O.: Introduction to Machine Learning. New Advances in Machine Learning. InTech, Rijeka (2010)
5. Yang, J., Payne, B., Hitz, M., Zhang, Y., Guo, P., Li, L.: Fair gain based dynamic channel allocation for cognitive radios in wireless mesh networks. J. Comput. (2014)
6. Yang, J., Fei, Z.: HDAR: hole detection and adaptive geographic routing for ad hoc networks. In: 2010 Proceedings of the 19th International Conference on Computer Communications and Networks (ICCCN), pp. 1–6. IEEE (2010)
7. Yang, J., Fei, Z.: Broadcasting with prediction and selective forwarding in vehicular networks. Int. J. distrib. sens. netw. (2013)
8. Yang, J., Fei, Z.: Bipartite graph based dynamic spectrum allocation for wireless mesh networks. In: 28th International Conference on Distributed Computing Systems Workshops, ICDCS 2008, pp. 96–101. IEEE (2008)
9. Shen, J., Sen-Ching, S.C.: Layer depth denoising and completion for structured-light RGB-D cameras. In: 2013 IEEE Conference on Computer Vision and Pattern Recognition (CVPR), pp. 1187–1194. IEEE (2013)
10. Shen, J., Su, P., Sen-Ching, S.C., Zhao, J.: Virtual mirror rendering with stationary RGB-D cameras and stored 3-D background. IEEE Trans. Image Process. **22**(9), 3433–3448 (2013)
11. Shen, J., Tan, W.: Image-based indoor place-finder using image to plane matching. In: 2013 IEEE International Conference on Multimedia and Expo (2013)
12. Wang, Y., Martonosi, M., Peh, L.S.: A supervised learning approach for routing optimizations in wireless sensor networks. In: Proceedings of the 2nd International Workshop on Multi-Hop Ad Hoc Networks: From Theory to Reality, pp. 79–86 (2006)
13. Sawhney, A., Bhatia, R., Mahajan, P.: Congestion control in wireless communication network using fuzzy logic and machine learning techniques. Int. J. Adv. Res. Electr. Electron. Instrum. Eng. **3**(11) (2014)
14. Kincaid, D., Cheney, W.: Numerical Analysis: Mathematics of Scientific Computing, 3rd (edn.) (2005)

Machine Learning and Information
Processing in Wireless Sensor Networks

Energy-Balanced Routing Algorithm in Wireless Sensor Networks Using Cauchy Operator

Feng Li[1][(✉)], Li Wang[1], Jiangxin Zhang[1], and Xin Liu[2]

[1] College of Information Engineering, Zhejiang University of Technology,
Hangzhou 310023, China
{fenglzj,liwang2002,zjx}@zjut.edu.cn
[2] School of Information and Communication Engineering,
Dalian University of Technology, Dalian 116024, China
liuxinstar1984@dlut.edu.cn

Abstract. Efficient and reliable routing is a critical issue in wireless sensor networks in which network availability and node lifetime involved in the course of routing design need to be deliberately considered. In this study, we propose an energy-aware routing scheme using Cauchy operator and considering the factors of node's residual energy adjacent to current transmit node along with the routing distance. Based on Cauchy inequation, we achieved the relationships between the single-hop distance and path energy consumption as well as the overall routing distance and energy usage of the whole networks. By fixing a relay selection parameter and identifying the transmission hops appropriately, we obtain the balancing energy-aware routing algorithm. Numerical results are provided to testify the lifetime and equilibrium of the network energy by compared with traditional approaches.

Keywords: Wireless sensor networks · Energy-aware · Router · Optimization

1 Introduction

Wireless sensor networks (WSNs) are event-driven network systems which have attracted sustained research interest because of their wide applications in environmental and habitat monitoring, medical diagnostics and healthcare etc. [1,2]. Due to usually employed in harsh and distant circumstances, it is envisioned that the wireless sensor nodes cannot be repowered and cared frequently, therefore the efficient usage of limited energy is clearly of great crucial to maintain the overall network's stability.

Many approaches such as multi-hop cooperative transmission, cluster management and sleep mechanism in WSNs have been deeply studied for lifetime elongation of the sensor networks [3–5]. How to design suitable strategies of energy allocation, working mode as well as node deployment for the relay transmission networks attracts lots of attention where various of technique tools, including topology, game theory and intelligent algorithms etc. [6] are applied.

© ICST Institute for Computer Sciences, Social Informatics and Telecommunications Engineering 2017
X.-L. Huang (Ed.): MLICOM 2016, LNICST 183, pp. 237–246, 2017.
DOI: 10.1007/978-3-319-52730-7_24

Specifically, reference [3] reveals the shortcomings of currently evaluative standard and current cluster-based routing protocol by HEED, and proposes a clustering patch hierarchical routing protocol with the purposes of improving network coverage rate and effective network lifetime. The authors in [4] consider the problem of optimal deployment of numbers of sink nodes in a WSN for minimizing average hop distance between sensors and its nearest sink with maximizing degree of each sink node which can solve hot spot problem which is another critical issue of WSNs design. Reference [5] designs a mobile sensor node platform to achieve a highly accurate localization mechanism by using ultrasonic, dead reckoning, and radio frequency information which is processed through a particle filter algorithm. In [6], the authors propose a hub-spoke network topology that is adaptively formed according to the resources of its members. A protocol named resource oriented protocol which divides the network operation into two phases is developed to build the network topology.

The main challenges in a WSN include the energy, bandwidth constraint and complicated topology structure of the sensor nodes which triggers many innovations of software algorithms, hardware solutions and transmission protocols. One of the most important innovations of an energy efficient solution lies in the network layer of the WSNs, the routing protocol and route node finding to elongate nodes energy and network's lifetime. In this work, we mainly focus on the optimal route node searching for multi-hop relay transmission in sensor networks in light of the energy-balancing-aware of the whole system. In the course of searching next optimal relay node, a dynamic route rather than static route need to be figured out ceaselessly so as to avoid the paralysis of the hot transmit path with the results of the interrupt of overall networks. With the development of various applications of WSNs, new routing approaches are attracting plenty of attention in recent years [7–9]. In [7], the authors propose a routing algorithm which intends for WSNs that needs a period and event-driven approach and can adapt to the situation the sensor faces. Reference [8] develops a routing solution off-network control processing that achieves control scalability in large sensor networks by shifting certain amount of routing functions to an off-network server. A tiered routing approach, consisting of coarse grain server based global routing, and distributed fine grain local routing is designed for achieving scalability by avoiding network wide control message dissemination. In [9], the authors propose an energy-aware trust derivation scheme using game theoretic approach, which manages overhead while maintaining adequate security of WSNs. A risk strategy model is first presented to stimulate WSNs node's cooperation.

In this paper, we propose a dynamic routing algorithm for WSNs to elongate network lifetime by searching optimal relay node uninterruptedly and balancing energy consumption over whole networks. To extend the survival cycle of wireless sensor nodes and avoid network paralysis due to regional usage overheating, source sensor nodes need to select different transmission paths according to node's residual energy and holistic energy consumption in whole route. In every single cycle, the source node is supposed to start data transmission after sending detecting signal package to acknowledge the transmit route. Based on Cauchy

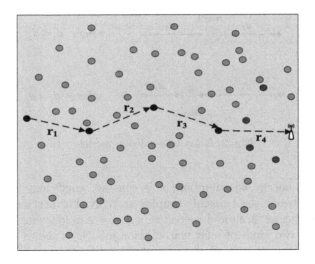

Fig. 1. Multi-hop relay transmission in WSNs

operator, we design a relay selection parameter to ascertain the next transmit node dynamically. Since the transmit tasks require multiple hops rather than single hop, we pursue to find the optimal routing by taking the expenditure of transmit energy on the whole router into account rather than only next hop. In order to evaluate the network performance and energy consumption effects, we perform several simulated network tests to testify the consumption of node's energy and the number of the death node. A uniform distribution mapping of residual network energy along with relative few death nodes are achieved. Also, comparison experiments are also performed to evaluate the performance of our proposed algorithm.

2   System Model

In this paper, we consider a model of WSNs where numbers of wireless sensor nodes and a fusion center locate in the networks as shown in Fig. 1. Because of wide distribution area and limited battery capacity within the sensor nodes, multi-hop relay transmission should be frequently applied. Relay communication was originally encountered in bent-pipe satellites where the primary function of the spacecraft is to relay the uplink carrier into a downlink [10]. In general, relay cooperative transmission can benefit sensor transmitter in terms of power saving, degressive outage probability and improved system capacity. We further assume the relay node works in regenerative mode, and suppose the distribution of the sensor nodes is relatively static and the channel conditions are changing not much over time.

The energy consumption of transmit sensor node mainly depends on the transmit distance, thus we can consider the following equation

$$E = \kappa d^n \tag{1}$$

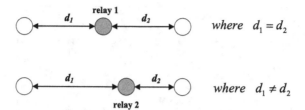

Fig. 2. Relay transmission models

where E is the energy consumption, κ is constant coefficient and d denotes the distance between the transmit couples in WSN. Furthermore, n is power coefficient locating at $[2, 4]$, and we can choose $n = 2$ in following deductions.

We consider two kinds of relay transmission models as shown in Fig. 2. Suppose a source sensor needs to select a relay station to finish its signal transmission as shown in Fig. 2. In the first case, the relay station locates at the middle position between the source and the destination nodes. According to (1), we can obtain the energy consumption as

$$E = \kappa(d_1^2 + d_2^2) \tag{2}$$

As there is $a^2 + b^2 \geq \frac{(a+b)^2}{2} = \frac{m^2}{2}$, and the equation holds only when $a = b = \frac{m}{2}$, thus we have

$$E_1 = 2\kappa d_1^2 < E_2 = \kappa(d_1^2 + d_2^2) \tag{3}$$

In WSNs, it always requires multiple hops to finish the signal transmission through source node to sensor fusion. Then, the total energy consumption over the multi-hop relay transmission can be expressed as

$$E = \sum_{j=1}^{n} \kappa d_i^2 \tag{4}$$

Then the optimization objective function can be given as

$$min \sum_{j=1}^{n} \kappa d_i^2 \tag{5}$$

Also, according to Cauchy inequation, there is

$$\sum_{i=1}^{n} a_i^2 \sum_{i=1}^{n} b_i^2 \geq \{\sum_{i=1}^{n} a_i b_i\}^2 \tag{6}$$

Let $b_i = 1$, we have

$$(a_1^2 + a_2^2 + \cdots + a_n^2) \geq \frac{1}{n}(a_1 + a_2 + \cdots + a_n)^2 = \frac{C^2}{n} \tag{7}$$

where $a_1 + a_2 + \cdots + a_n = C$, and the equation holds if and only if $a_1 = a_2 = \cdots = a_n = \frac{C}{n}$. It means the entire energy consumption can be minimal when every relay distance is equal in condition of fixed total route length. In real monitoring environment, the distribution of wireless sensor nodes is usually random, and the distance between source node and sensor fusion is inconstant. Whereas, we can control the distances of every hop to make every d_i similar or even equal. In condition of same SINR communication effects, shorter transmission distance brings evident energy benefits. Therefore, we can decrease the total energy consumption by suitably choose the senor node in every hop.

In this study, we want to investigate the optimal relay selection of every single hop in precondition of settled total router length. According to the analyses above, we can obtain that the energy consumption of whole sensor networks will be optimized and the network lifetime can be prolonged by carefully control the distance between adjacent relay hops when the router length is fixed. Furthermore, we need to balance the energy usage overall the networks by preferring the node with more residual energy. Hence, we consider the following equation as our proposed method of relay selection

$$\gamma_i = S(i).E - \frac{A\{d(i-1,i) - d(i-2,i-1)\}}{\log_2(r+1)} \tag{8}$$

$$where\ r \geq 1, S(i).E \geq 0, i \geq 2$$

where $S(i).E$ denotes the residual energy of the next hop, γ_0 means the source node station, $d(i,j)$ is the distance between relay i and relay j, A is a correlation coefficient and r is the cycle number which also means the data transmit times. In this relay selection strategy, a sensor node will be chosen in priority of more usable energy. Besides, we can receive a rational energy efficiency when the transmit length of every hop approaches to the previous hop. When $i = 1$ which means the source node is looking for the first relay, the objective function returns to be $\gamma_i = S(i).E$. On the other hand, when the monitoring sensor networks have carried out many rounds of signal transmission, it can be concluded the available energy within the wireless sensor nodes will be very limited. Hence, we introduce a parameter denoted as the cycle number which means the emphasis of the relay selection will be put on current residual energy. We set the parameter on denominator to balance the impacts of residual energy and relay distance. With the increase of transmit times, the available energy $S(i).E$ becomes smaller and the distribution of node's energy overall the networks gets proportional. We add a parameter on the latter section to balance the effects between the two parts. Besides, we need to confirm every new hop is closer to the fusion center. Otherwise, the relay transmission gets meaningless. Also, when no available sensor node can meet the selection conditions, the parameter $d(i,j)$ which denotes the searching coverage will be increased. The thresholds of searching distance and residual energy are further settled to guarantee the rationality of the proposed algorithm. The diagram of this routing method is shown as Fig. 3.

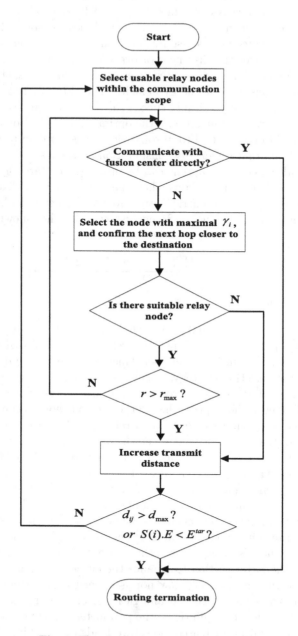

Fig. 3. Flowchart of our routing algorithm

3   Simulation Results and Performance Evaluation

In this section, we evaluate the performances of our proposed balancing energy-aware routing algorithm in Matlab simulation platform. As shown in Fig. 4, we randomly generate wireless sensor topologies with size $500 \times 500 \, m^2$, and node's initial energy is randomly settled with maximum $1500 \, J$. Initial transmission distance is $60 \, m$, and the communication range can be suitably increased when there is no available node within the initial distance. We ignore the effects from other cells, and suppose the usable bands is enough.

In the set of simulations, 360 sensor nodes randomly locate in the networks as shown in Fig. 4. The QoS threshold of every relay hop is identical. Sensor nodes select next hop according to Eq. (8). The energy consumption in every single transmission is identified by (1) where we set $\kappa = 2$. Here, we suppose the maximal transmit-distance of sensor node to be $140 \, m$ [11]. If the current sensor node cannot find suitable hop with enough energy within the scope, the routing will be terminated. Also, the number of death nodes is upgrading with the increase of the transmit-cycle, and we set a minimal threshold of active sensor nodes which is 3% total node's quantity.

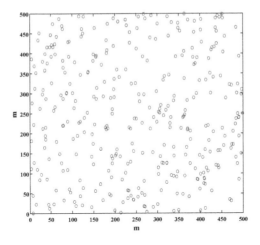

Fig. 4. User distribution in WSNs

As shown from Figs. 5, 6 and 7, we give the performance of node's residual energy. In Fig. 5, an initial energy distribution is presented. Due to random distribution of node's residual energy, we can obtain from the figure that the usable energy of every node is asymmetric, therefore the source node should carefully select the router to balance network energy according to our proposal. In this test, we fix the initial source node at the coordinate (0,250) and the fusion center at (500,250). In the following tests, we also evaluate the performance in variant source nodes. In Fig. 6, the energy distribution of all the sensor nodes at 400 transmission cycles is shown. In fact, the sensor node in this WSNs does not know the residual energy of its adjacent nodes. It needs to send detecting signal to acquire the energy information of the other nodes within the transmit-scope firstly. In our proposed method, more residual energy node will likely be chosen so that the energy distribution in this network becomes smooth over time. Furthermore, in Fig. 7, after 2000 transmission cycles, the energy figure gets flat except for several heaves in four corners where the sensor nodes are very distant to the fusion center which will lead to a selection discarding as a result. In the routing algorithm, we suppose the next hop must be closer than the prior one, so very isolated sensor nodes will not often be selected to assist the cooperative transmission.

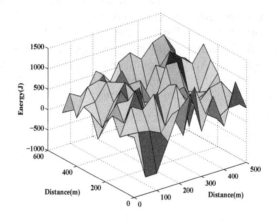

Fig. 5. Residual energy of sensor nodes in initial status

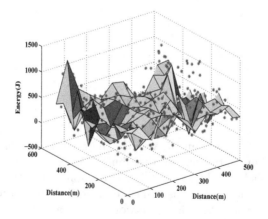

Fig. 6. Residual energy of sensor nodes with 400 transmission cycles

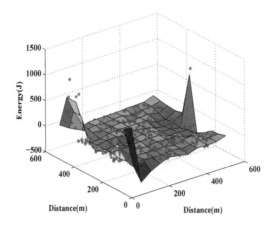

Fig. 7. Residual energy of sensor nodes with 2000 transmission cycles

4   Conclusion

In this paper, we develop a balancing energy-aware routing algorithm for WSNs. In pursuing the goal of energy consumption minimum, we take care of energy equilibrium over the networks. According to Cauchy operator, we achieve the energy equation by investigating the distances between every relay hop. In given total router length, minimal energy consumption can be attained in condition of equal hop length. Furthermore, a corresponding routing program is provided to testify our proposal. In the progress, we identify the routing nodes by considering the node's residual energy as well as the distance compared with prior hop. Numerical results show the algorithm's equilibrium is rational, and the energy efficiency is evaluated by a comparison tests.

Acknowledgment. This work was supported by the National Natural Science Foundation of China (No. 51404211) and Natural Science Foundation of Zhejiang Province (No. LY14F010009 and LQ14H180001) along with Foundation of Zhejiang Educational Committee (No. Y201431820), Zhejiang Open Foundation of the Most Important Subjects (No. 20150705) and the Fundamental Research Funds for the Central Universities under Grant No. DUT16RC(3)045.

References

1. Mascarenas, D., Flynn, E., Farrar, C., Park, G., Todd, M.: A mobile host approach for wireless powering and interrogation of structural health monitoring sensor networks. IEEE Sens. J. **9**, 1719–1726 (2009)
2. Choi, B., Lee, J.: Sensor network based localization algorithm using fusion sensor-agent for indoor service robot. IEEE Trans. Consum. Electron. **56**, 1457–1464 (2010)
3. Poonguzhali, P.K.: Energy efficient realization of clustering patch routing protocol in wireless sensors network. In: International Conference on Computer Communication and Informatics, Coimbatore, India, pp. 1–6 (2012)

4. Dandekar, D.R., Deshmukh, P.R.: Energy balancing multiple sink optimal deployment in multi-hop wireless sensor networks. In: IEEE International Advance Computing Conference, pp. 408–412 (2013)

5. Luo, R.C., Chen, O.: Mobile sensor node deployment and asynchronous power management for wireless sensor networks. IEEE Trans. Industr. Electron. **59**, 2377–2385 (2012)

6. Ma, Y., Aylor, J.H.: System lifetime optimization for heterogeneous sensor networks with a hub-spoke topology. IEEE Trans. Mob. Comput. **3**, 286–294 (2004)

7. Masruroh, S.U., Sabran, K.U.: Emergency-aware and QoS based routing protocol in wireless sensor network. In: International Conference on Intelligent Autonomous Agents, Networks and Systems, Bandung, Indonesia, pp. 47–51 (2014)

8. Wu, T., Biswas, S.: Off-network control for scalable routing in very large sensor networks. In: IEEE ICC, pp. 3357–3363 (2007)

9. Duan, J., Gao, D., Yang, D., Foh, C.H., Chen, H.: An energy-aware trust derivation scheme with game theoretic approach in wireless sensor networks for iot applications. IEEE Internet Things J. **1**, 58–69 (2014)

10. Li, F., Tan, X., Wang, L.: Power scheme and time-division bargaining for cooperative transmission in cognitive radio. Wirel. Commun. Mob. Comput. **15**, 379–388 (2015)

11. Sun, B., Gao, S., Lu, Q.: Study on optimization of communication radii of nodes in hierarchically clustered wireless sensor networks. J. Graduate School Chin. Acad. Sci. **27**, 818–823 (2010)

Coverage Improvement Strategy Based on Voronoi for Directional Sensor Networks

Shan You[1,2], Guanglin Zhang[1,2(✉)], and Demin Li[1,2]

[1] College of Information Science and Technology, Donghua University,
Shanghai 201620, People's Republic of China
shanyou@mail.dhu.edu.cn, {glzhang,deminli}@dhu.edu.cn
[2] Engineering Research Center of Digitized Textile and Apparel Technology, Ministry
of Education, Shanghai 201620, People's Republic of China

Abstract. Nowadays, directional sensor networks (DSNs) have drawn a lot of attentions, which are made up of a large number of tiny directional sensors that are different from traditional omnidirectional sensors. Directional sensor is characterized by working direction and angle of view (AOV). In this paper we study area coverage of DSNs. We exploit Voronoi theory to divide sensors into polygons, by optimizing the local coverage in each polygon to achieve the overall coverage. We take full use of Voronoi vertexes and edges to judge whether a sensor gets full coverage inside in current polygon, if not, then the sensor calls Move Inside Cell Algorithm (MIC) and Rotate Working Direction Algorithm (RWD) algorithms we have designed. Compared to the similar methods to solve this question our algorithms are relatively simple and moving distance is shorter. Simulation results reveal that our algorithms outperform some existing methods in term of the area coverage.

Keywords: Coverage · Voronoi · DSNs · Sensor network

1 Introduction

With the development of the current technology and wide applications of wireless sensor networks (WSNs), WSNs are playing an increasingly important role in the daily life and have gained wide attentions. Coverage is the fundamental problem in WSNs, previous works most focus on the omnidirectional sensor network that means the working coverage area is a circle. In recent years, attentions have been focused on the DSNs which are different from the general WSNs. Directional sensor is limited to a certain working direction and AOV, such as camera sensor, infrared sensor and so on. In the WSNs, sensing coverage is mainly related to R_s(sensing radius) and the location of sensor. Whereas in the DSNs, sensing coverage is related not only to the location and R_s but also to the working direction and AOV.

G. Zhang—College of Information Science and Technology, Donghua University Shanghai, 201620, P. R. China. Engineering Research Center of Digitized Textile and Apparel Technology, Ministry of Education

Currently, there are some studies related to the DSNs area coverage. For example, some studies centered on using virtual force between adjacent sensors [4,7], others adopted the Voronoi [9] principle to reduce the complexity of sensor to make decisions [2,3]. In this paper we adopt Voronoi diagram mainly based on the following points: (1) Voronoi can be constructed by the randomly deployed sensors, every sensor lies exactly in the Voronoi cell; (2) by the information of voronoi vertexes, sensor can make a decision to move and rotate; (3) the sensor's moving trajectory is limited to the cell of current sensor. In this paper, we raise coverage enhancement algorithm MIC and RWD. Sensor by adjusting working direction and updating location to get full coverage in each Voronoi polygon. Our algorithm can get the better coverage and RWD needs sensor to move the shorter distance compared to DVSA proposed in [2].

The rest of the paper can be organized as follows. In Sect. 2, we summarize the related work. In Sect. 3, we state the problem clearly and introduce some preliminaries. Then, we analyse the theoretical framework and provide coverage increment algorithms in Sect. 4. Performance evaluation and analysis are presented in Sect. 5. Finally, we conclude conclusions in Sect. 6.

2   Related Work

For the DSNs and WSNs area coverage, some scholars conducted extensive researches in [6]. Wang et al. [1] and Ghosh [8] studied hybrid WSNs. They employed Voronoi diagram to solve coverage problem. Liang et al. [5] presented two distributed self-deployment schemes of mobile sensors, namely Circumcenter-based and Incenter-based. The former constructs Voronoi diagram by the circumcenter of sensor's sensing sector, if sensor does not get the optimal coverage, then sensor moves until the circumcenter coincides with the centroid of current cell. The later also adopts this pattern, difference is that it employs inceneter of sensor's sensing sector to construct Voronoi diagram.

In [4,7], authors put forward coverage enhance algorithms based on virtual force between sensors. Dan et al. [7] presented potential field based coverage-enhancing algorithm (PFCEA). PFCEA mainly takes centroid of sensing sector area as stressed point, sensor adjusts its working direction by resultant force from neighbor sensors. Liang et al. [4] also employed virtual force, the difference is that they added Voronoi diagram on the basis of [7] and proposed a scheme consists of four different forces caused by neighboring sensors and uncovered regions in the field, namely: the Centroid Push Auxiliary point Force (CPAF), the Centroid Push Centroid Force (CPCF), the Voronoi point Pull Centroid Force (VPCF), and the Neighbor Repulsive Force (NRF). Their core idea is to adjust working direction based on resultant force of sensor.

Sung and Yang [2,3] mainly adopted Voronoi theory and Delaunay triangulation to design distributed self-redeployment coverage enhancement algorithm. In DVSA [2], main idea is to calculate corresponding field angle and side length of each vertex in each polygon, and then compare AOV and R_s with field angle and side length. If result is equivalent, then sensor moves to that vertex and take the longer side as one boundary of sensing sector. In [3], they presented several coverage increment algorithms, Vertex-based adjustment with Voronoi diagram(V-VD), Edge-based adjustment with Voronoi diagram(E-VD), Edge-based adjustment with Delaunay triangulation (E-DT) and Angle-based adjustment with Delaunay triangulation (A-DT). In E-VD sensor chooses the midpoint of the farthest edge of its own cell as working direction.

In this paper we study the area coverage of DSNs in the case of random deployment. Sensor makes a decision on adjusting working direction and updating location by the vertex information of Voronoi diagram. The final goal for the sensor is to get local maximum coverage in current cell.

3   Preliminaries

3.1   DSN Sensing Model

Compared with the traditional WSNs, directional sensor has some differences as shown in Fig. 1(a), which is characterized by directional working direction $\omega(-\pi \leq \omega \leq \pi)$ and AOV α. $\overrightarrow{wd}$ splits angle α and is defined as working direction vector which is unit vector. The effective coverage field of sensor is the sector area $\alpha R_s^2/2$. Two auxiliary points a_l, a_r we introduce are to help make decision later in our algorithm. Given a point $p(x,y)$ which is covered by the sensor $s(x_s, y_s)$, the following two conditions must be met:

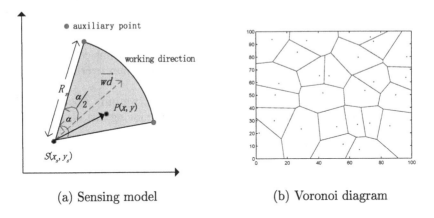

(a) Sensing model   (b) Voronoi diagram

Fig. 1. Sensing model of DSN and voronoi diagram.

(1) Euclidean distance between p and s must be smaller than sensing radius R_s.

$$d(s,p) = \sqrt{(x - x_s)^2 + (y - y_s)^2} \leq R_s \qquad (1)$$

(2) Absolute included angle between $\overrightarrow{wd}$ and $\overrightarrow{sp}$ must be smaller then $\alpha/2$.

$$\phi = \arccos \frac{\overrightarrow{wd}\overrightarrow{sp}}{\|sp\|} \leq \frac{\alpha}{2} \qquad (2)$$

3.2   Voronoi Diagram and Some Assumptions

Voronoi diagram is an important data structure in computational geometry which is widely used in many fields. The main properties we need to use in this paper are that (1) each sensor s_i lies exactly in the current cell and any point p within the current cell has the shortest distance between sensor s_i and point p compares to other sensor i.e. $d(s_i, p) \leq d(s_j, p)$; (2) point in a shared edge of two sensors has equivalent distance from two sensors. Here, as a example we choose 25 random nodes to generate Voronoi diagram as Fig. 1(b) shows. In order to refine the question we study and make the key researched point stand out, some assumptions are made as follow:

(1) All of directional sensors are homogeneous, that is to say every sensor has the same sensing radius R_s, viewing angle α, rotation ability and mobility.
(2) For every sensor we can acquire accurate coordinate by GPS or other localization algorithm such as DV-hop, Amorphous and so on.
(3) All sensors have strong transmission ability to ensure the network connected and Voronoi diagram constructed successfully.

3.3   Problem Statement

Our ultimate aim is to reduce overlap and enlarge effective coverage in the target area. By using the rotating ability and mobility of sensor coupled with Voronoi information to make sensor get full coverage in every cell. Mobility and working direction adjustment must meet three principles:

(1) For every sensor, moving range is restricted at the current cell.
(2) To get maximal coverage in current cell and minimum overlap with other cell's sensor.
(3) Although we do not consider energy consuming, we only take the mobile distance as a measure of standards i.e. minimum moving distance meanwhile maximum coverage.

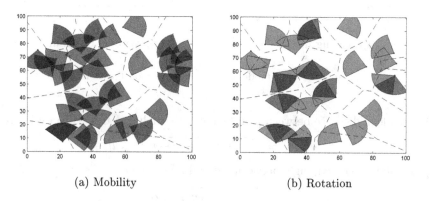

(a) Mobility   (b) Rotation

Fig. 2. Example of results before and after execution of algorithms.

Here, we give a simple exhibition of mobility and motility as shown in Fig. 2. Figure 2(a) is mobility deployment, Fig. 2(b) is rotation deployment, grey area is the status after mobility and rotation.

4   Theoretical Analysis

4.1   Judge Whether Working Area is Wrapped by Polygon

If the sensor s and auxiliary points a_l, a_r all inside in polygon as Fig. 3(b) and (c) show, then we approximately think that sensor s gets full coverage, which means sensor does not need to move and rotate.

To prove the sensors's coverage field is wrapped in the cell, we can prove points s, a_l and a_r are all in the convex polygon. To prove whether one point (assuming p) is in the convex polygon, we exploit triangle segmentation to compute the convex polygon area S_{V_s} and the area S_{pv_s} constructed by p and V as Fig. 3(a) demonstrates. V_s denotes vertex set of sensor S.

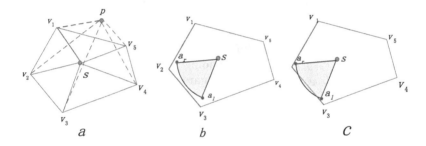

Fig. 3. Proof point in convex polygon and sector wrapped in convex polygon

$$S_{v_s} = \sum_{v_i=1}^{V_s-1} S_{\Delta sv_iv_{i+1}} \quad S_{pv_s} = \sum_{v_i=1}^{V_s-1} S_{\Delta pv_iv_{i+1}} \tag{3}$$

If $S_{v_s} = S_{pv_s}$, than we can judge that point p is inside in polygon. If two auxiliary points a_l, a_r of sensor s are all in the polygon, we can infer that sensor get relatively full coverage although not completely as shown in Fig. 3(c). If $S_{v_s} \neq S_{pv_s}$, which means sensor does not get full coverage in current polygon, then sensor will move or adjust $\overrightarrow{wd}$.

4.2   Move and Rotate Based on Vertex

If auxiliary points a_l, a_r are not all inside in current polygon, location and working direction will adjust. We choose the farthest vertex (assume v_1) of polygon as the $s's$ initial working direction mainly consider that original coverage may be relatively full, which can help to move by the shorter moving distance or make

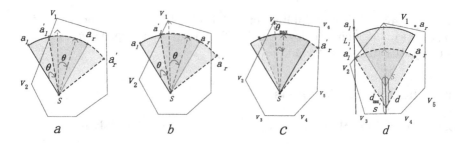

Fig. 4. Move and rotate inside cell

the smaller adjustment of working direction. Let we see Fig. 4(a), select v_1 as working direction $\overrightarrow{sv_1}$, there are two situations here:

Case 1: Rotate Working Direction Algorithm (RWD). $R_s \leq d(s, v_1)$ and auxiliary point a_r is inside of polygon the other auxiliary point a_l is not.

For this situation we can rotate the working direction by θ to get the aim of letting the $a_l's$ position update to a_l'. While rotating θ may lead a_r' to be out of polygon as Fig. 4(b) shows, therefore we should primarily figure out θ_{max} (see in Fig. 4(c)) which is maximal rotation angle. If $\theta \leq \theta_{max}$, rotate working direction by θ, if not, rotate by θ_{max}. To compute θ and θ_{max}, as follow elaboration:

According to $s(x_s, y_s)$ and corresponding angle ω of $\overrightarrow{wd}$, we can get the included angle β of $L_{sa_l'}$ relative to coordinate axis and coordinate of a_l'.

$$\psi = \arctan(\frac{y_{v_1} - y_s}{x_{v_1} - x_s}) \tag{4}$$

if $\psi \leq 0$.

$$\begin{cases} \omega = \psi & x_{v_1} \geq x_s, y_{v_1} \leq y_s \\ \omega = \psi + \pi & x_{v_1} \leq x_s, y_{v_1} \geq y_s \end{cases} \tag{5}$$

if $\psi \geq 0$.

$$\begin{cases} \omega = \psi & x_{v_1} \geq x_s, y_{v_1} \geq y_s \\ \omega = \psi + \pi & x_{v_1} \leq x_s, y_{v_1} \leq y_s \end{cases} \tag{6}$$

if rotate to right

$$\begin{cases} \beta = \omega + \frac{\alpha}{2} - \theta & \omega \geq 0. \\ \beta = \omega - \frac{\alpha}{2} + \theta & \omega \leq 0. \end{cases} \tag{7}$$

if rotate to left

$$\begin{cases} \beta = \omega - \frac{\alpha}{2} + \theta & \omega \geq 0. \\ \beta = \omega + \frac{\alpha}{2} - \theta & \omega \leq 0. \end{cases} \tag{8}$$

We can get the θ for point $a_l'(x_{a_1'}, y_{a_1'})$ is at line $L_{v_1 v_2}$.

$$\begin{cases} x_{a_1'} = x_s + R_s \cos(\beta) \\ y_{a_1'} = y_s + R_s \sin(\beta) \\ L_{v_1 v_2} = \arctan(\frac{y_{v_1} - y_{v_2}}{x_{v_1} - x_{v_2}})(x - x_{v_1}) + y_{y_1} \end{cases} \tag{9}$$

In a similar way see in Fig. 4(c), we can solve the θ_{max} according to the above method, the only difference is that we combine a'_r with line $L_{v_5 v_6}$.

Case 2: Move Inside Cell Algorithm (MIC). Two auxiliary points of sensor s are all outside of polygon, Fig. 4(d).

For this situation, sensor should move, moving direction and moving distance are primarily under consideration. We choose the reverse direction of $\overrightarrow{sv_1}$ as moving direction and take a_l which is far from polygon as reference point for reason that if a_1 is moved into polygon then a_r is also in polygon. Sensor moves until a_1 is exactly in the edge of polygon, so moving distance $d = d(a_l, a'_l) = d(s, s')$. If $d \leq d_{max}$, moving distance by d, if not move distance by d_{max}. a'_l is intersection of line L_1 and $L_{v_1 v_2}$, then if we can find out L_1 and $L_{v_1 v_2}$, problem will be solved. Constructed calculation of d and d_{max} is as below:

$$
\begin{cases}
x_{a_1} = x_s + R_s cos(\omega + \frac{\alpha}{2}) \\
y_{a_1} = y_s + R_s sin(\omega + \frac{\alpha}{2}) \\
L_1 = \psi(x - x_{a_1}) + y_{a_1} \\
L_{v_1 v_2} = \frac{y_{v_1} - y_{v_2}}{x_{v_1} - x_{v_2}}(x - x_{v_1}) + y_{v_1}
\end{cases}
\tag{10}
$$

we can get the coordinate of a'_1 by combining L_1 with $L_{v_1 v_2}$. Similarly d_{max} can be worked out by combining $L_{ss'}$ with $L_{v_3 v_4}$. Updated sensor location (x'_s, y'_s):

$$
x'_s = \begin{cases}
x_s - d_{max} * cos(\omega) & d_{max} \leq d \\
x_s - d * cos(\omega) & d_{max} > d
\end{cases}
\tag{11}
$$

$$
y'_s = \begin{cases}
y_s - d_{max} * sin(\omega) & d_{max} \leq d \\
y_s - d * sin(\omega) & d_{max} > d
\end{cases}
\tag{12}
$$

5   Performance Evaluation

In this section, we conduct corresponding simulations about the algorithms we have designed. In order to make a remarkable contrast, we compare RWD and MIC with random deployment and DVSA [2]. Some essential parameters in the simulation are listed at following Table 1.

Table 1. Main notations.

The number of sensors N	N = 30, 60, 90, 120, 150, 180, 210, 240, 270, 300
Sensing radius	$R_s = 6\,m, 8\,m, 10\,m, 12\,m$
Angle of view AOV	$\alpha = \frac{\pi}{2}$
Size of monitoring area	$Area = 100\,m \times 100\,m$

5.1 Sensing Coverage

We chose $R_s = 12$ m, AOV is $\frac{\pi}{2}$ and the number of sensor is 30 as the initial conditions. We compare our algorithm with Random Deployment and DVSA [2]. Figure 5(a) is the random deployment and Fig. 5(b) is the DVSA deployment, Fig. 5(c) is the RWD deployment, Fig. 5(d) is MIC deplyment, the last one Fig. 5(e) is RWD coupled with MIC.

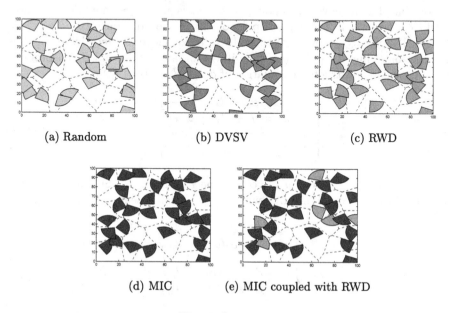

(a) Random   (b) DVSV   (c) RWD

(d) MIC   (e) MIC coupled with RWD

Fig. 5. Coverage.

5.2 Coverage Ratio

Coverage ratio is basic measure of DSNs. Figure 6 shows five kinds of coverage ratio. From the simulation results we can see that MIC coupled with RWD get the optimal result. when the number of sensors N is less, coverage increment ratio is not so remarkable and when N is relatively big, coverage increment ratio is obvious, if N is big enough, then advantage disappears. However, this situation is foreseeable and understandable.

5.3 Moving Track

Although in this paper we do not consider energy consuming, we only take moving distance as the measure of energy consumption. In DVSA, sensor moves to the vertex with side length and field angle similar to R_s and AOV, while in MIC sensor makes a decision to move based on whether the auxiliary points

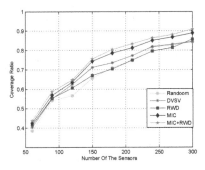

Fig. 6. Coverage ratio.

are in current polygon. Obviously, the former needs all sensors to move to the vertex, the latter does not. Figure 7(a) is the DVSA moving track and Fig. 7(b) is the MIC moving track, blue point is initial position and red point is updated position. Clearly, we can see MIC moving distance is the shorter than DVSA, besides, not all sensors in MIC need to move.

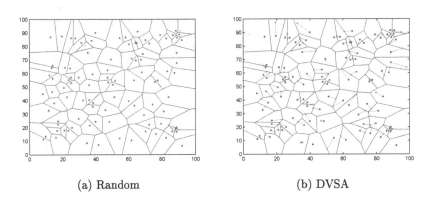

(a) Random   (b) DVSA

Fig. 7. Moving track. (Color figure online)

6 Conclusions and Future Work

In this paper, we study area coverage problem in DSNs and propose MIC and RWD coverage increment algorithms based on Voronoi. In order to have a measure for mobility and rotation, we introduce two auxiliary points a_l and a_r. By comparing the simulation results of MIC and RWD with DVSA and random deployment in sensing coverage, coverage ratio and moving track aspects, our algorithms outperform others. As to our future work, we are prepared to embark on researching energy consumption and networks life-time problems.

Acknowledgement. This work is supported by the NSF of China under Grant No. 61301118; the Innovation Program of Shanghai Municipal Education Commission under Grant No. 14YZ130; the International S&T Cooperation Program of Shanghai Science and Technology Commission under Grant No. 15220710600; and the Fundamental Research Funds for the Central Universities under Grant No. 16D210403.

References

1. Wang, G., Cao, G., Berman, P., Porta, T.F.L.: Bidding protocols for deploying mobile sensors. IEEE Trans. Mob. Comput. **6**(5), 515–528 (2007)
2. Sung, T.-W., Yang, C.-S.: Distributed Voronoi-based self-redeployment for coverage enhancement in a mobile directional sensor network. Int. J. Distrib. Sens. Netw. **2013**, 1456–1459 (2013)
3. Sungn, T.-W., Yang, C.-S.: Localised sensor direction adjustments with geometric structures of voronoi diagram and delaunay triangulation for directional sensor networks. Int. J. Ad Hoc Ubiquit. Comput. Arch. **20**(2), 91–106 (2015)
4. Liang, C.-K., Chung, C.-Y., Li, C.-F.: A virtual force based movement scheme for area coverage in directional sensor networks. In: 2014 Tenth International Conference on Intelligent Information Hiding and Multimedia Signal Processing, pp. 718–722 (2014)
5. Liang, C.-K., He, M.-C., Tsai, C.-H.: Movement assisted sensor deployment in directional sensor networks. In: 2010 Sixth International Conference on Mobile Ad-hoc Sensor Networks, pp. 226–230 (2010)
6. Guvensan, M.A., Yavuz, A.G.: On coverage issues in directional sensor networks: a survey. Ad Hoc Netw. **9**(7), 1238–1255 (2011)
7. Tao, D., Ma, H.-D., Liu, L.: A virtual potential field based coverage-enhancing algorithm for directional sensor networks. J. Softw. **18**(5), 1152–1163 (2007)
8. Ghosh, A.: Estimating coverage holes and enhancing coverage in mixed sensor networks. In: IEEE International Conference on Local Computer Networks, pp. 68–76 (2004)
9. Aurenhammer, F.: Voronoi diagrams - a survey of a fundamental geometric data structure. ACM Comput. Surv. **23**(3), 345–405 (1991)

Text Detection in Natural Scene Image:
A Survey

Shupeng Wang$^{(\boxtimes)}$, Chenglin Fu, and Qi Li

College of Communication and Information Engineering,
Xi'an University of Science and Technology, Xi'an 710054, China
1013366723@qq.com

Abstract. Text detection in natural scene image is the extraction of the text regions from a natural scene image. The extraction information can be used in the system of text recognition. The texts in natural scene image contain important information. Text detection is an important prerequisite for many computer vision applications, such as license plate recognitions system, information filtering system, automatic navigation and so on. Text detection as a real-life application has to quickly and successfully process the texts in different fonts and under different environmental conditions. It should also be generalized to process texts in different languages and directions. We categorize different text detection techniques according to the methods used for each stage, and compare them in terms of merits, demerits and performance. Feature forecasts of text detection in natural scene image are given at the end.

Keywords: Text detection · Natural scene image · Text information

1 Introduction

Along with the popularization of smart phone, tablets and smart wearable equipment, it is more and more convenient to capture high-quality scene images. The natural scene images contain wealth semantic information, such as road signs, posters, license plate and signboards. Text recognition from the natural scene image is the important prerequisite for many computer vision applications, such as automatic image understanding and content-based image analysis tasks. Efficient text detection is the foundation of semantic information extraction. To acquire the text regions in a natural scene image, a lot of techniques, for instance, object detection, image processing and pattern recognition, will be used. The variations of the images and text cause challenges in text detection.

The text detection system, which extracts the regions of text from a given natural scene image, can be composed of five stages. The first stage is to acquire the natural scene image using a camera. The view of the camera, the illumination and the image quantity should be considered. The second stage is to process the images for the follow-up work. The third stage is to extract the candidate character regions based on some text features. The fourth stage is to ensure the character regions using classifiers. The last stage is to extract the text regions. As the characters in natural scene image are

© ICST Institute for Computer Sciences, Social Informatics and Telecommunications Engineering 2017
X.-L. Huang (Ed.): MLICOM 2016, LNICST 183, pp. 257–264, 2017.
DOI: 10.1007/978-3-319-52730-7_26

always exit in words. The characters are grouped into words based on some features. Figure 1 shows the structure of text detection process.

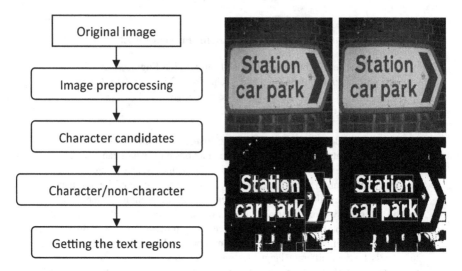

Fig. 1. On the left is the flowchart of the common text detection systems. On the right is the results of the main stages, including the original image, one of the channels after image preprocessing, the character regions and the text regions.

The reminder of this paper is organized as follows. In Sect. 2, the features of character are introduced. In Sect. 3, text extraction methods are classified with a detailed review. In Sect. 4, the protocols of the performance analysis of the Text Detection system are introduced. In Sect. 5, we summarize this paper and discuss areas for future research.

2   Character Features

The texts in images can be classified into two groups, the artificial texts and the scene texts. The artificial texts, such as the subtitle in the video, are artificially added. The scene texts, such as road signs and posters, are parts of the image. At the third stage of a text detection system, the character regions are extracted based on the features. The optional features are summarized as follows.

2.1   Morphological Features

(a) Color: the texts in natural scene image always exit as words, and the colors of the characters in the same words are always similar.

(b) Size: the characters in a same text line are always have the similar size. Though the size of the characters in an image may be various, the size may not be too big or too small.

(c) Distance: the characters in a same words may have similar distances, and the distances between adjacent words may have a certain ratio to the character distance.

(d) Shape: the character may have some specific shapes, such as aspect ratio and some holes. The area and perimeter can also be used.

2.2 Some Advanced Features

(e) Euler number: as the character regions may always have some holes, this feature means the difference between the number of connected components and the number of holes.

(f) Edge: the character regions may have difference with the background and wealth edge information.

(g) Horizontal crossings: the number of transitions between pixels belonging the region and not belonging to the region can computed to exclude the non-character regions.

(h) There are some more complex features, such as hole area ratio and convex hull ratio.

With the development of pattern recognition and digital image-processing, more and more features will be designed to improve the accuracy and efficiency of character regions detection.

3 The Techniques of Text Detection

According to the way of candidate character regions extraction, existing methods for scene text detection can roughly be categorized into two groups, texture-based methods and connected components-based methods.

3.1 Texture-Based Methods

The methods in this group exploit a sliding window and compute the texture features to search for the possible characters in the image. Then classifiers are used to identify the candidate character regions. At last group the regions into words.

In [2], the color clustering is used to extract the candidate character regions using the feature of pixel value. Then a support vector machine (SVM) is designed to remove the non-character regions. The adaptive mean shift algorithm (CAMSHIFT) is used to group the character regions into text regions. The method presented a detection rate of 87% using 50 images of different size, fonts and formats.

In [3], a special feature, combined the feature of HOG and the feature LBP, is designed to locate the characters in the image. Then cascade adaptive boosting (AdaBoost) classifier is adopted to ensure the character regions. To get the text regions, a window grouping method is used to generate text lines. At last a Markov Random Fields (MRF) model is used to filtered out the non-text regions. The method presented a recall rate of 67%, and the precision rate of 68% using the ICDAR 2003 Dateset.

In [4], Taking advantage of the desirable characteristic of gray-scale invariance of local binary patterns (LBP), a modified LBP operator is designed to extract the features of the characters. Then the classifier for is made by a polynomial neural network (PNN) to get the character regions. At last a post-processing procedure including verification and fusion is used to produce text regions. The method presented a recall rate of 87.7%, and the precision rate of 68% using the ICDAR 2003 Dateset.

In [5], six different classes features are used to extract the character regions. Then Modest AdaBoost with multi-scale sequential search is designed to get the text regions. The method use some complex features to improve the accuracy rate. However the complexity of the algorithm is also high. The method presented a recall rate of 75%, and the precision rate of 66% using the ICDAR 2003 Dateset.

In [6], AdaBoost is combined with Haar-like features to obtain cascade classifiers for text regions extraction. The method presented a recall rate of 79.9%, and the precision rate of 72.6% using 128 street view images.

Different from the above methods, in [7], wavelet transform is applied to the image and the distribution of high-frequency wavelet coefficients is considered to statistically characterize text and non-text areas. Then the k-means algorithm and projection analysis are used to detect and refine the text regions. The method presented a recall rate of 90%, and the precision rate of 87% using a set of video frames taken from the MPEG-7 video test set. Based on this methods, in [8], use a new Fourier-Statistical Features (FSF) in RGB space to detect texts of different fonts, size and scripts. The experimental results show that the method has made some improvement in terms of recall rate and precision rate.

The methods in this group use a sliding window to localize individual characters, or the whole words. Strengths of such method include robustness to noise and blur. The main drawback is how to define the size of the window. Since the too big windows may result in too much noise, and the too small windows may give rise to the difficulty of computing the features.

3.2 Connect Component Based Methods

The Connect component based methods extract candidate character regions by connect component. Then group the characters regions into text. And some additional checks may be used to refine the detection results.

As text in the natural scene images always have closely spaced edges, the edge feature can be used to detect the character regions. In [9], the Sobel operator is used to get edges. Then local thresholding and hysteresis edge recovery are applied to get the character regions. The projection analysis is used to group the character regions into text regions. The method can process multilingual text characteristics, including English and Chinese.

In [10], the edges are detected by the wavelet transform and scanned into patches by a sliding window. Then a simple classification procedure with two learned discriminative dictionaries is applied to get candidate text areas. At last, adaptive run-length smoothing algorithm and projection analysis are used to refine the candidate text areas.

As text in the natural scene images always have special color, intensity and stroke width, these features can be used to detect some special connect component as the character regions. In [11], firstly the Stroke Width Transform (SWT) is applied to the image. Then detect the connect components with similar stroke widths as the candidate character regions. At last the text lines are built with the features of shape and distance. In [12], after getting the character regions by SWT, the color and shape of the regions are used to build multi-direction text lines.

After the concept of maximally stable extremal regions (MSER) is presented in [13], the feature region has been widely used in the system of image retrieval and object detection. In [1], the MSERs are detected as the candidate character regions. Then the connected component analysis (CCA) is used to get the text regions.

In [14], after detecting the MSERs, an efficiently pruned exhaustive search algorithm is used to filter out the nesting or duplicate regions. Then the morphological features and Single-link algorithm are used to group the character regions into text regions. The posterior probabilities of text candidates corresponding to non-text are estimated with a SVM classifier. And the Bayesian decision rule is used to refine the detection results. The method presented the best performance in ICDAR 2015 [15].

In [16], the extremal regions (ER), with a more simple calculation procedure, are detected as the candidate character regions. Then a two-stage algorithm is designed to pruning the non-character regions. In the first stage, SVM and five features are used to estimate the class-conditional probabilities of ERs. And in the second stage, some more complex features and AdaBoost classifier are used to refine the results. Finally, a clustering-based method is used to group the character regions into text regions.

The Connect component based methods, recently more popular approach, can detect most text regions in the natural scene images. And these methods are robustness to incline, rotation and blur. However, there are still drawbacks. One problem is that the number of the feature regions will be large. Another problem is the absence of an effective text candidates construction algorithm.

4   Evaluation Protocols

With so many approaches and datasets, reproducing all of them and comparing them with each dataset are problematic. For text detection, the ICDAR protocols are most commonly adopted. The ICDAR 03 [17] and ICDAR 11/13 [15] datasets are prepared for scene text, covering tasks of text location, character segmentation and word recognition. And the Street View Text (SVT), a more complex dataset, is another commonly used dataset.

In [18], Wolf and Jolion proposed the DteEval protocol that comprise the area overlap [17] and the object level evaluation. As shown in Fig. 2, it supports one-to-one, many-to-one and one-to-many matches among the ground truth and detections. This protocol was adopted in ICDAR 2011 and ICDAR 2013 "Robust Reading" competitions.

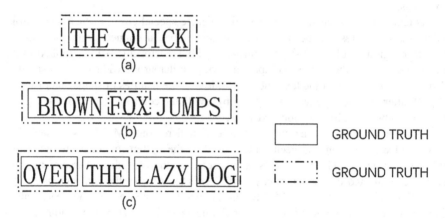

Fig. 2. Different match types between ground truth rectangles and detected rectangles: (a) one-to-one match; (b) a split: a one-to-many match with one ground truth rectangle; (c) a merge: a one-to-many match with one detected rectangle

The precision and the recall are used to measure the performance of the approaches. The precision is defined as the ratio between the area of intersection regions and that of detected text regions. Recall is defined as the ratio between the area of intersection regions and that of ground truth regions. In case of multiple images or a single image with multiple text rectangles, a natural way has been proposed to get the results. With saving the detection results and ground truth in XML format, there has been an available system to evaluation the algorithms.

$$Recall\ rate = \frac{N.o.correctly\ detected\ rectangles}{N.o.rectangles\ in\ the\ database} \tag{1}$$

$$Precision\ rate = \frac{N.o.correctly\ detected\ rectangles}{N.o.rectangles\ in\ the\ database} \tag{2}$$

5 Conclusion

This paper presented a comprehensive survey on existing techniques of text detection in natural scene images. Although significant process of text detection has been made in the last few decades, there is still a lot of work to be done since a robust system should work effectively under a variety of environmental conditions and text formats.

In most text detection systems, extracting the character regions and grouping them to text regions are always two independent stage. Taking advantage of the affiliation of characters and texts may provide some new methods for this problem. With the rapid expansion of machine learning technique, design more effective classifiers can be an important task in this filed.

References

1. Neumann, L., Matas, J.: A method for text localization and recognition in real-world images. In: Kimmel, R., Klette, R., Sugimoto, A. (eds.) ACCV 2010. LNCS, vol. 6494, pp. 770–783. Springer, Heidelberg (2011). doi:10.1007/978-3-642-19318-7_60
2. Kim, K.I., Jung, K., Jin, H.K.: Texture-based approach for text detection in images using support vector machines and continuously adaptive mean shift algorithm. IEEE Trans. Pattern Anal. Mach. Intell. **25**(12), 1631–1639 (2003)
3. Pan, Y.F., Hou, X., Liu, C.L.: A robust system to detect and localize texts in natural scene images. In: The Eighth IAPR International Workshop on Document Analysis Systems, pp. 35–42. IEEE (2008)
4. Ye, J., Huang, L.L., Hao, X.: Neural network based text detection in videos using local binary patterns. In: Chinese Conference on Pattern Recognition, CCPR 2009, pp. 1–5 (2009)
5. Lee, J., Lee, P.H., Lee, S.W., et al.: AdaBoost for text detection in natural scene. In: International Conference on Document Analysis and Recognition, pp. 429–434. IEEE Computer Society (2011)
6. Song, Y., He, Y., Li, Q., et al.: Reading text in street views using Adaboost: towards a system for searching target places. In: 2009 IEEE Intelligent Vehicles Symposium, pp. 227–232. IEEE (2009)
7. Gllavata, J., Ewerth, R., Freisleben, B.: Text detection in images based on unsupervised classification of high-frequency wavelet coefficients. Proc. Int. Conf. Pattern Recogn. **1**(3), 425–428 (2004)
8. Shivakumara, P., Phan, T.Q., Tan, C.L.: New fourier-statistical features in RGB space for video text detection. IEEE Trans. Circ. Syst. Video Technol. **20**(11), 1520–1532 (2010)
9. Lyu, M.R., Song, J., Cai, M.: A comprehensive method for multilingual video text detection, localization, and extraction. IEEE Trans. Circ. Syst. Video Technol. **15**(2), 243–255 (2005)
10. Zhao, M., Li, S., Kwok, J.: Text detection in images using sparse representation with discriminative dictionaries. Image Vis. Comput. **28**(12), 1590–1599 (2010)
11. Epshtein, B., Ofek, E., Wexler, Y.: Detecting text in natural scenes with stroke width transform. In: 2013 IEEE Conference on Computer Vision and Pattern Recognition, pp. 2963–2970. IEEE (2010)
12. Yao, C., Bai, X., Liu, W., et al.: Detecting texts of arbitrary orientations in natural images. In: IEEE Conference on Computer Vision & Pattern Recognition, pp. 1083–1090 (2012)
13. Matas, J., Chum, O., Urban, M., et al.: Robust wide-baseline stereo from maximally stable extremal regions. Image Vis. Comput. **22**(10), 761–767 (2004)
14. Yin, X.C., Yin, X., Huang, K., et al.: Robust text detection in natural scene images. IEEE Trans. Pattern Anal. Mach. Intell. **36**(5), 970–983 (2013)
15. Karatzas, D., Gomezbigorda, L., Nicolaou, A., et al.: ICDAR 2015 Competition on Robust Reading. International Conference on Document Analysis and Recognition (2015)
16. Neumann, L., Matas, J.: Real-time lexicon-free scene text localization and recognition. 1 (2015)

17. Lucas, S.M., Panaretos, A., Sosa, L., et al.: ICDAR 2003 robust reading competitions. In: International Conference on Document Analysis and Recognition, p. 682. IEEE Computer Society (2003)
18. Wolf, C., Jolion, J.M.: Object count/area graphs for the evaluation of object detection and segmentation algorithms. Doc. Anal. Recogn. **8**(4), 280–296 (2006)

Machine Learning for Multimedia

Research of Speech Amplitude Distribution Based on Hadamard Transformation

Jingxue Tu[1], Jingyun Xu[1(✉)], and Xiaoqun Zhao[2]

[1] School of Engineering, Huzhou University, Huzhou 313000, China
happy_865709718@qq.com, xujingyunsh@gmail.com
[2] School of Electronic and Information Engineering,
Tongji University, Shanghai 201804, China
Zhao_Xiaoqun@tongji.edu.cn

Abstract. In view of PCM (Pulse Code Modulation) of speech signal, this paper puts forward a method of speech processing based on hadamard matrix transformation to change the amplitude distribution of speech signal, which can reduce the standard deviation of speech signal. Experiments show that the hadamard matrix transformation algorithm can obviously reduce the amplitude range of speech signal. Speech signal standard deviation is reduced by 20% after the transformation. At the same time, speech quality after decoding is not decreased according to listening experimenter. The algorithm reduces amplitude range and standard deviation of the speech signal, which can code the speech signal with less bits, and compression efficiency can be further improved.

Keywords: Speech · Amplitude distribution · Hadamard matrix · Standard deviation

1 Introduction

Speech are transmitted and processed in analogy manner in early stage of research. Since the PCM method has been proposed, Speech can be stored and transmitted as a digital data. However, A huge amount of data is a big issue, therefore it is necessary to compress speech signal. From the original 64 KB/s standard PCM waveform encoder to the at or below 4 KB/s parameter coding vocoder now, speech compression coding is gaining steam rapidly for decades. Speech coding and compression are extensively used in many applicants such as GSM system and IP telephone system [1–4]. GSM mobile communication use wireless channel transmission, since the frequency of the wireless channel resources are limited, the utilization rate of channel using speech compression technology is improved. Telephone system usually adopts linear PCM, which is sample rate of 8 kHz and quantization number is 11 bit, Coding rate reaches up to 88 kbit/s if using 8 bit non-uniform quantization, the coding rate reaches up to 64 kbit/s [5].

The most important business is speech business in mobile communication, precious wireless spectrum resources requires each user to take up the narrow spectrum as soon

XU. Jingyun—Project supported by the National Nature Science Foundation of China (No. 61271248), Natural Science Foundation of Huzhou City (No. 2015YZ04).

© ICST Institute for Computer Sciences, Social Informatics and Telecommunications Engineering 2017
X.-L. Huang (Ed.): MLICOM 2016, LNICST 183, pp. 267–273, 2017.
DOI: 10.1007/978-3-319-52730-7_27

as possible. The size of the spectrum is directly related in the compression ratio of speech. We need to compress the speech signal in order to save storage media. The purpose of the speech compression coding is to meet the demand of the narrowband channel low bit rate transmission and realize speech storage efficiently on the premise of guarantee the quality of speech [5].

At the same time, due to the special nature and construction method of hadamard matrix, which has a wide range of USES in communication [6, 7]. Thus, the paper puts forward a method of speech processing based on hadamard matrix transformation to change the amplitude distribution of speech signal, which can reduce the standard deviation of speech signal. Experiments show that speech compression efficiency can be further improved by the hadamard matrix transformation algorithm.

2 Speech Signal Amplitude Distribution

Suppose that the sample size of speech signals is $K = M \times N$ point, denoted as matrix $X = (X_1, X_2, ..., X_K)$, among K, M, N is a positive integer. The matrix form of the data points can be expressed as

$$X_{N \times M} = \begin{bmatrix} X_1 & \cdots & X_M \\ \vdots & \ddots & \vdots \\ X_{N-1 \times M+1} & \cdots & X_{N \times M} \end{bmatrix} \tag{1}$$

The duration of speech signal is 10 min, the sampling rate of speech signal is 8 kHz, the quantitation precision and range is 8 bit and between $128 \sim 128$ respectively in the experiment.

The most part of speech signal sample data points fall within the scope of the $(0.2 \sim 0.2)$ from Fig. 1, so we make the scope of the $(0.2 \sim 0.2)$ as the benchmark for comparisons in this paper. The experimental results show that the probability of data point fall within the scope is 91.78%.

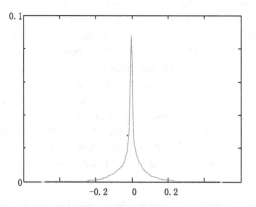

Fig. 1. Amplitude distribution of speech

Speech signal is non-uniform and has significant correlation from the time domain. Speech signal has strong unevenness and certain redundancy from the power spectral density. Due to the special nature and construction of hadamard matrix, We can take advantage of the hadamard matrix of speech signal to eliminate the correlation operation, which reduces the dynamic range of speech signal and improve the coding efficiency.

3 Hadamard Matrix and Its Properties

Hadamard matrix [4] is made up of element +1 and 1 and nonsingular N order phalanx, 2 order hadamard matrix can be defined as

$$H_2 = \begin{bmatrix} 1 & 1 \\ 1 & -1 \end{bmatrix} N \times N \tag{2}$$

4 order hadamard matrix can be defined as:

$$H_4 = \begin{bmatrix} H_2 & H_2 \\ H_2 & -H_2 \end{bmatrix} = \begin{bmatrix} 1 & 1 & 1 & 1 \\ 1 & -1 & 1 & -1 \\ 1 & 1 & -1 & -1 \\ 1 & -1 & -1 & 1 \end{bmatrix} \tag{3}$$

The general equation for hadamard matrix:

$$H_{2N} = \begin{bmatrix} H_N & H_N \\ H_N & -H_N \end{bmatrix} \tag{4}$$

Where H_N is a hadamard matrix of size $N \times N$, Hadamard matrix (H matrix) and its main properties:

(A) any two rows (columns) of the matrix are orthogonal.
(B) the square sum of all the elements in any row (column) is equal to the squares of order number.
(C) the hadamard matrix order number is 2 or 4 multiples.

4 Speech Signal Distribution Transformed

4.1 Mixed Speech Signal Distribution

We can get the matrix of speech X, N × M order. The matrix X is transformed by hadamard method to the mixed matrix Y:

$$Y = \mathrm{H} \times X' = \begin{bmatrix} 1 & \cdots & 1 \\ \vdots & \ddots & \vdots \\ 1 & \cdots & -1 \end{bmatrix} \times \begin{bmatrix} X_1 & \cdots & X_M \\ \vdots & \ddots & \vdots \\ X_{N-1 \times M + 1} & \cdots & X_{N \times M} \end{bmatrix} \quad (5)$$

The experimental results show that the probability of the mixed matrix Y data point fall within the scope is 94.11% (Fig. 2).

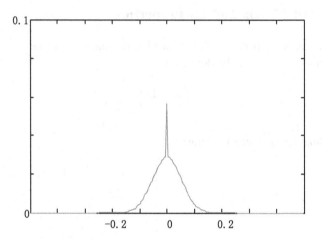

Fig. 2. Mixed speech signal distribution

4.2 Speech Signal Distribution Transposed

In order to further eliminate the correlation of speech signal among the samples, the speech signals are divided into a set of N samples, and carries on the hadamard transformation, the resulting speech signal is equivalent to transpose for mixed speech signals, so called transposed speech signal. The matrix form of the speech signal F is expressed as:

$$F_{M \times N} = \begin{bmatrix} X_1 & \cdots & X_N \\ \vdots & \ddots & \vdots \\ X_{M-1 \times N + 1} & \cdots & X_{M \times N} \end{bmatrix} \quad (6)$$

The correlation of speech signal mainly embodied in the $(X_1, X_2, \ldots X_N)$, ..., $(X_{M-N + 1}, X_{M-N + 2}, \ldots X_M)$, in order to take advantage of the nature of the hadamard matrix to remove the correlation of speech signal, to transpose for F, we can get F' are expressed as

$$F'_{N \times M} = \begin{bmatrix} X_1 & \cdots & X_{M-1 \times N + 1} \\ \vdots & \ddots & \vdots \\ X_N & \cdots & X_{M \times N} \end{bmatrix} \quad (7)$$

N order hadamard matrix H multiply matrix X', We can get the transposed matrix Z:

$$Z = HX' = \begin{bmatrix} 1 & \cdots & 1 \\ \vdots & \ddots & \vdots \\ 1 & \cdots & -1 \end{bmatrix} \begin{bmatrix} X_1 & \cdots & X_M \\ \vdots & \ddots & \vdots \\ X_{(N-1)M+1} & \cdots & X_{NM} \end{bmatrix} \tag{8}$$

The dynamic range of speech signal are compressed and the coding bits is reduced by the method, The experimental results show that the probability of the transposed matrix Z data point fall within the scope is 95.78% (Fig. 3).

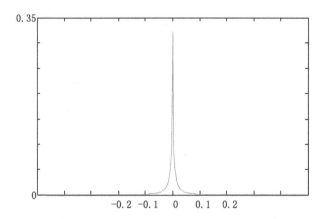

Fig. 3. Speech signal amplitude distribution transposed

5   Order Selection of Hadamard Matrix

The speech signal is transformed by using different order hadamard matrix and is to do its statistics of amplitude distribution with the increase of hadamard matrix of order, the speech signal compression ratio is higher from the simulation results. However, when the hadamard matrix has achieved a certain order, signal compression rate no longer increase, if we further increase the order number of hadamard matrix, compression effect decreases instead.

The experiments shows that N = 256 order of hadamard matrix has the optimal compression effect. So, this paper uses the hadamard matrix of 256 order.

6   Result and Analysis

The matrix of original speech signal X, the hadamard matrix H, the matrix Y is the results of transformation, Y is expressed as

$$Y = H \times X \tag{9}$$

In order to restore the original speech signal X, do the following operation for Y:

$$\frac{1}{N} \times H' \times Y = \frac{1}{N} \times H' \times H \times X = \frac{1}{N} \times N \times X = X \tag{10}$$

The original signal can be restored after the above operation, Fig. 4 shows the original speech and speech after two kinds of inverse transformation of time domain waveform, it can be seen that speech waveform recovered is no distortion.

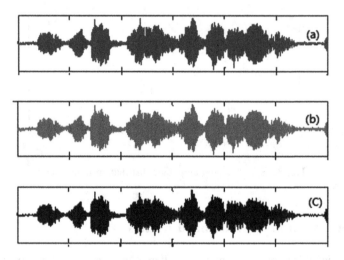

Fig. 4. Time domain waveform of speech signal: (a) original speech; (b) mixed speech; (c) transposed speech

The simulation data as shown in Table 1. The standard deviation is defined as the overall standard units with the square root of the arithmetic average of the mean square deviation. It reflects the degree of discrete between individuals in the group. In the experiment, the standard deviation is normalized, the results are shown in Table 2.

Table 1. Original signal standard deviation

	Original	Mixed	Transposed
Standard deviation	0.9178	0.9411	0.9578

Table 2. Signal standard deviation before DC eliminated

	Original	Mixed	Transposed
Standard deviation	1.00	0.974	0.952

Table 3. Signal standard deviation after DC eliminated

	Original	Mixed	Transposed
Standard deviation	1.00	0.862	0.805

After the original speech is eliminated the dc to, the standard deviation of the original signal and signal transformed are calculated, the result as shown in Table 3. From Table 3, the standard deviation of mixed signal is reduced by 14%, and standard deviation of the transposed signal is reduced by 20% relative to the original signal.

7 Conclusion

This paper first analyzes the correlation of speech signals between adjacent samples using hadamard transformation, the experimental results show that this method can significantly reduce the speech signal dynamic range, which can improve the compression ratio, at the same time greatly reduce the standard deviation of speech signals. The method in speech coding and wireless communication has certain reference value.

References

1. Ahmed, N., Rao, K.R.: Orthogonal Transforms for Digital Signal Processing. Springer Science & Business Media, Heidelberg (2012)
2. Kaur, H., Kaur, R.: Speech compression and decompression using DWT and DCT. Int. J. Comput. Technol. Appl. **3**(4), 1501–1503 (2012)
3. Hillebrand, F.: The creation of standards for global mobile communication: GSM and UMTS standardization from 1982 to 2000. IEEE Wirel. Commun. **20**(5), 24–33 (2013)
4. Amira, A., Bouridane, A., Milligan, P., et al.: Novel FPGA implementations of Walsh-Hadamard transforms for signal processing. IEE Proc. Vis. Image Sign. Process. **148**(6), 377–383 (2001). IET
5. Wang, G., Huang, H., Liu, Y., Zhang, X., Wang, Z.: Uncertainty estimation of reliability redundancy in complex systems based on the Cross-Entropy method. J. Mech. Sci. Technol. **23**(10), 2612–2623 (2009)
6. Fu, Z., Zhao, J.: Information Theory and Coding. Electronic Industry Press, Beijing (2008)
7. Kumar, A., Singh, G.K., Rajesh, G., et al.: The optimized wavelet filters for speech compression. Int. J. Speech Technol. **16**(2), 171–179 (2013)

A Pitch Estimation Method Robust to High Levels of Noise

Xu Jingyun[1,2(✉)], Zhao Xiaoqun[2], and Cai Zhiduan[1]

[1] School of Engineering, Huzhou University, Huzhou 313000, China
xujingyunsh@gmail.com, czddule@hutc.zj.cn
[2] School of Electronic and Information Engineering,
Tongji University, Shanghai 201804, China
Zhao_Xiaoqun@tongji.edu.cn

Abstract. Pitch is one of the most key parameter in speech coding, speech synthesis and so on, the traditional methods for pitch detection are prone to error at a low SNR at present. A pitch detection method based on pitch harmonic (PH) and the harmonic number based on PH is proposed in this paper. At first, the pitch harmonic is roughly estimated by pitch estimation filter with amplitude compression (PEFAC). Secondly, the weighted algorithm based on modified circular average magnitude difference function (MCAMDF) and pulse sequence is used to compute the pitch harmonic number. At last a pitch tracking method is applied to compute the pitch period candidates accurately. By simulation experiments, it is shown that the proposed pitch detection method has more accurate and more low algorithm complexity than the traditional methods at both high and low SNR.

Keywords: Pitch detection · Pitch estimation filter with amplitude compression · MCAMDF · Dynamic programming

1 Introduction

Pitch is a key important characteristic parameter of speech signal processing, Pitch detection has vital significance in speech synthesis, speech coding and speech recognition and so on. Since the 1960s, a variety of effective pitch detection method is proposed in the time and frequency domain [1, 2]. In time domain, waveform similarity is used to extract the pitch period and the harmonic peaks location is identified and located to extract the pitch period in frequency domain. Most of them have fine performance for clean speech [2].

Due to speech signal is derived from the real environment, speech signal is prone to pollute by different types of noise (white noise, cars noise and so on.) and signal to noise ratio (−20 db – +20 db), the cycle time domain and frequency of the speech in different extent was distorted, thus conventional methods will become unreliable or

XU. Jingyun—Project supported by the National Nature Science Foundation of China (No. 61271248), Natural Science Foundation of Huzhou City (No. 2015YZ04).

X.-L. Huang (Ed.): MLICOM 2016, LNICST 183, pp. 274–280, 2017.
DOI: 10.1007/978-3-319-52730-7_28

even completely ineffective. At present, performance improvement in noisy environments is still desired [2, 3]. Pitch detection in the real environment gradually become the focus of research, people put forward a lot of methods for this purpose.

Paper [4] extracted some candidate pitch in time domain, and each of them was weighted in the frequency domain, dynamic programming (DP) was then utilized to select the pitch candidates. HSAC-SIM method estimated a PH based on HSAC and estimated the pitch from the harmonic number based on impulse-train weighted SAMDF. The methods in [4, 5] show good performance by utilizing the current frame and the adjacent frame of acoustic characteristics in time domain, frequency domain, however, are not adapt to severe noisy conditions. Paper [6] discussed a method which eliminates the noise of the pitch period harmonic characteristics by calculating spectral peaks. Paper [2] use the PEFAC method which attenuate strong noise components, extract three pitch candidate value and determine the most optimal pitch by dynamic programming. The methods in [2, 6], especially in [2], could extract pitch under severe noisy conditions by de-noising the pitch harmonic feature. PEFAC treats directly the max amplitude point as the highest probability pitch frequency in the log-frequency; however, the max amplitude point usually is not pitch frequency but PH.

According to the above the advantages and disadvantages of the paper [2–6], we put forward a pitch estimation method referred to as PH-SIM, we firstly extract a PH based on PEFAC, and then we determine the harmonic number based on MCAMDF impulse-train method. Experiments results show that the PEF-SIM estimate pitch more accurate than the HSAC-SIM and PEFAC method in real environment.

2 Extraction of PH

The noisy speech is eliminated DC component, normalized and segmented. We can get noisy speech frame $s(k)$ which can be expressed in time-domain as

$$s(k) = a(k) + b(k) \tag{1}$$

Here, $a(k)$ is denote the clean speech frame and $b(k)$ is the noise signal.

We can estimate rough a PH from $s(n)$ base PEFAC, the complete algorithm comprises the following steps[9]:

(1) Calculating power spectral density of $s(n)$ in the log-frequency

$$R(p) = X(p) + E(p) = \sum_{i=1}^{I} b_i \delta(p - \log(f_0 i)) + E(p) \tag{2}$$

Where $x(p)$ *and* $E(p)$ is the spectral density of power for the clean speech and noise respectively, $p = \log f$, b_i represents the power of the ith harmonic, I the number of harmonics and δ the Dirac delta function.

(2) Calculating the normalized period gram of $R(p)$

$R_t(p)$ is the period gram of the log-frequency power spectral density $R(p)$ at tth frame, $R'_t(p)$ is the normalized period gram of $R(p)$, which can be written as

$$R'_t(p) = \frac{R_t(p)}{\tilde{R}_t(p)} L(p) \tag{3}$$

where $R'_t(p) = R_t(p) * o(t,p)$; $o(t,p)$ is the moving average filter, $o(t,p) = 1$ for $|t| < T_0$, $|p| < Q_0$, otherwise $o(t,p) = 0$. $L(p)$ denotes the LTASS spectrum;

(3) Matched filter for $R'_t(p)$

we can get $Z_t(p)$ by matching filter for $R'_t(p)$, which is expressed as

$$Z_t(p) = R'_t(p) * h(-p) \tag{4}$$

here, the matched filter is defined as

$$h(p) = \begin{cases} 1/[\lambda - \cos(2\pi e^p)] - v & when \quad \log(1/2) < x < \log(I + 1/2), \\ 0 & otherwise \end{cases} \tag{5}$$

where the parameter v is introduced to determine $\int T(p)dp = 0$ and the parameter λ controls the pitch peak width while I the number of pitch peaks, it has the big number to include all harmonics with significant energy.

(4) Estimating the PH

The pitch frequency maximum probability candidate corresponding to the maximum peak of $Z_t(p)$ ranging from 60 Hz to 1250 Hz denote the exact pitch position of a PH ω_q.

3   Estimation of PH Number

The pitch harmonic number p_{opt} from the PH ω_{popt} is estimated in this section. Thus we maximize an function defined as an symmetrical impulse-sequence weighted MCAMDF (SIM) to estimate q_{opt} in time domain.

The MCAMDF is defined as

$$\varepsilon(\tau) = \sum_{n=0}^{\beta} |s(\mod(n+\tau, N+\tau_{\max})) - s(n)|, \tau = 0, 1, 2, \cdots \beta \tag{6}$$

Here, $\beta = N + \tau_{\max} - 1$, $\tau_{\max}$ is the maximum possible pitch of speech signal. $\varepsilon(\tau)$ with symmetrical features in $\tau_s = (\beta + 1)/2$, so $\varepsilon(\tau)$ is only calculated in the range $\tau \in [0, \tau_s]$. $\varepsilon(\tau)$ has the most possibility of having deep-valleys at $\tau = \rho T$ with $0 \le \rho T \le \tau_s (\rho = 0, 1, 2, \cdots)$. However, the pitch peaks features can be utilized to estimate pitch [3]. In order to change valleys to peaks, so the following function is defined as

$$\eta(\tau) = \frac{\phi_{\max}}{N - \mu_{\max}} - \varepsilon(\tau), \tau = 0, 1, 2, \cdots, \tau_s \tag{7}$$

Here, $\phi_{\max}$ is the maximum value of $\phi(\tau)$ in the range of $0 < \tau \le \tau_s$ and $\mu_{\max} \le \tau_s$ is the index of $\eta_{\max}$. The function of MCAMDF $\eta(\tau)$ reverses the peaks when $\varepsilon(\tau)$ is the valleys.

The harmonic number p_{opt} can be determined by the function MCAMDF $\eta(\tau)$

$$p_{opt} = \arg\max \sum_{\tau=0}^{\tau_s} J(\tau, p)\eta(\tau) \tag{8}$$

Where the impulse-training is represented as

$$I(n, p) = \sum_{\mu=0}^{q-1} \delta(\tau - 2\theta p\pi / \omega_{popt}), \tau = 0, 1, 2, \cdots, \tau_s \tag{9}$$

Here, ω_{popt} is the maximum probability of pitch harmonic. θ is the number of unit impulses. For the (8), we can get the p_{opt}, thus, we can get the optimum value of pitch $F_0 = F_s\omega_{opt}/2\pi p_{opt}$, here, F_s is sampling frequency.

4 Experimental Results and Analysis

The speech library is derived from the Keele pitch detection reference in this experiment. This library contains 10 speakers, five women and five men which read the same paragraph of English each speech file is about 30 s, all speech is sampled for 20 kHz and quantified for 16 bits and the library provides the reference pitch value of every frame, frame length is 512 sampling point and the frame shift is 200 sampling point. The input speech is sampled at 8 kHz, frame length 200 points and frame moving 80 points in this paper, so, speech files is down-sampled to 8 kHz in this library, and speech frame of pitch reference is multiplied by 0.4 as a reference for the final test.

Experiment parameter Settings are as follows: the sampling frequency is 8 kHz, speech frame length is 200 points, the frame shift is 80 points, Hamming window by zero padding from 200 points to 1600 points, the frequency resolution of 5 Hz, logarithmic frequency range of 40–4000 Hz; $L(p)$ data is from the literature [7] see Table 2, $T_0 = 1.5 s$, $p_0 = 10f_0$; $\lambda = 1.8$, $v = 0.6700$, $i = 10$.

In order to compare quantitatively HSAC-SIM and PEF-SIM method extracted pitch harmonic performance, we randomly selected from a group of 400 frames voiced speech signal respectively in different SNR (−20, 10, 0, 10 and 20 db) and different noise (white and so on) and combined 15 groups which per group 400 frames. Average execution time (AET) and average total degree of the fundamental frequency offset (Gross Pitch Harmonic Offset Degree, GPD) of extract pitch is computed for the two kinds of algorithm respectively.

GPD is defined as

$$GPD = \sum_{i=1}^{N} \frac{|f_e(i)/h - f_r(i)|}{f_r(i)} \tag{10}$$

Where, $f_r(i)$ represents the real pitch frequency of the frame, $f_e(i)$ represents the I frame extraction pitch harmonic, h is the pitch harmonic number (manually determination), N is the number of frames, the smaller the total GPD is the more accurate pitch harmonic estimation.

Two methods of quantitative performance comparison results are seen in Table 1, it shows that the GPD of PEF-SIM is less than the HSAC-SIM method, It is shown that the PH-MIM of pitch harmonic estimation is more accuracy than that of HSAC-SIM; The AET of PEF-SIM method is 0.2 fold of the HSAC-SIM method. Overall, PEF-SIM method is better performance than HSAC-SIM method, which is more advantageous to the subsequent pitch estimation.

Table 1. Two methods of quantitative performance comparison

	PEF-SIM	HSAC
AET (s)	1.6	8.5
GPD	0.93	0.77 (Rough estimation)
		0.84 (Fine estimation)

We use gross pitch error (GPE) to evaluation the PEF-SIM method. Dynamic programming algorithm in [3] is introduced. The pitch estimation effect of the RAPT [5], PEFAC, HSAC-SIM and the proposed PEF-MIM select the pitch. Pitch estimation is considered as correct if its GPE is the range [− 5%, +5%] of the correct value. It shows the performance of the algorithms in Fig. 1 at +20 dB SNR, four algorithms have a good effect. Due to RAPT is not specially designed for noise robustness; When SNR is lower than 0 dB, the performance of RAPT falls quickly for all noise types; The HSAC-SIM method has much better performance at low SNR of −5 dB in the white and car noise. However, the HSAC-SIM give relatively bigger values of GPE for babble noise and at lower than SNR of −5 dB; the PEFAC is prone to produce the double and half error; The proposed PEF-SIM method provides much better results from 20 db to −20 dB SNR for different noise types.

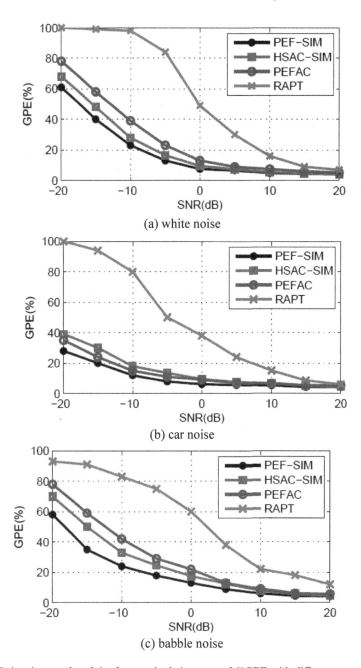

(a) white noise

(b) car noise

(c) babble noise

Fig. 1. Estimation results of the four methods in terms of %GPE with different types of noise

5 Conclusion

The paper is proposed a PEF-SIM method to estimate pitch in real environment. First, we propose an algorithm to extract the PH based on PEFAC and the result shows that the extraction algorithm of PH can extract PH accurately. And then, we introduce the SIM method to extract the number of PH Finally, pitch is smoothed by dynamic programming. The GPE of PEF-SIM method is less than the RAPT, PEFAC and HSAC-SIM method, the method of PEF-SIM has high performance especially for babble noise. The AET of PEF-SIM method is 0.2-fold of the HSAC-SIM method. The results shows that the proposed method is superior to PEFAC and HSAC-SIM under low SNR.

References

1. Hong, W.: Low Bit Rate Speech Coding. National Defense Industry Press, Beijing (2005)
2. Gonzalez, S., Brookes, M.: PEFAC-a pitch estimation algorithm robust to high levels of noise. IEEE Trans. Audio Speech Lang. Process. **22**(2), 518–530 (2014)
3. Jingyun, X., Xiaoqun, Z.: Voiced/unvoiced classification and pitch estimation based on amplitude compression filter. J. Electron. Inf. Technol. **38**(3), 586–593 (2016)
4. Xu, J.D., Chang, L., Cui, H.J., et al.: A pitch period detection algorithm using time and frequency analyses. J. Tsinghua Univ. **52**(3), 413–415, 420 (2012)
5. Shahnaz, C., Zhu, W.P., Omair, M.: Pitch estimation based on a harmonic sinusoidal autocorrelation model and a time-domain matching scheme. IEEE Trans. Acoust. Speech Sig. Process. **20**(1), 322–335 (2012)
6. Huang, F., Lee, T.: Pitch estimation in noisy speech using accumulated peak spectrum and sparse estimation technique. IEEE Trans. Audio Speech Lang. Process. **21**(1), 99–109 (2013)
7. Byrne, D., Dillon, H., Tran, K., et al.: An international comparison of long term average speech spectra. J. Acoust. Soc. Am. **96**(4), 2108–2120 (1994)

Main Track

An Redundant Networking Channel to Support Reliable Communications in the Internet of Things Applications

Michael Ortiz[1], Yu Sun[1(✉)], Gilbert S. Young[1], and Qingquan Sun[2]

[1] Computer Science Department, California State Polytechnic University, Pomona,
Pomona, USA
{mdortiz,yusun}@cpp.edu
[2] School of Computer Science and Computer Engineering,
California State University, San Bernardino, USA
qsun@csusb.edu

Abstract. Within the context of the Internet-of-Things (IoT), the number of interconnected devices is increasing dramatically and allowing for access to physical data that was previously unimaginable. Physical data is rapidly changing which makes it important to keep networking connections active. Any drop in communication can lead to the loss of sensitive data. A redundant network connection is an attempt to utilize common networking solutions in order to decrease the likelihood of network downtime. It does this by adding a new level of abstraction to networking, allowing data to be sent over multiple networking solutions as if it were a single network, as well as an intelligent decision engine to determine the most optimized and reliable connection to use dynamically.

Keywords: IoT · Communication channel · Code generation

1 Introduction

The Internet-of-Things (IoT) is a rapidly growing area of technology that is impacting aspects of everyday life in both the working and domestic sectors. The basic idea of this concept is the pervasive presence around us of a variety of things or objects – such as Radio-Frequency Identification (RFID) tags, sensors, actuators, mobile phones, which through unique addressing schemes, are able to interact with each other and cooperate with their neighbors to reach common goals [1]. IoT is gaining attention because of its flexibility and cost-effective nature, allowing access to data that was previously unimaginable. From a statistical standpoint, IoT devices have overtaken the human population by reaching 11 billion devices in 2011 and this number is expected to reach 24 billion devices by 2020 [2].

IoT is powerful because of its flexibility and ability to connect to multiple devices. With connections to different components, there comes the need for secure and reliable communication channels. Digitization of physical objects means that those objects must perform the same operations without any extra

© ICST Institute for Computer Sciences, Social Informatics and Telecommunications Engineering 2017
X.-L. Huang (Ed.): MLICOM 2016, LNICST 183, pp. 283–292, 2017.
DOI: 10.1007/978-3-319-52730-7_29

complexity. This must occur while also keeping data safe as it travels through networks. Current communication systems such as Wi-Fi and Bluetooth have had years of work put into them in order to allow secure transfer of data and is active in numerous amounts of current mobile devices. Other communications methods such as RFID are not as prevalent and would not be able to connect to multiple devices.

Open Problem: Current networking solutions are not fit for the demand of consistent data transmission from IoT devices as high-level languages abstract I/O communications and expect network failures. Networking applications today can function with an interrupted connection for short periods of time to counter unreliable connections. However, if the IoT applications are programmed in the same fashion, it will inevitably undermine the full potential of IoT devices.

Interactions with the physical world are constant, thus, information can be gained or lost with any disconnection. General purpose networking techniques, such as TCP/IP, focus on reliably delivering packets rather than timing. More specialized networking techniques must come into development to use in IoT devices. Moreover, network time synchronization technology must be improved considerably. Networks must offer the possibility of timing coherency across multiple, distributed computations. Networking innovations will dramatically change the way distributed real-time software is designed.

Solution Approach ⇒ Model-based network management and the abstraction of communication interfaces. To address these challenges, this paper presents a network model that combines common communication protocols together to provide reliable connections across multiple devices. A model-based framework for communication between the IoT devices and the server. All IoT systems follow the same concept of sending data from devices to a server which processes the data and sends to other clients. Each device may differ based on its functionality, but the nature of communication using the Internet applies to all the IoT systems. The approach presented in this paper abstracts the common elements and entities used in the implementation of communication channels in order to both simplify communication to the necessary messages and manage the network to reduce possible downtime.

The remainder of this paper is organized as follows. Section 2, we provide a motivating example for this paper. We list down the challenges faced while developing reliable communication system in Sect. 3 and present our solution to address these challenges in Sect. 4. We analyze the related work in Sect. 5 and conclude this paper as well as provide some insight on future work and scope of this framework, in Sect. 6.

2   Motivating Example

IoT brings the possibility of new devices to consumers and health systems are taking notice. Major academic research has gone into innovative solutions for mobile healthcare delivery and sensors. In particular, major advances were introduced in the mobile broadband and wireless internet m-health systems [4]. This widespread and unprecedented evolution of m-health systems and services in recent years has been reflected in a 2010 study by McKinsey estimated that the opportunities in the global mobile healthcare market are worth between 50 billion and 60 billion [5]. As healthcare data becomes increasingly profitable, so will the wireless technologies bundled with the devices. There will be multiple reasons to address the communications streams for health devices:

1. The ability to relay information constantly. It is expected that mobile healthcare devices continue to monitor the user at all times. Devices such as the Fitbit [6], Jawbone [7], Misfit [8], etc. need to monitor both active lifestyle activity as well as sleeping patters. These are devices that will consist of as little down time as possible.
2. Choosing the right mobile technology for price and communication. IoT devices must stay on for long periods of time. Some might be connected directly to an outlet, others will use batteries and some might even utilize passive radio transmission (RFID). The goal for any of these devices is to relay information to both the user and the management system for the data.
3. Dealing with communication failure. In the event that a device goes offline, it is imperative that the device reconnect in the simplest way possible for the user. Down time will reduce the usability of healthcare devices and can cause profit loss for businesses.
4. Security of wireless technologies. A common problem for modern devices is adapting to modern technologies while also keeping data safe. Sensitive health data need to be kept secure wherever the user might go which means that the mobile communication must be able to connect to all other IoT devices securely.
5. Standardization to allow adaptable communication. A standard protocol for sending and receiving health data will allow devices to adapt to the rapid advancement of technology. If a new devices comes along with better or changed hardware, standardizing the exchange of data will allow older devices to keep in communication.

With advancements in these areas, mobile health will be able to create a more focused effort at improving livelihood. By allowing devices constant information to a user's heartrate, food consumption, exercise routine, medical regiment, and more, we will be able to decrease the duration of hospital stays along with improve a doctor's knowledge of his or her patients. This comes too with a need to improve security. The amount of data that will be available needs to be kept safe so users will not be used maliciously. Trust that mobile health devices keep a user's information private will be just as important as the direct service. Users will feel more comfortable using communication devices that they are familiar with such as Wi-Fi or Bluetooth or adapt to newer technologies like RFID as long as they do not complicate the application. By keeping a developmental standard in current and future healthcare devices, businesses will be able to gain quicker market adaptation and access to other medical information.

3 Challenges

Abstraction of any computer system is never without complication which goes double for areas that require synchronization such as networks. Current IoT connection problems are as follows:

1. *The inconvenience of networking channel setup in IoT development.* Research development for IoT devices becomes a challenge as most devices do not keep common developer interfaces while active. Communication settings such as Wi-Fi and Bluetooth are hosted in locations that are not convenient to edit initially (e.g., using the Raspberry Pi development board, you need to connect to a monitor, use a keyboard and mouse to edit and control Bluetooth devices).
2. *The challenge of implementing the communication channel.* For instance, Bluetooth has the advantage of easy setup for clients, but programming Bluetooth is challenging, particularly for different types of devices and protocols. Creating a client to find Bluetooth devices must be synchronous as to not interrupt other device communications.
3. *For mission critical IoT applications, there lacks a reliable communication channel to ensure data integrity.* Relying on a single communication channel and protocol is not reliable enough. Most communication systems are expected to fail [8]. This means that allowing only one pathway for communication is yet truly reliable.
4. *The lack of an intelligent decision engine on choosing the most optimized communication channel.* With multiple communication, understanding which portal to send data through is imperative but IoT devices do not have decent space for large scale, dynamic efficiency scaling.

For this project, the goal was to help pave the way to reduce some of these challenges as well as demonstrate the areas that still need work.

4 Solution

To reduce the challenge of network connectivity and problems with multiple connections on a single device, a generic framework has been developed - ReliableConnection, that allows multiple communication services to be unified to perform a singular function. As previously mentioned, the program works by the abstraction of network communication services, such as Bluetooth and Wi-Fi, and then managed autonomously in order to decrease the chance of downtime for multiple devices.

The two main components for the framework is the Network class and the Protocol class. The Network consists of a linked list of Protocols and manages which Protocol will send or receive data. The Protocol is an interface that abstracts TCP/IP and Bluetooth to a simpler functionality. This allows the network to observe the casted protocol rather than deal with specific complications in the communication services. To allow these classes to work with minimal

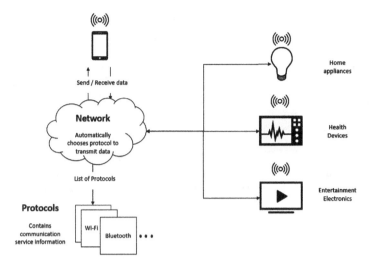

Fig. 1. Overview of ReliableConnection framework

supervision, an Observer pattern is implemented to both classes. The Network observes Protocols so that any changes to a Protocol will quickly notify the Network (Figs. 1, 2, 3 and 4. Addressed below are some of the ways the ReliableConnection framework addresses challenges mentioned in Sect. 3:

1. *Simplify the networking channel setup in IoT development with the default connection manager.* The connection manager is a built-in component in the development framework that handles the channel setup and initialization process. Allowing a framework to manage connection data will allow developers more time to focus on the logic of their code rather than reliability. Keeping a major factor in IoT development under stricter guidelines reduces the learning curve and invites increased innovation.
2. *Abstract the common communication channel implementations.* By creating a single interface for multiple communication channels, complexity of the communication is reduced while the benefits can be manipulated. For communication such as Bluetooth and TCP/IP, IoT devices can now switch between them without disruption. On the back-end, energy efficiency, distance from other devices and more can be monitored to allow the best communication possible for IoT devices.
3. *Enhance the data integrity by applying an intelligent and redundant connection channel.* Allowing ReliableConnection to manage networking activity provides standardization for network activity. By sending data over multiple communication services, critical IoT applications can increase reliability and chain devices with different hardware configurations.
4. *Support a optimized communication channel by runtime checking and machine learning techinques.* ReliableConnection's Network class provides developers a way to weigh different communication services against each other. By giving

such measurements, data can pass through the different services depending on the developer's needs without complex code. In addition, the framework has a built-in verification engine to periodically check the performance using the two different connections, which enables a dynamic decision on which type of connection to use. We are also collecting the decision data, aligned with feature factors such as time, location, protocol, and data type, so that we will be able to apply machine learning to predict the best type of connection to use based on the actual application scenario.

With this framework, new communication services can be quickly added to the Network and contribute to the reliability without disrupting other Protocol behavior. A more descriptive discussion about the library follows.

4.1 Protocol

The Protocol class is a superclass that is modeled after the Observational design pattern and adaptable for future additions. As an Observable class, it contains a method to notify observers of any changes to the class, notifyObservers(). This method will pass along a reference of the Protocol to an Observer, the Network class.

The rest of the Protocol is defined in its interface implementation. The interface methods will be status(), getOutputStream(), getInputStream(), connect() and close(). The status() method is used to check if the Protocol is actively connected to a network. getOutputStream() has a return object of

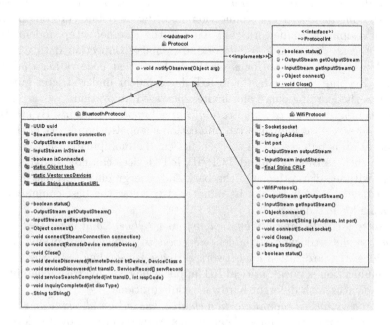

Fig. 2. UML diagram of the Protocol framework

java.io.OutputStream. for the communication standard contained in a Protocol object, Java's Output stream is a general way to send bytes of information across a network and will work with most networking devices. Similarly, getInputStream() is of java.io.InputStream. This allows the Protocol to have an input and output interface for the Network object to manipulate. The connect() method is used to implement any necessary function calls for connecting a network device and returns type object. The return for connect() is expected to be the stream for both getInputStream() and getOutputStream to utilize for that Protocol. Finally, close() is a standard method for closing the stream within the protocol for the Network or the user to utilize. The intention for the close() method is so that the Network object can close or reconnect to a Protocol automatically, for any debugging reasons.

4.2   Network

The Network is a class focused on the observation and management of Protocols for the client device. The way the Network class discovers and retrieves an active Protocol is through Java's Observer design pattern. Java's utility library has a simple Observer interface with an update method. This method, update(), is called when an Observable object's notifyObservers method is activated. The Observable method in this case is a Protocol. Once a Protocol is confirmed to be active, it will notify the network through notifyObservers() that it is a viable candidate for communication. The update method in Network then adds that Protocol to a Linked List.

When a user would like to send or receive data to another client they will go through the Network object. Network contains BufferedReader and PrintWriter objects for sending and receiving data. Network only allows users to use PrintWriter's println() and BufferedReader's readln() method. This abstraction from a stream formatter allows for dynamic allocation of IO streams. The Network class scans through Protocols and uses their IO streams

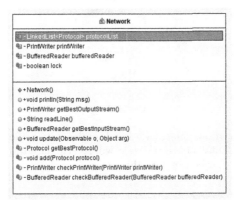

Fig. 3. UML diagram of the Network framework

interchangeably within BufferedReader and PrintWriter. What the user will see is Network.println(String msg) and Network.readln() functions. The Network.println(String msg) function will send a string to all output streams available while the Network.readln() function will return a string from the best input stream available.

5   ReliableConnection in Action

To demonstrate the functionality of this framework, a multi-client chat application was created and hosted on Windows, Android and the Raspberry Pi microprocessor board. The client is given both the IP address of a server as well as discovers the Bluetooth hardware affiliated with the server. The Server will then relay messages from one client to any other client connected with handler threads. This application helps demonstrate how the ReliableConnection framework will make data transfer simpler for IoT devices. Each chat client can efficiently transfer data to a server using either TCP/IP or Bluetooth without the developer having to directly send data over each protocol. The clients will not have the challenge of complex networking design since the Network class in ReliableConnection will manage it. At the same time, the connection reliability for each client is increased since it can communicate over Bluetooth and TCP/IP interchangeably. Now that the framework manages each protocol in one area, information about each protocol can be compared to one another. With this data, intelligent IoT design is realized, dynamic communication is now simple and efficient. A more in-depth look of the client and the server is provided below.

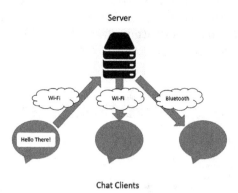

Fig. 4. The chat application. Clients send and receive data while the Network handles communication. The Chat client to the right will receive data through Bluetooth, since it was deemed best to service the data that way.

5.1   Client

The ChatClient class is a chat GUI for observing and receiving messages from other clients. On Windows, the chat client is a JFrame with a text field and a message area. For the applications on Windows and Raspberry PI, Bluetooth is done with the BlueCove Java library [9]. This gives a lightweight Java API for working with Bluetooth stack calls. From BlueCove, the ChatClient can detect available Bluetooth devices. Once all of the network information is added, the ChatClient will finalize its connection to the server. If the connection is successful, the server will request a name for the client. That name will be a unique identifier for when other clients are added. As long as the name is unique, the server will accept the information and allow a thread for communication. Now, the Network handles both reading and printing information on the Protocols available. The user now has both Wi-Fi and Bluetooth to send data over but will not interface with the complex information associated with those communication services. Instead, the user simply instantiates and provides connection information to the classes BluetoothProtocol and WifiProtocol. Then, those objects add the Network as an observer. The ChatClient's network now observes behaviors of the communication services without having the user worry about connectivity issues. All that is passed to the network will be strings of information displayed on the ChatClient's message area.

5.2   Server

For the server side of this application, a ChatServer class was created to seamlessly broadcast client information on either Bluetooth or Wi-Fi. The server is run only on Raspberry Pi or Windows as Android support was not necessary to test the application. Once run, the ChatServer first opens up a dialog for developers to choose what kind of connections should it accept. It can accept Wi-Fi and Bluetooth at the same time or each one individually. Once a client is accepted, a new thread will handle data sent to and from that client. Simply put, the Handler thread is a stripped down chat client. The constructor will get the socket information for the client and create a WifiProtocol and Bluetooth-Protocol object to manage it. The Network then observes the Protocols to keep them active, same as the ChatClient class. This makes the ChatServer simpler to understand and modify, not having to worry about the network information.

6   Related Work

Communication is a layer that all devices have to deal with but now, more than ever, the security of data from household devices and health systems makes it increasingly important [10]. This communication layer is one that has to work well with both the IoT cloud and the local embedded hardware systems. Cloud services take the work out of storing and manipulating data leaving the Software Developer to figure out how to both create IoT hardware and securely upload the data to the cloud. This framework, ReliableConnection, was created to bridge the gap between the hardware and the cloud and help developers focus on the logic of their system rather than the reliability of data flow. This frameworks also allows for new applications to develop since data can be extracted in new locations. If one protocol such as Wi-Fi is out of reach, possibly Bluetooth can find local devices to hop over in a sort of P2P style system. This system provides new possibilities as well as enhances IoT communication by offering a framework to make IoT a better, more connected reality.

7   Conclusion and Future Work

The Network and Protocol framework allow IoT applications a standardized way of sending data over a network, utilizing the advantages of multiple communication methods. This helps create applications that have increased trust in messages going through since a single Protocol failure will not halt communication.

For the future of this framework, more protocols such as RFID or IR could be added to allow new communication features or triggers. The Network could also be refined to monitor more details about each Protocol. Monitoring the speed and efficiency of each Protocol could assist in delivering information while reducing energy consumption and computational load.

References

1. Giusto, D., Lera, A., Morabito, G., Atzori, L. (eds.): The Internet of Things (2010)
2. Gubbi, J., Buyya, R., Marusic, S., Palaniswami, M.: Internet of Things (IoT): a vision, architectural elements, and future directions. Future Gener. Comput. Syst. **29**(7), 1645–1660 (2013)
3. World Health Organization: mHealth: New Horizon for Health Through Mobile Technologies (Global Observatory for e-health Services), vol. 3. WHO, Geneva (2011)
4. Niyato, D., Hossain, E., Diamond, J.: IEEE 802.16/WiMAX-based broadband wireless access and its application for telemedicine and e-health services. IEEE Wirel. Commun. Mag. **14**(1), 104–111 (2010)
5. Istepanaian, R.S., Zhang, Y.-T.: Guest editorial introduction to the special section: 4G health-the long-term evolution of m-health. IEEE Trans. Inf. Technol. Biomed. **16**(1), 1–5 (2012)
6. "FitBit." Fitbit Official Site for Activity Trackers & More. N.p., n.d. Web: 30 May 2016
7. "UP by Jawbone—Fitness Trackers for a Healthier You." Jawbone. N.p., n.d. Web: 30 May 2016
8. "Misfit." Misfit—Wearables, Activity Trackers, Fitness and Sleep Monitors. N.p., n.d. Web: 30 May 2016
9. Desmedt, Y.: Man-in-the-middle attack. In: Tilborg, H.C.A., Jajodia, S. (eds.) Encyclopedia of Cryptography and Security, pp. 759–759. Springer, US (2011)
10. Wortmann, F., Flüchter, K.: Internet of things. Bus. Inf. Syst. Eng. **57**(3), 221–224 (2015)

Research on Decentralized Group Replication Strategy Based on Correlated Patterns Mining in Data Grids

Danyang Qin[1(✉)], Ruixue Liu[2], Jiaqi Zhen[1], Songxiang Yang[1],
and Erfu Wang[1]

[1] Key Laboratory of Electronics Engineering, Heilongjiang University,
Harbin 150080, People's Republic of China
{qindanyang,zhenjiaqi,yangsongxiang,
wangerfu}@hlju.edu.cn
[2] Harbin Institute of Technology Shenzhen Graduate School,
Shenzhen 518055, People's Republic of China
liuruixue@hlju.edu.cn

Abstract. Aiming at the problem that most of the existing data mining based replication strategies cannot extract correlations between files effectively, a new decentralized replication strategy based on maximal frequent correlated patterns mining, called RSMFCP, is proposed. By translating the files access history to the binary access history, applying maximal frequent correlated patterns mining and performing replication, RSMFCP can extremely eliminate redundancy and optimize the replication performance. Data analysis and simulation results show that, comparing with other strategies like no replication, PRA, DR2 and PDDRA, RSMFCP can extract correlations more effectively and gain lower mean job execute time under different access patterns, which will provide a new option to reduce transmission delay in data grid.

Keywords: Data mining · Correlated patterns · Data replication · Distributed groups

1 Introduction

Data grid is a kind of integrated architecture to manage plenty of distributed data generated in some scientific, financial and medical fields [1]. In data grids, using data replication strategies can greatly reduce the bandwidth cost, improve the response time and maintain the reliability of the system. However, only single files are considered as the replicating object in most of the existing replication strategies, and the relationships between files are neglected. Because of the fact that nowadays many intensive

D. Qin—This work is supported by the National Natural Science Foundation of China (61302074, 61501176, 61571181), Natural Science Foundation of Heilongjiang Province (QC2013C061), Modern Sensor Technology Research and Innovation Team Foundation of Heilongjiang Province (2012TD007), and Postdoctoral Research Foundation of Heilongjiang Province (LBH-Q15121).

© ICST Institute for Computer Sciences, Social Informatics and Telecommunications Engineering 2017
X.-L. Huang (Ed.): MLICOM 2016, LNICST 183, pp. 293–302, 2017.
DOI: 10.1007/978-3-319-52730-7_30

applications need to discover the relationships between files, it is important to extract file correlations more effectively in the related research fields. Data mining can help extract valuable information from the large data sets. Using data mining in data grids can effectively find hidden correlations between files, thus achieve the goal to optimize the replicas management module.

Two measures can be concluded to discover the hidden correlations between files, which are frequent sequence mining and correlated patterns mining. Typical strategies like PRA [2] and PDDRA [3] are mainly based on frequent sequence mining. In order to predict future requested files, when execute the strategies aforementioned, the process of frequent sequence mining will be constantly running, which will increase the number of replicas and greatly impact the value of response time and occupied storage percentage in data grids. As one of the traditional correlated patterns mining based strategies, Apriori [4] can identify the frequent item sets from the large-scale data sets and produce strong correlated patterns. Apriori is a kind of sophisticated data mining algorithm, whose optimized and derived mining measures can be applied in many different industries and fields [5–7]. However, most of the common correlated patterns mining based strategies are redundant and cannot reflect the true relationships between files to some extent [8, 9]. Therefore, based on the previous research, define the groups of associated files distributed in different sites as the distributed groups and propose a Replication Strategy based on Maximal Frequent Correlated Patterns (RSMFCP). By optimizing period parameter and designing a Maximal Frequent Correlated Patterns Miner (MFCPM), RSMFCP can be periodically invoked in the real data grids, which will help realize the goals to reduce the network delay and quickly access the valuable remote files.

2   Maximal Frequent Correlated Patterns Mining

2.1   Basic Definitions

Item is a kind of binary attribute, using logical value 0 or 1 to indicate whether the given job can access the corresponding files or not. Suppose $\mathcal{I} = \{i_1, i_2, \ldots, i_n\}$ as a set of n items and the transaction associated with a unique identifier as a subset of $\mathcal{I}$. In this paper, items are defined as the accessed target files, the transaction is defined as a set of files accessed by the given job and the pattern is defined as an item set. With regard to an item set $X \subseteq \mathcal{I}$, $Supp(X)$ represents the support of X, which is calculated as the ratio of the number of transactions including X to the number of all the transactions. In fact, $Supp(X)$ is the probability of the emergence of the transactions including X. If the support of the pattern is not less than the minimum support threshold specified by users, then the pattern is frequent. Moreover, if the correlation measure of the pattern, which is denoted as $Corr(X)$ is not less than the minimum correlation measure threshold, then the pattern is correlated.

Definition 1 All-Confidence. All-confidence is a kind of correlation measure used to estimate the correlated degree for the patterns. The all-confidence of the item set $X \subseteq \mathcal{I}$ can be calculated by Eq. (1).

$$all - confidence(X) = \frac{Supp(\wedge X)}{\max\{Supp(\wedge i)|i \in X\}} \tag{1}$$

where $\max\{Supp(\wedge i)|i \in X\}$ represents the maximum support of the items in X, $Supp(\wedge X)$ represents the support of X, and i represents an item in X. All-confidence simultaneously possesses the anti-monotone, cross-support and null-invariant properties:

(1) Anti-monotone property. For any item set $I \subseteq \mathcal{I}$, $I_1 \subseteq I$, if the fact that I satisfies the constraint Q can infer that I_1 also satisfies Q, then the constraint Q is considered to be anti-monotone.

(2) Cross-support property. Given the threshold $t \in [0,1]$ and the item set $I \subseteq \mathcal{I}$ that contains item x and y, if $(Supp(\wedge x)/Supp(\wedge y)) < t$, then I is considered to be cross-support with respect to the threshold t.

(3) Null-invariant property. When it comes to the correlation of the pattern, the null-invariant property can make sure that only the transactions including the specific pattern are analyzed [10]. For the pattern $I \subseteq \mathcal{I}$, the transactions that do not contain I is deemed as the null transactions. It makes no sense to deduce the correlation of I according to the number of null transactions, which will also help avoid the bad influence of the null transactions.

Definition 2 Frequent correlated pattern. Support and all-confidence are the measures respectively corresponding to the frequency and correlation of the pattern. Given the minimum support threshold *minsupp* and the minimum correlation measure threshold *mincorr*, if $Supp(X) \geq minsupp$ and $Corr(X) \geq mincorr$, then the pattern X is considered to be a frequent correlated pattern.

Definition 3 Maximal frequent correlated pattern. If X is a frequent correlated pattern, and the superset of X is definitely not a correlated frequent pattern, then X is deemed to be a maximal frequent correlated pattern. The definition of the maximal frequent correlated pattern can contribute to extremely decrease the number of distributed groups to replicate, decrease the occupied storage in data grids and optimize the replicating process.

2.2   Maximal Frequent Correlated Patterns Miner

In order to mine the distributed groups in data grids, it is necessary to extract the maximal frequent correlated pattern defined before. In this section, a maximal frequent correlated patterns miner, called MFCPM, is designed by Algorithm 1. The notations used in MFCPM are defined in Table 1.

Table 1. Notations used in MFCPM.

Notation	Meaning	Notation	Meaning
X_k	A pattern X of k items	$\mathcal{FCP}_k$	A frequent correlated pattern of k items
C_k	A candidate set of k items	$\mathcal{MFCP}$	A set of maximal frequent correlated patterns

Algorithm 1.   MFCPM algorithm.

Input: A binary access history, $minsupp$, $min-all-confidence$
Output: $\mathcal{MFCP}$
Begin
   $k := 1$;
   $\mathcal{FCP}_1 := \{i \in \mathcal{I} \mid Supp(i) \geq minsupp\}$;   %determine the fcp
   $\mathcal{MFCP} := \mathcal{FCP}_1$;
  While $\mathcal{FCP}_k \neq \phi$ **do**
      $\mathcal{FCP}_{k+1} := \text{GENERATE\_NEXT\_}\mathcal{FCP}\left(\mathcal{FCP}_k, minsupp, min\text{-}all\text{-}confidence\right)$;
     Foreach $\left(X_{k+1} \in \mathcal{FCP}_{k+1}\right)$ **do**
        If $\left(\exists X_k \subset X_{k+1} \mid \left(X_k \in \mathcal{MFCP}\right)\right)$ **then**
          remove X_k from $\mathcal{MFCP}$
     $\mathcal{MFCP} := \mathcal{MFCP} \cup \mathcal{FCP}_{k+1}$; %make sure no rp exist
     $k := k+1$;
  Return $\mathcal{MFCP}$
End

---

3   Distributed Groups Replication Strategy Based on MFCPM

3.1   The Procedures of RSMFCP

Based on the MFCPM module proposed in the last section, the RSMFCP strategy is proposed aiming at P2P data grid topology and consists of 4 phases.

Extract the files access history. To require local and remote files, the current site should locally record the files access history during every executing period and the job access order is determined by the access patterns.

Translate the files access history to a binary access history. The binary access history is essentially a logical table consists of the accessed object files and jobs.

Generate the $\mathcal{MFCP}$ pattern. Design the MFCPM module to find the hidden correlations between the distributed groups and simplify the later replication process.

Replicate and replace. Choose $\mathcal{MFCP}$ as the input of this phase, and select to retain or replace the files primarily by calculating the average weight of the files to replicate and delete.

3.2   The Translation of the Binary Access History

Each site should maintain its files access history. The files access history of the site S_i is defined as a matrix **A** of $n \times m$, while n represents the total number of jobs running in the given period, m represents the sum of accessed files and $\mathbf{A}_{j,k} = \#request(F_j, J_k)$ represents the number of times that the job J_k accesses the file F_j. Before data mining, it is necessary to translate the files access history into the binary access history including logical value 0 or 1. In order to translate more quickly, the popularity of the file is introduced. If the jobs executed in S_i frequently access F_j, then F_j is considered to be popular within the scope of S_i. The average file accessed times $AvgAccess(F_j)$ is introduced to make it more convenient to evaluate the popularity of F_j in S_i. $AvgAccess(F_j)$ is calculated by Eq. (2).

$$AvgAccess(F_j) = \frac{\sum_{k=1}^{n} \#request(F_j, J_k)}{n_j} \tag{2}$$

where n_j represents the total number of jobs that access F_j.

3.3 The Replication Process of RSMFCP

In this section, the replication process of the RSMFCP strategy will be elaborated. In order to replicate the distributed groups, choose the $\mathcal{MFCP}$ pattern as the input of the replication process. Suppose $\mathcal{MFCP} = \{\alpha_1, \alpha_2, \ldots, \alpha_n\}$ and any element $\alpha_i \in \mathcal{MFCP}$ is the set of the files frequently accessed by the jobs. The specific steps of the replication process of RSMFCP are as follow: (1) For each $\alpha_i \in \mathcal{MFCP}$, sort the elements in $\mathcal{MFCP}$ according to a descending order of the number of the patterns contained in α_i. (2) For each $\alpha_i \in \mathcal{MFCP}$, if the storage space in S_i is enough to store all the files in α_i, then replicate all the files in α_i to S_i. (3) Otherwise, select candidate files to delete by calculating the weight of F_j in S_i according to Eq. (3).

$$FileWeight(F_j) = \frac{size(F_j) \times \#request(F_j, S_i)}{Bandwidth(S_i, S_r)} \tag{3}$$

where $size(F_j)$ represents the size of F_j and $Bandwidth(S_i, S_r)$ represents the bandwidth between the site S_i and the site S_r that contain the best replica of F_j. (4) Calculate the average weight of the files which will be replicated and deleted respectively according to Eqs. (4) and (5).

$$AvgGroupRepWeight = \frac{1}{|ToReplicate|} \times \sum_{f \in ToReplicate} FileWeight(f) \tag{4}$$

where $AvgGroupRepWeight$ represents the average weight of the files which will be replicated, the last item represents the total weight of the files which will be replicated and $|ToReplicate|$ represents the total number of the files which will be replicated.

$$AvgCandidateDelWeight = \frac{1}{|CandidateDel|} \times \sum_{f \in CandidateDel} FileWeight(f) \tag{5}$$

where *AvgCandidateDelWeight* represents the average weight of the file which will be deleted, the last item represents the total weight of the candidate file which will be deleted and |*CandidateDel*| represents the total number of the candidate files to delete. (5) Compare the two types of average weight values aforementioned. The candidate files will be replaced to delete with the files to replicate, or give up replicating if there is *AvgGroupRepWeight* > *AvgCandidateDelWeight*.

4 Performance Analysis and Simulation Evaluation

4.1 Simulation Environment

In this paper, the OptorSim [11, 12] simulator is used to test the job scheduling and replicating strategies and simulate the actual data grid topology. OptorSim is a simulation package wrote by Java, which consists of the users, resource agent and many sites. Each site consists of the Computing Element (CE), Replica Management (RM) and Storage Element (SE). The simulation environment in this paper is CMS testbed grid. The CMS testbed grid consists of 20 imitative sites in Europe and America. Except for the sites in CERN and FNAL own the storage of 100 Gb, the other sites all own the storage of 50 Gb and a CE. At the beginning, the initial size of the files in the distributed groups is 1 Gb, the total number of the files is 97, the total number of the jobs is 1000 and all stored in the SEs. In addition, the sequential access pattern is selected to access the files, and the current and queued jobs access cost scheduling algorithm is applied to schedule the jobs.

4.2 Impact of the Executing Period on Strategy Performance

Considering that the given number of jobs is 1000, *minsupp* and *min-all-confidence* are both fixed, analyze the impact of different executing periods on the mean job execute time of RSMFCP. Mean job execute time is defined as the total individual executing time of every job divided by the total number of jobs executed. The mean job execute time is shorter, the performance of RSMFCP is better. The simulation result is shown in Fig. 1. When 1000 jobs are executed, it is not hard to deduce that, invoking RSMFCP

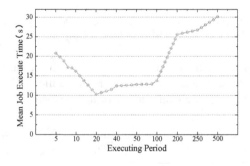

Fig. 1. Mean job execute time of RSMFCP for different periods.

after executing every 20 jobs (2%) can obtain the minimum mean job execute time. Whether the period is shorter or longer will lead to frequently accessing the remote files, which will cause the mean job execute time increase and the replicating efficiency decrease.

4.3   Impact of the Threshold to Strategy Performance

Given that the number of jobs is 1000, the executing period is 2%, *min-all-confidence* and *minsupp* respectively equal to 0.2, 0.4 and 0.6, analyze the impact of the related *minsupp* and *min-all-confidence* on mean job execute time of RFMFCP. The simulation results are shown in Fig. 2.

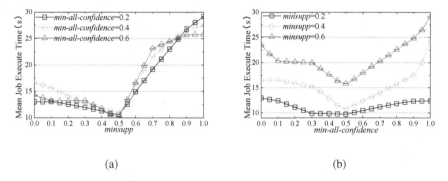

(a)   (b)

Fig. 2. Mean job execute time of RSMFCP for different thresholds (a) Mean job execute time of RSMFCP for different *minsupp* thresholds; (b) Mean job execute time of RSMFCP for different *min-all-confidence* thresholds.

It can be inferred from Fig. 2 that the mean job execute time will slowly decay when the threshold is between 0 and 0.5. In addition, the mean job execute time will rapidly grow up when the threshold exceeds 0.5, which means that the strategy performance begins to deteriorate. The simulation results show that the increase of the threshold value can result in the deterioration of the strategy. So it can be concluded that when *minsupp* and *min-all-confidence* both equal to 0.5, the mean job execute time is the smallest and the strategy performance is optimal.

4.4   Impact of the Access Patterns to Strategy Performance

Given that the number of jobs is 1000, the period is 2% and *minsupp* and *min-all-confidence* both equal to 0.5, compare the performance of the proposed RSMFCP strategy with the other four replication strategies under different access patterns. The four strategies are no replication. The five different access patterns are random access pattern, sequential access pattern, random Zipf access pattern and random walk Gaussian access pattern. Each comparison process repeats at least 10 times, after which calculate the mean values.

Mean job execute time and Effective Network Usage (ENU). The mean job execute time and ENU of the five strategies under different access patterns are shown in Fig. 3. ENU ranging from 0 to 1 is a specific ratio of the transformed files to the accessed files, which is calculated by Eq. (6). Apparently, ENU is lower, and the strategy performance is better.

$$ENU \doteq \frac{N_{remotefileaccesses} + N_{filereplications}}{N_{remotefileaccesses} + N_{localfileaccesses}} \tag{6}$$

where $N_{remotefileaccesses}$ its the number of the accessed remote files, $N_{filereplications}$ is the number of the replicas and $N_{localfileaccesses}$ is the number of the accessed local files. The simulation results show that comparing with no replication, DR2, PRA and PDDRA strategy, the mean job execute time of RSMFCP can respectively decrease 80%, 60%, 20% and 15% at most for different access patterns. One of the main goals of the research is to minimize the bandwidth cost and decrease the network traffic, to achieve that the performance of RSMFCP is better compared with other strategies.

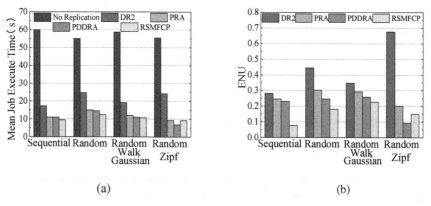

Fig. 3. Mean job execute time and ENU with different access patterns. (a) Mean job execute time with different access patterns; (b) ENU with different access patterns.

Amount of replications and occupied storage percentage. This is the number of the replicating times. Obviously, when the amount of replications is big, it indicates that most of the files required are stored in the remote sites. Besides, the occupied storage percentage is the average usage of the SEs in the grid sites. The usage of SE is the ratio of the storage resource used by files to the SE capacity. The amount of replications and occupied storage percentage of the five strategies are shown in Fig. 4. It is easy to deduce from the simulation results that the amount of replications of RSMFCP can decrease apparently with different access patterns, but it can still guarantee the availability of the files in the data grid. The amount of replications is bigger, which implies that the number of transferred files is also bigger. Therefore, the strategies of the same kind only consume the reasonable network bandwidth.

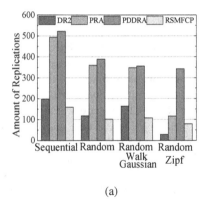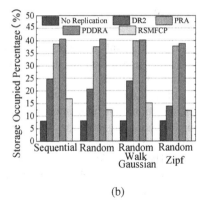

(a)   (b)

Fig. 4. Amount of replications and occupied storage percentage with different access patterns (a) Amount of replications with different access patterns; (b) Occupied storage percentage with different access patterns.

Hit Ratio (HR). HR is the ratio of the total number of times accessing the local files to the total number of times accessing all the files. HR can be calculated by Eq. (7) and the HR of the five strategies with different access patterns are shown in Fig. 5.

$$HR = \frac{N_{localfileaccess}}{N_{remotefileaccess} + N_{replications} + N_{localfileaccesses}}. \tag{7}$$

The simulation results show that compared with the same kind DR2, PRA and PDDRA strategy, the HR of RSMFCP can respectively increase 65%, 20% and 15% at most with all the access patterns.

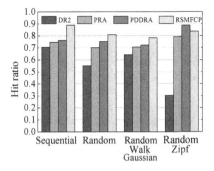

Fig. 5. Hit ratio with different access patterns.

5   Conclusion

Nowadays, the number of data generated in scientific and engineering fields gradually grows faster and faster, so the demand of computing and storing in each field is increasing. Therefore, data grid is generated as a reasonable solution. In this paper, taking the distributed groups of the sites in data grid as the mining object, MFCPM was added on traditional replication strategies and the RSMFCP strategy was proposed. Compared with the same kind of strategies, the mean job execute time and ENU of RSMFCP can decrease 80% at most, meanwhile the HR of RSMFCP can increase 65% at most. The simulation results showed that RSMFCP takes the distributed groups as the object of the research, which can reduce the number of files to replicate. Thus, RSMFCP can improve the grid performance and have some certain superiority and better application prospect. The future work will aim at optimizing the files access history in each site and applying multidimensional dynamic data mining technologies, in order to further improve the replication process and make the strategy more suitable for the realistic data grid environment.

References

1. Amornsinlaphachai, P.: Efficiency of data mining models to predict academic performance and a cooperative learning model. In: 8th International Conference on Knowledge and Smart Technology (KST), pp. 66–71 (2016)
2. Lee, M.C., Leu, F.Y., Chen, Y.P.: PFRF: An adaptive data replication algorithm based on star-topology data grids. Future Gener. Comput. Syst. 28(7), 1045–1057 (2012)
3. Saadat, N., Rahmani, A.M.: PDDRA: a new pre-fetching based dynamic data replication algorithm in data grids. Future Gener. J. Comput. Syst. 28(4), 666–681 (2012)
4. Agrawal, R., Imielinski, T., Swami, A.: A mining association rules between sets of items in large databases. In: Proceedings of the 1993 ACM SIGMOD Conference, pp. 207–216 (1993)
5. Taheri, J., Zomaya, A.Z., Bouvry, P., Khan, S.U.: Hopfield neural network for simultaneous job scheduling and data replication in grids. Future Gener. Comput. Syst. 29(8), 1885–1900 (2013)
6. Wei, H.: Correlation mining of multi-dimensional large data sets. Mod. Comput. 9(1), 3–8 (2012)
7. Shorfuzzaman, M., Graham, P.: Adaptive popularity-driven replica placement in hierarchical data grids. J. Supercomput. 51(3), 374–392 (2010)
8. Bellodi, E., Riguzzi, F., Lamma, E.: Statistical relational learning for workflow mining. Intell. Data Anal. 20(3), 515–541 (2016)
9. Jian, L., Wang, C., Liu, Y., Liang, S., Yi, W.: Parallel data mining techniques on graphics processing unit with compute unified device architecture (CUDA). J. Supercomput. 64(3), 942–967 (2013)
10. Wu, T., Chen, Y., Han, J.: Re-examination of interestingness measures in pattern Mining: a Unified framework. Data Min. Knowl. Discov. 21(3), 371–397 (2010)
11. Grace, R.K., Manimegalai, R.: Dynamic replica placement and selection strategies in data grids – a comprehensive survey. J. Parall. Distrib. Comput. 74(2), 2099–2108 (2014)
12. Ma, J., Liu, W., Glatard, T.: A classification of file placement and replication methods on grids. Future Gener. Comput. Syst. 29(6), 1395–1406 (2013)

Calibration Method of Gain-Phase Errors in Super-resolution Direction Finding for Wideband Signals

Jiaqi Zhen$^{(\boxtimes)}$, Danyang Qin, Jie Yang, and Yanchao Li

College of Electronic Engineering, Heilongjiang University,
Harbin 150080, China
zhenjiaqi2011@163.com

Abstract. Most super-resolution direction finding methods need to know the array manifold exactly, but there is usually gain and phase errors in the array, which directly lead to the discordance of the channels. The paper proposed a novel calibration method in super-resolution direction finding for wideband signals based on spatial domain sparse optimization when gain and phase errors exist. First, the optimization functions are founded by the signals of every frequency, then the functions are optimized iteratively, consequently the information of all frequencies is integrated for the calibration, thus, the actual directions of arrival (DOA) can be estimated. Simulations have proved the method is appropriate for low signal to noise ratio (SNR) and small samples.

Keywords: Super-resolution direction finding · Array calibration · Gain-phase errors · Wideband signals

1 Introduction

Super-resolution direction finding is one of the major research contents in array signal processing, it is widely used in radio monitoring [1–7] and internet of things [8, 9]. Most of the direction finding methods need to know the accurate array manifold, but there are often amplifiers in the channels, the gains of them are not consistent, and sometimes accompanied with discordant lengths of the channels in practical systems, which directly lead to the performance deteriorated of direction finding methods, and even failure, so they are necessary to be calibrated.

Gain-phase errors have no relation with DOA of the signal, they are caused by the different responses of the channels. Srinath [10] analyzed the effect of the gain and phase errors on traditional multiple signal classification (MUSIC) [11] algorithm, he proved that they have a great influence on the estimation, even lead to the failure. Most of the calibration methods are based on eigenstructure and lack adaptation to the

---

J. Zhen—This work was supported by the National Natural Science Foundation of China under Grant No. 61501176 and 61302074, Specialized Research Fund for the Doctoral Program of Higher Education under Grant No. 20122301120004, Natural Science Foundation of Heilongjiang Province under Grant No. QC2013C061.

© ICST Institute for Computer Sciences, Social Informatics and Telecommunications Engineering 2017
X.-L. Huang (Ed.): MLICOM 2016, LNICST 183, pp. 303–313, 2017.
DOI: 10.1007/978-3-319-52730-7_31

background of low signal to noise ratio (SNR) and small samples. Wang [12] proposed a simple and fast calibration algorithm that does not require any prior knowledge of the DOA along with sensor gain and phase uncertainties based on Toeplize characteristic; Jiang [13] provided the conventional and improved data models, then correct the array, the estimation accuracy is not affected regardless of how large the phase errors are; Xu [14] estimated DOA of strong and weak signals in the presence of array gain and phase mismatch; Cao and Ye [15] proposed a calibration method for channel gain and phase uncertainties based on fourth-order cumulant technique, it adapts to the background of non-Gaussian signals and Gaussian noise. All the methods above only adapt to narrowband signals, and need many samples, but there are rare published literatures of gain and phase errors calibration for wideband signals.

The paper proposed a novel array error calibration method in super-resolution direction finding for wideband signals based on spatial domain sparse optimization when gain-phase errors exist in the array, the corresponding optimization functions are founded by the signal of every frequency, then the functions are optimized iteratively, at last, the information of all frequencies is integrated to calibrated the errors, consequently the actual DOA can be acquired.

2 Signal Model

2.1 Ideal Signal Model

It is seen from Fig. 1, suppose there are K far-field wideband signals $s_k(t)$ $(k = 1, 2, \cdots, K)$ impinging on the uniform linear array composed of M omnidirectional sensors, the space of them is d, it is equal to half of the wavelength of the center frequency, DOAs of them are $\boldsymbol{\alpha} = [\alpha_1, \cdots, \alpha_k, \cdots, \alpha_K]$, the first sensor is defined as the reference, then output of the mth sensor can be written as

$$x_m(t) = \sum_{k=1}^{K} s_k(t - \tau_m(\alpha_k)) + n_m(t), m = 1, 2, \cdots, M \tag{1}$$

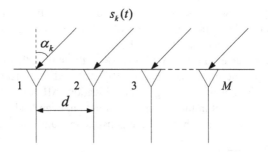

Fig. 1. Array signal Model

Where $\tau_m(\alpha_k) = (m-1)\frac{d}{c}\sin\alpha_k$ is the propagation delay for the kth signal arriving at the mth sensor with respect to the reference of the array, c is the propagating speed of the signal, $n_m(t)$ is the Gaussian white noise on the mth sensor.

Assume that the range of the frequency band of all signals is $[f_{\text{Low}}, f_{\text{High}}]$, before the processing, we divide the output vector into J nonoverlapping components, Discrete Fourier Transform(DFT) is performed on (1) and the array outputs of J frequencies can be represented as

$$X(f_i) = A(f_i, \alpha)S(f_i) + N(f_i) \quad i = 1, 2, \cdots, J \tag{2}$$

Where $f_{\text{Low}} \le f_i \le f_{\text{High}}$ $(i = 1, 2, \cdots, J)$, KP snapshots are collected at every frequency, then we have

$$X(f_i) = [X_1(f_i), \cdots, X_m(f_i), \cdots, X_M(f_i)]^{\text{T}} \tag{3}$$

Where

$$X_m(f_i) = [X_m(f_i, 1), \cdots, X_m(f_i, kp), \cdots, X_m(f_i, KP)] \tag{4}$$

$A(f_i, \alpha)$ is a $M \times K$ dimensional steering vector

$$A(f_i, \alpha) = [a(f_i, \alpha_1), \cdots, a(f_i, \alpha_k), \cdots, a(f_i, \alpha_K)] \tag{5}$$

$$a(f_i, \alpha_k) = \left[1, \exp(-\mathrm{j}2\pi f_i \frac{d}{c}\sin\alpha_k), \cdots, \exp\left(-\mathrm{j}(M-1)2\pi f_i \frac{d}{c}\sin\alpha_k\right)\right]^{\text{T}} \tag{6}$$

And

$$S(f_i) = [S_1(f_i), \cdots, S_k(f_i), \cdots, S_K(f_i)]^{\text{T}} \tag{7}$$

is the signal vector matrix after DFT to $s_k(t)$ $(k = 1, 2, \cdots, K)$, where

$$S_k(f_i) = [S_k(f_i, 1), \cdots S_k(f_i, kp), \cdots, S_k(f_i, KP)] \tag{8}$$

Here, $S_k(f_i, kp)$ is the kpth snapshots of the kth signal at f_i, then

$$N(f_i) = [N_1(f_i), \cdots, N_m(f_i), \cdots, N_M(f_i)]^{\text{T}} \tag{9}$$

$$N_m(f_i) = [N_m(f_i, 1), \cdots, N_m(f_i, kp), \cdots, N_m(f_i, KP)] \tag{10}$$

is the noise vector after performing DFT on $n_m(t)$ $(m = 1, 2, \cdots, M)$ with mean 0 and variance $\mu^2(f_i)$.

2.2   Gain-Phase Errors Model

For convenience, we only discuss the information at frequency f_i for the moment. When there is only gain and phase errors in the array, $W(f_i)$ is defined as perturbation matrix, it is

$$W(f_i) = \mathrm{diag}\left([1, W_2(f_i), \cdots, W_m(f_i), \cdots, W_M(f_i)]^T\right) \tag{11}$$

Here

$$W_m(f_i) = \rho_m(f_i)e^{j\varphi_m(f_i)}, m = 1, 2, \cdots, M \tag{12}$$

is the gain and phase perturbation of mth sensor, $\rho_m(f_i)$, $\varphi_m(f_i)$ are respectively the gain and phase of the mth sensor with respect to the reference sensor, so the perturbed steering vector is

$$
\begin{aligned}
a'(f_i, \alpha_k) &= [1, W_2(f_i)e^{j2\pi f_i \tau_2(\alpha_k)}, \cdots, W_m(f_i)e^{j2\pi f_i \tau_m(\alpha_k)}, \cdots, W_M(f_i)e^{j2\pi f_i \tau_M(\alpha_k)}]^T \\
&= \mathrm{diag}\left([1, W_2(f_i), \cdots, W_m(f_i), \cdots, W_M(f_i)]^T\right) a(f_i, \alpha_k) \\
&= W(f_i)a(f_i, \alpha_k) \quad (k = 1, 2, \cdots, K)
\end{aligned} \tag{13}
$$

So the corresponding array manifold matrix is

$$A'(f_i, \boldsymbol{\alpha}) = [a'(f_i, \alpha_1), \cdots, a'(f_i, \alpha_k), \cdots, a'(f_i, \alpha_K)] = W(f_i)A(f_i, \boldsymbol{\alpha}) \tag{14}$$

For the sake of simplicity, we also define the gain/phase uncertainty vector among sensors as $w(f_i) = [\rho_2(f_i)e^{j\varphi_2(f_i)}, \cdots, \rho_m(f_i)e^{j\varphi_m(f_i)}, \cdots, \rho_M(f_i)e^{j\varphi_M(f_i)}]^T$, so the output of the array at frequency f_i can be expressed as

$$
\begin{aligned}
X'(f_i) &= A'(f_i, \boldsymbol{\alpha})S(f_i) + N(f_i) = W(f_i)A(f_i, \boldsymbol{\alpha})S(f_i) + N(f_i) \\
&= A(f_i, \boldsymbol{\alpha})S(f_i) + \Lambda(f_i)w(f_i) + N(f_i)
\end{aligned} \tag{15}
$$

Where $\Lambda(f_i)$ is the vector related to the signal along with gain and phase errors.

3   Estimation Theory

We divide the searching area into some grids $\boldsymbol{\Omega} = [\bar{\alpha}_1, \cdots, \bar{\alpha}_l, \cdots, \bar{\alpha}_L]$, here $K \ll L$, take $\boldsymbol{\Omega}$ into (2)

$$\bar{X}'(f_i) = A'(f_i, \boldsymbol{\Omega})\bar{S}(f_i) + N(f_i) \ (i = 1, 2, \cdots, J) \tag{16}$$

The covariance matrix is

$$\bar{R}'(f_i) = E\left\{\bar{X}'(f_i)(\bar{X}'(f_i))^H\right\} \ (i = 1, 2, \cdots, J) \tag{17}$$

In (16), $\bar{S}(f_i) = [\bar{S}(f_i,1),\cdots,\bar{S}(f_i,kp),\cdots,\bar{S}(f_i,KP)]$, where $\bar{S}(f_i,kp) = [\bar{S}_1(f_i,kp),\cdots,\bar{S}_l(f_i,kp),\cdots,\bar{S}_L(f_i,kp)]^T$ is a sparse matrix, it only contains K non-zero elements, they are non-zero if and only if $\bar{\alpha}_l = \alpha_k$ and $\bar{S}_l(f_i,kp) = S_k(f_i,kp)$ ($l = 1, 2, \cdots, L$; $k = 1, 2, \cdots, K$), so $\bar{S}(f_i)$ can be regarded as $S(f_i)$ jointed many zero elements.

Define $\delta(f_i) = [\delta_1(f_i),\cdots,\delta_l(f_i),\cdots,\delta_L(f_i)]^T$ as the vector formed by variances of the elements in $\bar{S}(f_i)$, it reflects the energy of the signal, that is

$$\bar{S}(f_i) \sim N(0, \Sigma(f_i)) \tag{18}$$

Where $\Sigma(f_i) = \text{diag}(\delta(f_i))$, as $\bar{S}(f_i)$ is $S(f_i)$ jointed many zero elements, $\delta(f_i)$ contains K non-zero elements too.

It can be seen from (16) and (18), probability density of the output signal at f_i along with the error is

$$\begin{aligned}P(\bar{X}'(f_i)|\bar{S}(f_i);w(f_i),\mu^2(f_i)) &= \left|\pi\mu^2(f_i)I_M\right|^{-KP}\exp\left\{-\mu^2(f_i)\|\bar{X}'(f_i) - A'(f_i,\Omega)\bar{S}(f_i)\|_2^2\right\} \\ &= \left|\pi\mu^2(f_i)I_M\right|^{-KP}\exp\left\{-\mu^2(f_i)\times\|\bar{X}'(f_i) - W(f_i)A(f_i,\Omega)\bar{S}(f_i)\|_2^2\right\}\end{aligned} \tag{19}$$

Combining (16), (18) and (19), probability density of $\bar{X}'(f_i)$ is

$$\begin{aligned}&P(\bar{X}'(f_i);\delta(f_i),w(f_i),\mu^2(f_i)) \\ &= \int P(\bar{X}'(f_i)|\bar{S}(f_i);w(f_i),\mu^2(f_i))P(\bar{S}(f_i);\delta(f_i))\mathrm{d}\bar{S}(f_i) \\ &= \left|\pi(\mu^2(f_i)I_M + A'(f_i,\Omega)\Sigma(f_i)(A'(f_i,\Omega))^H)\right|^{-KP} \\ &\quad\exp\left\{-KP\times\text{tr}\left(\left(\mu^2(f_i)I_M + A'(f_i,\Omega)\Sigma(f_i)(A'(f_i,\Omega))^H\right)^{-1}\bar{R}'(f_i)\right)\right\}\end{aligned} \tag{20}$$

Then Expectation Maximization (EM) method [16] can be employed to estimate each parameter, compute distribution function of $P(\bar{X}'(f_i),\bar{S}(f_i);\delta(f_i),w(f_i),\mu^2(f_i))$, in the E-step:

$$\begin{aligned}&F(\bar{X}'(f_i),\bar{S}(f_i);\delta(f_i),w(f_i),\mu^2(f_i)) \\ &=\langle\text{In}P(\bar{X}'(f_i),\bar{S}(f_i);\delta(f_i),w(f_i),\mu^2(f_i))\rangle \\ &=\langle\text{In}P(\bar{X}'(f_i)|\bar{S}(f_i);w(f_i),\mu^2(f_i)) + \text{In}P(\bar{S}(f_i);\delta(f_i))\rangle \\ &=\left\langle -M\times KP\times\text{In}\mu^2(f_i) - \mu^{-2}(f_i)\|\bar{X}'(f_i) - W(f_i)A(f_i,\Omega)\bar{S}(f_i)\|_2^2 \right. \\ &\quad\left. -\sum_{l=1}^{L}\left(KP\times\text{In}\delta_l(f_i) + \frac{\left(\sum_{kp=1}^{KP}|\bar{S}_l(f_i,kp)|^2\right)}{\delta_l(f_i)}\right)\right\rangle\end{aligned} \tag{21}$$

In the M-step, solve derivatives of $F(\bar{X}'(f_i), \bar{S}(f_i); \delta(f_i), w(f_i), \mu^2(f_i))$ for each parameter, that is

$$\frac{\partial F(\bar{X}'(f_i), \bar{S}(f_i); \delta(f_i), w(f_i), \mu^2(f_i))}{\partial w(f_i)} \tag{22}$$
$$= -2\mu^{-2}(f_i)\left[\langle \Lambda^H(f_i)\Lambda(f_i)\rangle w(f_i) - \langle \Lambda^H(f_i)(\bar{X}'(f_i) - A(f_i, \mathbf{\Omega})\bar{S}(f_i))\rangle\right]$$

$$\frac{\partial F(\bar{X}'(f_i), \bar{S}(f_i); \delta(f_i), w(f_i), \mu^2(f_i))}{\partial \mu^2(f_i)} \tag{23}$$
$$= -\frac{M \times KP}{\mu^2(f_i)} + \frac{1}{(\mu^2(f_i))^2}\left\langle \left\|\bar{X}'(f_i) - A'(f_i, \mathbf{\Omega})\bar{S}(f_i)\right\|_2^2\right\rangle$$

$$\frac{\partial F(\bar{X}'(f_i), \bar{S}(f_i); \delta(f_i), w(f_i), \mu^2(f_i))}{\partial \delta_l(f_i)} = -\frac{KP}{\delta_l(f_i)} + \frac{1}{\delta_l^2(f_i)}\left\langle \sum_{kp=1}^{KP} |\bar{S}_l(f_i, kp)|^2\right\rangle \tag{24}$$

Set them to be 0 respectively, the estimation of every parameter of the pth iteration is

$$w^{(p)}(f_i) = \langle \Lambda^H(f_i)\Lambda(f_i)\rangle^{-1}\langle \Lambda^H(f_i)(\bar{X}'(f_i) - A(f_i, \mathbf{\Omega})\bar{S}(f_i))\rangle \tag{25}$$

$$(\mu^2(f_i))^{(p)} = \frac{1}{M \times KP}\left\langle \left\|\bar{X}'(f_i) - (A'(f_i, \mathbf{\Omega}))^{(p)}\bar{S}(f_i)\right\|_2^2\right\rangle \tag{26}$$

$$\delta_l^{(p)}(f_i) = \frac{1}{KP}\left\langle \sum_{kp=1}^{KP} |\bar{S}_l(f_i, kp)|^2\right\rangle \tag{27}$$

Here (p) denotes number of iterations, after several times, $w(f_i)$, $\mu^2(f_i)$ and $\delta_l(f_i)$ tend to be zero, then they are deemed to be convergent, we can acquire their final estimation: $\hat{w}(f_i)$, $\hat{\mu}^2(f_i)$ and $\hat{\delta}_l(f_i)$. We can use them for array calibration, define X as the vector composed by sum of signal of all frequencies, as the signal of every frequency is independent of one another, the joint probability density of X is

$$P(X) = \prod_{i=1}^{J} P\left(\bar{X}'(f_i); \hat{\delta}(f_i), \hat{w}(f_i), \hat{\mu}^2(f_i)\right)$$

$$= |\pi|^{-J \times KP}\prod_{i=1}^{J}\left|\left(\hat{\mu}^2(f_i)I_M + A'(f_i, \mathbf{\Omega})\hat{\Sigma}(f_i)(A'(f_i, \mathbf{\Omega}))^H\right)\right|^{-KP} \tag{28}$$

$$\times \exp\left\{-KP \times \sum_{i=1}^{J} \mathrm{tr}\left(\left(\begin{array}{c}\hat{\mu}^2(f_i)I_M + A'(f_i, \mathbf{\Omega}) \times \\ \hat{\Sigma}(f_i)(A'(f_i, \mathbf{\Omega}))^H\end{array}\right)^{-1}\bar{R}'(f_i)\right)\right\}$$

Solve logarithm operation on (28), we have

$\ln(P(\mathbf{X}))$

$$= -J \times KP \times \ln\pi - KP \times \left(\sum_{i=1}^{J} \ln\left| \hat{\mu}^2(f_i)\mathbf{I}_M + \mathbf{A}'(f_i, \mathbf{\Omega})\hat{\Sigma}(f_i)(\mathbf{A}'(f_i, \mathbf{\Omega}))^{\mathrm{H}} \right| \right) \tag{29}$$

$$- KP \times \sum_{i=1}^{J} \mathrm{tr}\left(\left(\begin{matrix} \hat{\mu}^2(f_i)\mathbf{I}_M + \mathbf{A}'(f_i, \mathbf{\Omega}) \times \\ \hat{\Sigma}(f_i)(\mathbf{A}'(f_i, \mathbf{\Omega}))^{\mathrm{H}} \end{matrix} \right)^{-1} \bar{R}'(f_i) \right)$$

Solve the partial differentiation of $\ln(P(X))$ with regard to $\boldsymbol{\alpha}$

$$\frac{\partial \ln(P(X))}{\partial \boldsymbol{\alpha}} = 0 \tag{30}$$

Combing (29) with (30), we have

$$\hat{\alpha}_k = \arg\max_{\alpha_k} \mathrm{Re}\left\{ \left| \begin{matrix} \left[\sum_{i=1}^{J} \left[(\boldsymbol{a}'(f_i, \alpha_k))^{\mathrm{H}} \times \left(\begin{matrix} \hat{\mu}^2(f_i)\mathbf{I}_M + \mathbf{A}'(f_i, \mathbf{\Omega}_{-k}) \times \\ \hat{\Sigma}_{-k}(f_i)(\mathbf{A}'(f_i, \mathbf{\Omega}_{-k}))^{\mathrm{H}} \end{matrix} \right)^{-1} \right) \right] \right] \\ \times \left[\sum_{i=1}^{J} \left(\left(\begin{matrix} \boldsymbol{a}'(f_i, \alpha_k)(\boldsymbol{a}'(f_i, \alpha_k))^{\mathrm{H}} \times \\ \left(\hat{\mu}^2(f_i)\mathbf{I}_M + \mathbf{A}'(f_i, \mathbf{\Omega}_{-k}) \times \\ \hat{\Sigma}_{-k}(f_i)(\mathbf{A}'(f_i, \mathbf{\Omega}_{-k}))^{\mathrm{H}} \end{matrix} \right)^{-1} \bar{R}'(f_i) \right) \right) - \sum_{i=1}^{J} \left(\bar{R}'(f_i) \left(\begin{matrix} \hat{\mu}^2(f_i)\mathbf{I}_M + \\ \mathbf{A}'(f_i, \mathbf{\Omega}_{-k})\hat{\Sigma}_{-k}(f_i) \times \\ (\mathbf{A}'(f_i, \mathbf{\Omega}_{-k}))^{\mathrm{H}} \end{matrix} \right)^{-1} \times \\ \boldsymbol{a}'(f_i, \alpha_k)(\boldsymbol{a}'(f_i, \alpha_k))^{\mathrm{H}} \right) \right) \right] \\ \times \sum_{i=1}^{J} \left[\left(\begin{matrix} \hat{\mu}^2(f_i)\mathbf{I}_M + \mathbf{A}'(f_i, \mathbf{\Omega}_{-k}) \times \\ \hat{\Sigma}_{-k}(f_i)(\mathbf{A}'(f_i, \mathbf{\Omega}_{-k}))^{\mathrm{H}} \end{matrix} \right)^{-1} \times \frac{\partial \boldsymbol{a}'(f_i, \alpha_k)}{\partial \alpha_k} \right] \end{matrix} \right|^{-1} \right\} \tag{31}$$

Thus, the DOA can be estimated.

We will obtain $\rho_2(f_i)e^{j\varphi_2(f_i)}, \cdots, \rho_M(f_i)e^{j\varphi_M(f_i)}$ according to $\hat{w}(f_i)$, thus $\mathbf{W}(f_i)$ can be calculated by (11) and (12), then $\boldsymbol{a}'(f_i, \alpha_k)$ and $\mathbf{A}'(f_i, \mathbf{\Omega}_{-k})$ can be acquired, we will get the accurate estimation based on (31) and the parameters above.

The method is used for wideband signal, and has employed spatial domain sparse optimization for gain and phase errors, so we can call it WSGP for short.

4 Simulations

Here, some simulations are presented for the method, consider some wideband chirp signals impinge on a uniform linear array with 8 omnidirectional sensors from $(16°, 28°, 35°)$, the center frequency of the signals is 2 GHz, width of the band is 20% of the center frequency, the band is divided into 10 frequencies, and spacing d between adjacent sensors is equal to half of the wavelength of the center frequency. Now we will simplify the generation process of the error, suppose the gain and phase uncertainties are respectively selected between $(0 \sim 2)$ and $(-45° \sim 45°)$ randomly.

Table 1. Gain and phase errors estimation

	Actual error at f_1	Estimated error at f_1	Actual error at f_2	Estimated error at f_2
gp_2	−0.185+j0.159	−0.246+j0.106	0.585+j0.079	0.527+j0.031
gp_3	0.189−j0.015	0.122−j0.073	0.635+j0.951	0.584+j1.104
gp_4	0.209+j0.197	0.270+j0.261	−0.451+j0.606	−0.513+j0.661
gp_5	0.665−j0.681	0.719−j0.617	0.855+j0.376	0.798+j0.422
gp_6	0.823+j0.267	0.761+j0.195	0.377+j0.167	0.329+j0.224
gp_7	0.506+j0.343	0.562+j0.402	−0.559−j0.403	−0.615−j0.358
gp_8	0.637+j0.203	0.795+j0.151	0.351+0.195	0.406+j0.249
	Actual error at f_3	Estimated error at f_3	Actual error at f_4	Estimated error at f_4
gp_2	0.529+j0.357	0.486+j0.322	−0.805+j0.227	−0.772+j0.205
gp_3	0.833+j0.228	0.797+j0.262	0.517+j0.201	0.488+j0.183
gp_4	−0.918+j0.219	−0.884+j0.265	0.252+j0.192	0.226+j0.221
gp_5	−0.663−j0.135	−0.616−j0.102	−0.478−j0.220	−0.501−j0.256
gp_6	0.388+j0.276	0.425+j0.238	0.804+j0.184	0.768+j0.204
gp_7	−0.489+j0.535	−0.443+j0.496	−0.566+j0.380	−0.542+j0.346
gp_8	0.742+j0.048	0.708+j0.005	0.309+j0.148	0.279+j0.117
	Actual error at f_5	Estimated error at f_5	Actual error at f_6	Estimated error at f_6
gp_2	0.391+j0.742	0.372+j0.763	0.703+j0.443	0.726+j0.429
gp_3	−0.836−j0.568	−0.821−j0.589	0.492+j0.562	0.478+j0.539
gp_4	0.185+j0.477	0.206+j0.455	0.678+j0.360	0.658+j0.345
gp_5	−0.516−j0.344	−0.541−j0.357	0.291+j0.124	0.267+j0.143
gp_6	−0.348−j0.342	−0.352−j0.291	0.599+j0.549	0.623+j0.537
gp_7	0.571+j0.464	0.543+j0.447	0.410+j0.166	0.392+j0.183
gp_8	0.293+j0.255	0.274+j0.273	0.231+j0.197	0.219+j0.215
	Actual error at f_7	Estimated error at f_7	Actual error at f_8	Estimated error at f_8
gp_2	0.331+j0.290	0.348+j0.317	0.663+j0.991	0.625+j0.959
gp_3	0.669+j0.212	0.643+j0.186	0.888+j0.612	0.846+j0.570
gp_4	−0.578+j0.619	−0.605+j0.645	0.292+j0.157	0.321+j0.206
gp_5	0.243+j0.517	0.272+j0.546	−0.701+j0.447	−0.660+j0.409
gp_6	0.490+j0.318	0.518+j0.353	0.479+0.686	0.426+j0.649
gp_7	0.547+j0.202	0.512+j0.233	−0.147+j0.413	−0.186+j0.372
gp_8	0.479+j0.114	0.511+j0.148	0.958+j0.391	0.906+j0.369
	Actual error at f_9	Estimated error at f_9	Actual error at f_{10}	Estimated error at f_{10}
gp_2	−0.421+j0.879	−0.372+j0.937	0.471-j0.763	0.525-j0.707
gp_3	0.597+j0.430	0.548+j0.382	0.751+j0.116	0.697+j0.053
gp_4	0.554+j0.169	0.605+j0.123	−0.585-j0.224	−0.642-j0.166
gp_5	0.667+j0.297	0.624+j0.232	−0.195-j0.524	−0.266-j0.463
gp_6	−0.234+j0.212	−0.285+j0.266	0.461+j0.288	0.409+j0.343
gp_7	0.716+0.330	0.658+0.287	0.352+j0.662	0.427+j0.703
gp_8	−0.502-j0.249	−0.456-j0.305	0.727+j0.165	0.668+j0.225

4.1 Gain and Phase Errors Estimation

Suppose SNR is 10dB, the number of samples at every frequency is 30, WSGP is employed for estimating gain and phase errors, 200 Monte-Carlo simulations are repeated, their average is deemed as the final results, the estimation errors of every frequency are shown in Table 1.

Table 1 shows the method can effectively estimate the gain and phase errors existing in the array, especially when the frequency is near to the center bin, we can use these results to calibrate the array and obtain the DOA.

4.2 DOA Estimation

First, traditional two-sided correlation transformation (TCT) [17] and WSGP methods are employed for estimating DOA of wideband signals along with the gain and phase errors above, here, TCT is performed without correction, the estimation error of DOA is defined as $\sum_{k=1}^{K} |\alpha_k - \hat{\alpha}_k|$. 200 Monte-Carlo simulations are repeated, their average values are deemed as the final results. Suppose samples of every frequency is 30, other conditions are the same with 4.1, estimation error versus SNR are shown in Fig. 2; then suppose SNR is 10dB, that versus number of samples are shown in Fig. 3.

Figures 2 and 3 show that WSGP can effectively estimate the DOA of wideband signals along with the gain and phase errors, the estimation error approximately converges to 0.7° at last, but that of the traditional TCT method without correction converges to 1.6° under the same condition.

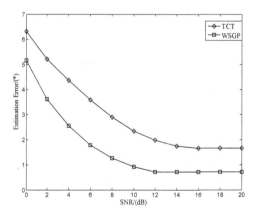

Fig. 2. Calibration Accuracy versus SNR

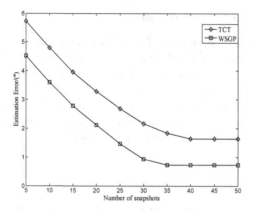

Fig. 3. Calibration Accuracy versus number of snapshots

5 Conclusion

The paper proposed a novel array calibration method in super-resolution direction finding for wideband signals based on spatial domain sparse optimization to the gain and phase errors existing in the array, it can calibrate the array and estimate the DOA relatively accurately.

Acknowledgments. I would like to thank Heilongjiang province ordinary college electronic engineering laboratory and post doctoral mobile stations of Heilongjiang University.

References

1. Muhammad, M.R., Iftekharuddin, K.M., Ernest, M.: Autonomous wireless radar sensor mote for target material classification. J. Digit. Sig. Process. **23**, 722–735 (2013)
2. Soh, P.J., Vanden, B.B., Xu, H.T.: A smart wearable textile array system for biomedical telemetry applications. J. IEEE Trans. Microw. Theory Tech. **61**, 2253–2261 (2013)
3. William, R.O., Aniruddha, G., Schmidt, D.C.: Efficient and deterministic application deployment in component-based enterprise distributed real-time and embedded systems. J. Inform. Softw. Technol. **55**, 475–488 (2013)
4. Mehmet, B., Guldogan, O.A.: A new technique for direction of arrival estimation for ionospheric multipath channels. J. Adv. Space Res. **44**, 653–662 (2009)
5. Muhammad, S., Ghulam, A.H.: Partial discharge diagnostic system for smart distribution networks using directionally calibrated induction sensors. J. Electr. Power Syst. Res. **119**, 447–461 (2015)
6. Li, J., Zhao, Y.J., Li, D.H.: Accurate single-observer passive coherent location estimation based on TDOA and DOA. J. Chin. J. Aeronaut. **27**, 913–923 (2014)
7. Giuseppe, F., Andrew, H.: A multipath-driven approach to HF geolocation. J. Sig. Process. **93**, 3487–3503 (2013)
8. Luis, S., Luis, M., Jose, A.G.: SmartSantander: IoT experimentation over a smart city testbed. J. Comput. Netw. **61**, 217–238 (2014)

9. Verdouw, C.N., Beulens, A.J.M., van der Vorst, J.G.A.J.: Virtualisation of floricultural supply chains: a review from an Internet of things perspective. J. Comput. Electr. Agric. **99**, 160–175 (2013)

10. Srinath, H., Reddy, V.U.: Analysis of MUSIC algorithm with sensor gain and phase perturbations. J. Sig. Process. **23**, 245–256 (1991)

11. Schmidt, R.O.: Multiple emitter location and signal parameter estimation. J IEEE Trans. Antennas Propag. **34**, 276–280 (1986)

12. Wang, T.D., Liao, H.S., Li, L.P.: Fast DOA estimation with ULA in the presence of sensor gain and phase errors. In: 2009 International Conference on Communications, Circuits and Systems, pp. 395–397. Milpitas (2009)

13. Jiang, J.J., Duan, F.J., Chen, J.: Two new estimation algorithms for sensor gain and phase errors based on different data models. IEEE Sens. J. **13**, 1921–1930 (2012)

14. Xu, L., Zeng, C., Liao, G.S.: DOA estimation for strong and weak signals in the presence of array gain and phase mismatch. In: 2010 International Conference on Multimedia Technology, pp. 1–4. Ningbo (2010)

15. Cao, S.H., Ye, Z.F., Hu, N., Xu, X.: DOA estimation based on fourth-order cumulants in the presence of sensor gain-phase errors. J. Sig. Process. **93**, 2581–2585 (2013)

16. Dempster, A.P., Laird, N.M., Rubin, D.B.: Maximum likelihood from incomplete data via the EM algorithm. J. R. Stat. Soc. **39**, 1–38 (1977)

17. Valaee, S., Kabal, P.: Wideband array processing using a two-sided correlation transformation. J. IEEE Trans. Sig. Process. **43**, 160–172 (1995)

Invited Paper

A Research on Underwater Acoustic Channel Modeling and Simulation of Shallow Sea

Bo Li, Hong-juan Yang$^{(\boxtimes)}$, Gong-liang Liu, and Xi-yuan Peng

School of Information and Electrical Engineering,
Harbin Institute of Technology (Weihai),
Weihai 264209, People's Republic of China
{libol983,hjyang,liugl,pxy}@hit.edu.cn

Abstract. In order to research the shallow underwater communication system further and conveniently, the paper study the characteristics of underwater acoustic propagation and choose "Ray Model" as the simulation. Then we describe this model from the point of geometry and analyze some parameters of shallow underwater acoustic communication. According to these factors, the received signal delay and loss can be calculated and the received signal can also be expressed. This paper uses the technique of adaptive equalization to solve the serious Inter symbol interference (ISI) from the multi-path signals by giving principles and formulas of the least mean square (LMS) error algorithm. Finally, the shallow underwater acoustic channel model can be simulated by MATLAB in 2ASK modulation. In the results of simulation, the severe ISI can be observed in the received signal and can be eliminated by equalization. We also get the bit error rate curve successfully. This simulation can provide the foundation for other underwater acoustic communication researches and works.

Keywords: Shallow underwater acoustic channels · Channel modeling · Ray model · Adaptive equalization

1 Introduction

With the continuous development of science and technology, together with the increase of world's population, scientific researchers in the major research institutions and universities are increasingly concerned about the ocean, known as a valuable resource to be explored. To realize marine survey, resource exploitation and utilization is one of the most concerned problems in coastal countries. By constructing shallow underwater communication system, we can monitor and gather military intelligence, along with detect conditions of ports and coast to realize group management. In addition, the construction can also provide a large amount of data for the marine environment, such as the sea temperature, salinity, and so on. In conclusion, the high-speed reliable shallow underwater acoustic communication technology has become a heated research direction in the field of underwater acoustic communication. However, the underwater acoustic channel is an extremely complex channel.

Because the electromagnetic wave attenuation is serious in seawater, low frequency acoustic wave is the most widespread shallow sea communication technology.

© ICST Institute for Computer Sciences, Social Informatics and Telecommunications Engineering 2017
X.-L. Huang (Ed.): MLICOM 2016, LNICST 183, pp. 317–327, 2017.
DOI: 10.1007/978-3-319-52730-7_32

However, this has caused the limitation of bandwidth of the shallow water communication. Moreover, the low propagation velocity of the 1500 m/s and the multipath effects caused by the shallow water propagation are inevitable. Worse physical properties are imposed on shallow water acoustic channels, due to the Doppler frequency shift generated by transceiver movement together with the random Doppler frequency shift generated by the sea movement. In order to facilitate the further study of the shallow seawater communication system, the modeling of the underwater communication channel is required [1]. Since the study of underwater acoustic communication, several underwater sound propagation model were promoted through different approximation and ameliorate method, which are ray model, normal mode model, parabolic equation model, fast field model and so on.

Description of the sound ray model is by the sound line to transmit energy, from the sound source of the sound line to follow a certain path to receiver, the received acoustic energy is all reach the superposition of voice. The ray model can describe the transmission path between the transmitter and the receiver, and gives the propagation loss and the time of each path.

Therefore, when analyzing and simulating the propagation characteristics of the multipath channel, the ray model is a common tool for analysis. In this paper, the basic characteristics of acoustic wave propagation in shallow water environment are studied. Based on the wave propagation, the most commonly used simulation model of underwater acoustic channel ray model is applied. In the end, the reasonable equalization technique is adopted to eliminate the multipath interference, and realize the transmission of the shallow water acoustic signal.

2   The Analysis of Underwater Acoustic Propagation Characters

This article will analyze the characters of underwater acoustic propagation from the following aspects:

(1) The propagation velocity of the sound waves in the sea: The propagation velocity of the sound waves has a great effect on the time we need to get the information from the receivers, which may influence the signal we receive. Therefore, this coefficient is of great importance. The average velocity of the sound in the sea is 1500 m/s. In the acoustic researches, conclusions have been drawn that the propagation velocity can be affected by the temperature, salinity and static pressure. Usually we use empirical formulas to demonstrate the relationship of the coefficients because of the complicated influence of environment. If accuracy is not required to be very high, Ude empirical formula can be used:

$$c = 1450 + 4.21T - 0.037T^2 + 1.14(S - 35) + 0.175P \qquad (1)$$

where T denotes temperature; S is salinity, P is static pressure.

(2) Sea surface: The sound waves can be reflected and scattered on the surface of sea. The wind and sea waves make the reflection various. Because of the regularity and the randomness of the waves on the surface, the researches of the waves' statistical characters will help us to analyze the surface's influence on the propagation. The reflection coefficients can be calculated in this way. The Bechmann-Spezzichino model, put forward by Coates, makes it possible to compute the reflection coefficients by the velocity of the wind and other variants. In addition, this model can be used perfectly in shallow sea underwater acoustic channel. Concrete details about calculation will be discussed in Sect. 3.2.

(3) Seabed: Different topographic texture and diverse sediments can influence the reflection coefficients and reflection loss. It affects the distance of the propagation in the sea. As the research show, the reflection coefficients are highly related to the seabed topographic texture. When it comes to shallow sea, the smooth silt and sand on the bottom provide little reflection loss. Combined with the other factors of the shallow sea, the grazing angle ψ is small on the reflecting surface. Based on the NUSC model, when ψ is smaller than $5°$, the frequency of sound signal is less than 50 kHz and the osmotic coefficient is smaller than 0.5 (fine sands, silts satisfy the condition), the reflection coefficient is close to 1, which means no loss.

(4) Noise: The noise of the sea environment has a great effect on the propagation of the signal. Severe noise may disturb the receivers' demodulation and decision. Usually, we take these sea noises into consideration: the noise from winds and waves, live beings, raindrops and environmental thermal noise. However, these factors are time-variable and also space-variable. Hence, we can only attain rough spectrum of the shallow sea noises. So the noises can be regarded as White Gaussian Noise when having theoretical analysis.

(5) Doppler shift: There are two kinds of Doppler shift. One is based on the relative motion of the receiver. Assumed the velocity of the transmitter is v_s, the velocity of the receiver is v_r, the frequency of the signal is f_s, the acoustic speed is c, and then the frequency of received signal under the effect of Doppler shift is $f_r = f_s(c - v_r)/(c - v_s)$.

Due to the velocity of the sound is far from that of the electromagnetic wave, and the frequency of the sound is too low, the Doppler shift can be severe.

Another situation is that the receiver does not have relative motion, and the Doppler Shift is caused by the waves on the surface and the current. This kind of Doppler shift may cause frequency expansion, which may lead to the time selective fading of the shallow water acoustic channels.

Doppler shift will cause great influence on the correct demodulation of the received signal. Hence, Doppler compensation is needed. If only frequency shift is considered, the compensation of the shift can be compensated easily. However, if frequency expansion occurs and it is frequency modulation, the interspace of the frequencies is required. When it comes to theoretical analysis, the Doppler frequency shift can be regarded as a compensated factor, and has no effect on the received signal.

3   Model Selection for Underwater Acoustic Channel

Received data from underwater sensor shows that the received signal of the underwater acoustic channel is a multipath signal, which is the superposition of the reflection of transmitted signal on the sea surface and seabed. Because of the small grazing angle and small boundary reflection loss caused by the shallow water environment, the received signal is a multipath signal. This has led to the amplitude of a number of non-diameter signals at the receiver are no less than the direct path, thus these signals cannot be ignored. In order to be able to consider the influence of environmental parameters and reflect multipath, "Ray Model" is used to describe and analyze the underwater acoustic channel. The impulse response of the channel can be attained based on the reflection and attenuation of multipath signals, which is convenient for the subsequent study of the received signal.

3.1   Description of the Ray Model

This model mainly analyzes more in the aspects of mathematics and physics, which is described by the method of geometry. In this way, it is more easily to reflect the propagation of rays, at the same time, more helpful for the further calculation and understanding of the parameters of the model. The model diagram is shown in Fig. 1.

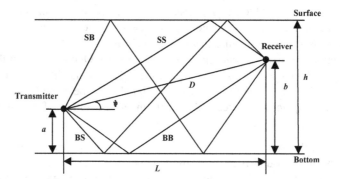

Fig. 1. Underwater ray model of shallow sea

In Fig. 1, a denotes the distance between transmitter and the bottom of the sea; b is the distance between receiver and the bottom of the sea; L stands for the horizontal distance between the transmitter and receiver; ψ is glancing angle (grazing angle).

The propagation paths can be divided into direct path D and multipath. Multipath signals can be divided into four categories according to the type of reflection: symbol SS denotes the signals which before reaching the receiver the first and last reflection is happened on the sea surface. Index is used to indicate the sequence of the signal. SB, BS, and BB are in the same way. These four types of signals are represented in Fig. 1 when $n = 1$. In order to facilitate the calculation, we assume that A < B.

In the shallow seawater acoustic environment ($h \ll L$), grazing angle ψ is very small. When the grazing angle ψ is less than the "boundary grazing angle (total internal reflection angle)", the boundary reflection can be regarded as specular reflection. This is a very important condition for the calculation of the model.

Each path propagation length can be calculated as $D \simeq L + (b - a)^2 / 2L$,

$$SS_n \simeq L + (2nh - a - b)^2 / 2L, \qquad\qquad SB_n \simeq L + (2nh - a + b)^2 / 2L,$$

$$BS_n \simeq L + (2nh + a - b)^2 / 2L, \text{ and } BB_n \simeq L + [2(n-1)h + a + b]^2 / 2L.$$

3.2 The Signal Loss

(1) Extension loss (spread loss): The extension loss is due to the amplitude (energy) attenuation caused by the propagation of sound waves in the sea. Generally speaking, the propagation loss caused by the extension can be indicated as $TL = n * 10 \lg r (\mathrm{dB})$.

According to different propagation conditions, n may take different values: $n = 0$: plane wave propagation, no expansion loss; $n = 1$: cylindrical wave propagation, the wave front is expanded according to the law of the cylinder, which is equivalent to the propagation condition of ideal waveguide, which is consisted of the total reflection sea surface and total reflection sea bottom; $n = 2$: spherical wave propagation, the wave front is extended by spherical surface.

In this study, we assume the acoustic wave is transmitted from the transmitter and expand in the sea in the form of spherical wave expansion, that is $n = 2$. Propagation loss (energy loss) is proportional to the square of the propagation length. Along the propagation direction of harmonic radius r of spherical wave, the pressure can be expressed as $p = \frac{p_0}{r} \exp[-i(wt - kr)]$, in which p_0/r is the sound pressure amplitude of the spherical wave, which is inversely proportional to the distance r.

In conclusion, as long as the propagation length of each path is known, the extension loss of this path can be expressed by the reciprocal of the length of the transmission.

(2) Absorption attenuation: The absorption attenuation is the sound intensity decrease caused by the absorption surface, which is related to the distance. Generally, α is used as a representation of the absorption coefficient, which unit is dB/m. In seawater, absorption attenuation and the loss caused by scattering presents at the same time, and it is difficult to separate them when measuring. However, when considering the absorption of homogeneous medium, it is found that the value of absorption coefficient of a majority of the liquid will be far greater than its theoretical value. We call this difference super absorption. As a result, we cannot use replace absorption attenuation with the theoretical value.

The structure relaxation theory proposed by Hall has explained the super absorption: in seawater, the propagation of acoustic waves is dissolved in the $MgSO_4$ of salt water. Because of its low solubility, the equilibrium of the original dissolution is destroyed, and a new equilibrium is reached. This process absorbs the energy of sound waves.

After a large amount of data testing, the empirical formula of the absorption coefficient [2] is $\alpha(f) = \frac{0.102f^2}{1+f^2} + \frac{40.7f^2}{4100+f^2}$ (dB/km).

(3) Reflection loss: Sea Surface: The reflection coefficient is dependent on wind speed. The empirical formula [3] has been given in the Bechmann-Spezzichino model. The application condition of the model is also a low grazing angle of the shallow sea underwater acoustic channel, thus this formula can be applied to this study. Empirical formula is $|r_s| = \sqrt{\left[1 + (f/f_1)^2\right] / \left[1 + (f/f_2)^2\right]}$, where $f_2 = 378\omega^{-2}$; $f_1 = \sqrt{10}f_2$.

In the above formula, the working frequency of the acoustic signal is f, and its unit is kHz; wind speed is expressed as ω, and its unit is knots. 1knots = 1.852 km/h = 0.514 m/s.

Sea Bottom: In order to simplify the model parameters, in the following calculation and simulation, it is assumed that the seabed reflection coefficient $|r_b| = 0.9$.

Considering each of the reflection of the sea surface and sea bottom is a specular reflection, as a result a 180° phase shift will be produced, so $r_s = -|r_s|$; $r_b = -|r_b|$. On the basis of the multipath classification, the reflection coefficient of each multipath signal in the sea surface and the sea bottom is different, and the reflection coefficient of each path can be obtained: $R_{SS_n} = r_s^n r_b^{n-1} = -r_s^n * 0.9^{n-1}$, $R_{SB_n} = r_s^n r_b^n = r_s^n * 0.9^n$, $R_{BS_n} = r_s^n r_b^n = r_s^n * 0.9^n$, where $n = 1, \ldots, \infty$.

3.3   The Signal Propagation Time Delay

Signal time delay is another important parameter in the received signal impulse response. According to the previous analysis, the condition of specular reflection can help to calculate the propagation length of each path. The propagation sound velocity in seawater can be regarded as constant c = 1500 m/s, so the propagation delay of each path can be expressed as: $t_D = D/c$, $t_{SS_n} = SS_n/c$, $t_{BS_n} = BS_n/c$, $t_{SB_n} = SB_n/c$, $t_{BB_n} = BB_n/c$.

3.4   The Joint Response of Received Signal

The multipath signals received by each path can be expressed by the sum of the sum of each path:

$$r(t) = \alpha \frac{e^{jw(t-t_D)}}{D} + \alpha \sum_{n=1}^{\infty} \left[\frac{R_{SS_n}}{SS_n} e^{jw(t-t_{SS_n})} + \frac{R_{SB_n}}{SB_n} e^{jw(t-t_{SB_n})} + \frac{R_{BS_n}}{BS_n} e^{jw(t-t_{BS_n})} + \frac{R_{BB_n}}{BB_n} e^{jw(t-t_{BB_n})} \right]$$

$$(2)$$

Because there are numerous multipath signals may be represented in this way, in this study, in order to limit the number of multipath, the multipath signal which amplitude is less than 1% of the signal of the direct path will be omitted. In this way, finite multipath signals are used to represent the received signal of a shallow sea underwater acoustic channel.

4 Underwater Acoustic Channels Equalization of Shallow Sea

4.1 The Work Mode and Structure of Adaptive Equalizer

Adaptive equalizer has the following two work modes [4–6]:

(1) Training Mode: The transmitter sends a fixed length set of training sequences (the typical training sequence is a binary pseudo-random sequence or a set of advance specified data). The equalizer makes compensation for the channel through recursive algorithm, which evaluates the channel characteristics, and constantly revises the filter coefficients. The filter coefficients tend to the optimal value after continuous recursive iterative process [7, 8]. (Recursive algorithm will be described in detail later.)

(2) Tracking Mode: After the most optimal filter coefficients are known, in order to eliminate ISI, we equalize the follow-up received signal.

According to the previous analysis, the environmental parameters of the underwater acoustic channel change slowly, that is, the channel is slowly time varying. So the channel characteristics can be regarded as the same for a time. Therefore, assume in this period, after continuous recursive iteration, the optimal filter coefficients of adaptive equalizer in the training mode can effectively eliminate inter symbol interference of multipath signal. For the next period, the adaptive equalizer will enter training mode again, which ensures the time varying characteristics of adaptive equalizer [9, 10].

The structure of the adaptive filter is shown in Fig. 2.

Figure 2 is a simple form of an adaptive filter: a linear transversal equalizer, in which there are a total of N tap coefficients. $y(n) = \sum_{i=0}^{N-1} w_i(n)x(n-i)$ is the equalized signal, which is obtained by the linear superposition of the current signal and the delay of the signal.

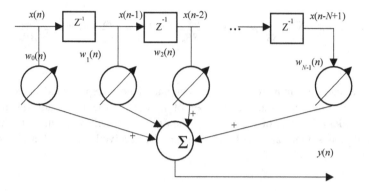

Fig. 2. The structure of adaptive filter

4.2 The Objective Function and Method of Iterative Recursive Adaptive Equalizer

(1) Objective Function of the Iteration: Determining the objective function iteration is to determine when the tap coefficient is best optimal. In this way, the tap coefficient can be used to equalize signal and ensure the output signal without inter-symbol interference.

Assuming that the expected signal (known training sequence) is $d(n)$, the error output sequence is $e(n) = d(n)-y(n)$. In the training mode, the tap coefficient $w(n)$ can be adjusted according to some algorithm, so that the cost of adaptive equalizer can be minimized. In this study, the least mean square error (LMSE) is adopted, which is the least mean square error between $d(n)$ and $y(n)$. This method is also referred to as the least mean square (LMS) algorithm.

(2) Iterative Method: The most widely used convergence principle, "steepest descent method" or said "gradient descent method" is adopted in this study to realize the iterative update of equalizer tap coefficients. Tap coefficients of next moment w $(n + 1)$ equal tap coefficients of this moment $w(n)$ add a negative square error gradient-$\nabla(n)$. Equation is expressed as $w(n + 1) = w(n)-\mu\nabla(n)$, in which μ is convergence factor for controlling the convergence rate.

Because the accurate calculation of the gradient is very complicated and difficult, the estimation of mean square error is used to approximate calculation, $\nabla(n) = -2e(n)x(n)$. Then, the iterative formula $w(n + 1) = w(n) + 2\mu e(n)x(n)$.

5 Simulation and Analysis

When simulating the system, the presumed channel and environmental parameters are: the depth of $h = 100$ m; horizontal distance between the transmitter and receiver $L = 1000$ m; distance between the transmitter and sea bottom $a = 20$ m; distance between the receiver and sea bed $b = 80$ m; working frequency of sound waves

f = 8 kHz; the speed of sound waves in the sea water c = 1500 m/s; the speed of wind on the surface of seawater speed_wind = 11.67 knots (6 m/s); sea bed reflection coefficient $|r_b|$ = 0.9; number of sent bit is 48 (modulation signal duration is 6 ms); modulation mode is 2ASK modulation; the channel signal-to-noise ratio SNR = 10; the training sequences are 8 pairs continuous 0,1 signals (the duration is 2 ms).

2ASK modulation is adopted. After bit 1 modulations the amplitude is 1, while after bit 0 modulations the amplitude is 0. In Fig. 3, the picture above is the entire received multipath signal, and the former blank is the time delay of the direct path from transmitter to receiver. The middle picture is the front part of the received signal. ISI can be found only in the latter part. This is because of the time delays difference between the second arrived path and the direct path. Even though the multipath signal is attenuated in the amplitude compared with the direct path, there is more than one path. As a result sometimes the maximum value of the received signal is greater than the direct path. The following picture is the whole the spectrum of the received signal. You can see the frequency is 8 kHz. The simulation of these instructions is successful.

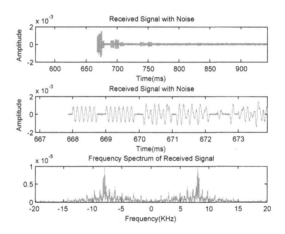

Fig. 3. The received multipath signal and its spectrum

In Fig. 4, the picture above is a modulated signal, and the following is an equalized signal of the received multipath signal. It can be discovered that the signal has a good recovery after the equalization. This is because the SNR of hypothetical channel is relatively high. As a result, the impact of noise on signal transmission is small.

According to Fig. 5, it can be seen that the error rate of the information transmission decreases with the increase of the channel SNR. There is no bit error in the channel of relatively high SNR. In this curve, I find that the bit error rate of the channel is relatively low when SNR is −1, for the equalizer can largely reduce the negative effect of the noise on the transmission.

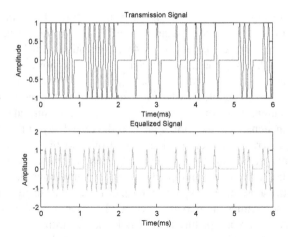

Fig. 4. The comparison between equalized signal and original signal

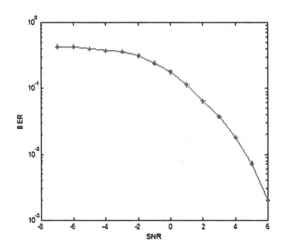

Fig. 5. The error rate curve of shallow seawater channel

6 Conclusions

(1) The characteristics of underwater acoustic communication are introduced, and the parameters of the underwater acoustic channel are analyzed from the perspective of physics. Then the "Ray Model" is chosen to simulate the channel, and the model is analyzed. After that, the model is described from the geometric point of view, and some parameters of the model are obtained. Then the delay and loss of the received signal are calculated. There are three kinds of loss, which is the extension attenuation, the absorption attenuation, and the attenuation. Finally, the expression of the received signal is given, which provides the basis for the follow-up work.

(2) In order to solve the inter symbol interference (ISI) caused by multipath effect of received signal, the adaptive equalization technique is adopted in this paper. After analyzing the reason of using the adaptive equalization technique, the working mode and structure of the adaptive equalizer is presented in this paper, and the principle and formula of iterative recursion is analyzed.

(3) MATLAB simulation of the underwater acoustic channel model is carried out using 2ASK modulation. In the simulation results, the severe Inter symbol interference can be observed, which indicates that the simulation of the channel can reflect the characteristics of the underwater acoustic channel. In the balanced simulation, the signal of the equalizer is almost the same as that before entering the channel, and the error rate curve is obtained, which indicates the equalizer has the ability to solve the severe Inter symbol interference.

Acknowledgments. This work is partly supported by National Natural Science Foundation of China (No. 61401118 and No. 61371100), Natural Science Foundation of Shandong Province (No. ZR2014FP016 and BS2012DX001), the Fundamental Research Funds for the Central Universities (No. HIT.NSRIF.2016100 and 201720), Subject Guide Foundation (No. WH20150109), and the Scientific Research Foundation of Harbin Institute of Technology at Weihai (No. HIT(WH)201409 and No. HIT(WH)201410).

References

1. Zielinski, A., Young-Hoon, Y., Lixue, W.: Performance analysis of digital acoustic communication in a shallow water channel. IEEE J. Ocean. Eng. **20**(4), 293–299 (1995)
2. Boshegn, L., Jiayu, L.: The principle of underwater acoustic, pp. 69–72. Harbin Eng. Univ. Press, Harbin (2010)
3. Coates, R.: An empirical formula for computing the beckmann-spizzichino surface reflection loss coefficient. IEEE Trans. Ultrason. Ferroelectr. Freq. Control **50**(4), 522–523 (1988)
4. Zhengkui, J., Teng, S.: Research on adaptive equalization algorithm of variable step size. Silicon Alley (2012)
5. Xiao-ling, N., Chen-liang, S., Zhong, L., Chen-liang, S.: Fast convergence adaptive equalization algorithm for underwater acoustic channels. Syst. Eng. Electron. **32**(12), 2524–2527 (2010)
6. Bei, Z.: Research on adaptive equalization technique in time varying multipath channel. Xidian University (2013)
7. Feng, T., Qin, P.: Experimental studies on shallow water acoustic channel equalization. J. Xiamen Univ. (Nat. Sci.) **50**(4), 724–728 (2014)
8. Xiang, C.: Research of adaptive equalizer system based on LMS algorithm. China New Telecommun. **5**, 60–63 (2010)
9. Mingyuan, X., Gongan, Q., Huafang, L.: Simulink experiment research on adaptive equalization system based on LMS algorithm. J. Syst. Simul. **15**(2), 176–178 (2003)
10. Peng, H., Cheng, L., Qin, S.: Analysis on adaptive equalization performance of LMS algorithm. Commun. Technol. **42**(11), 61–62 (2009)

Indoor WLAN Deployment Optimization Based on Error Bound of Neighbor Matching

Feng Qiu$^{(\boxtimes)}$, Mu Zhou, Zengshan Tian, Yunxia Tang, and Qiao Zhang

Chongqing Key Lab of Mobile Communications Technology,
Chongqing University of Posts and Telecommunications, Chongqing 400065, China
`qiufeng245@outlook.com`, {`zhoumu,tianzs`}`@cqupt.edu.cn`,
`13629735505@139.com`, `18716322725@139.com`

Abstract. In this paper, we propose a novel indoor Wireless Local Area
Network (WLAN) deployment optimization approach based on the error
bounds of Neighbor Matching Algorithms (NMAs). We derive out the
closed-form solution to the localization errors of NMAs with respect to the
environmental size, interval of Reference Points (RPs), number of neighbors, and locations of Access Points (APs). Based on the requirement of
localization precision, as well as networking overhead, we optimize the
networking parameters, like the interval of RPs, number of neighbors,
and locations of APs. Finally, the extensive experiments are conducted
to demonstrate that the proposed approach can effectively improve the
localization precision of NMAs in indoor WLAN environment.

Keywords: WLAN · Network optimization · Location fingerprinting ·
Neighbor matching · Error bound

1 Introduction

As the demand for the real-time location information increases remarkably, the
Location-based Services (LBSs) have attracted significant attention in recent
decade. The accurate localization in outdoor environment can be realized by
using the well-known Global Positioning System (GPS), whereas the localization accuracy decreases seriously in indoor environment since the signal from
the satellites is blocked by the buildings [1]. At the same time, there is growing interest in the indoor localization techniques which are based on the existed
indoor high-speed wireless access networks, like the Wireless Local Area Network (WLAN) [2], Zigbee, and Radio Frequency Identification (RFID). Due to
the consideration of the cost overhead and localization accuracy, the WLAN
technique is more favored by the current indoor localization systems.

Compared to the conventional trilateration based localization approach, the
location fingerprint based localization approach is preferred in WLAN localization. In the typical location fingerprint based localization system [3,4], the grids
of Reference Points (RPs) are first required to be calibrated. Second, the location
fingerprints which are typically the vectors of Received Signal Strength (RSS)

© ICST Institute for Computer Sciences, Social Informatics and Telecommunications Engineering 2017
X.-L. Huang (Ed.): MLICOM 2016, LNICST 183, pp. 328–335, 2017.
DOI: 10.1007/978-3-319-52730-7_33

mean from each hearable Access Point (AP) is collected at every RP. The set of location fingerprints is recognized as the radio map. Finally, when a location query occurs, the estimated location can be reported by matching the newly collected RSSs against the radio map.

Up to now, there is a batch of studies focusing on the design of localization algorithms, like the Nearest Neighbor (NN), K-nearest Neighbor (KNN) [5], weighted KNN (WKNN) [6], which are also known as the Neighbor Matching Algorithms (NMAs). NMAs are easily applied and featured with low computation overhead, practicability, and high-precision [7]. The KNN returns the location estimate as the average of the coordinates of the K neighbors corresponding to the smallest RSS distances to the newly collected RSSs. The NN is a special case of KNN as the number of neighbors equals to 1. The difference between the KNN and WKNN is that the latter one returns the location estimate as the weighted coordinates of the K neighbors, while the weights of neighbors are determined by the distances between the location fingerprints and newly collected RSSs. Since the NMAs are easily applied and featured with low computation overhead, practicability, and high-precision, we focus on deriving the error bounds of NMAs to investigate the theoretical relation between the localization error and networking parameters.

The remainder of the paper is organized as follows. The theoretical analysis for the error bound of NMAs is presented in Sect. 2. The analytical results are provided in Sect. 3. Finally, Sect. 4 concludes the paper.

2 Error Bound

In this paper, we focus on the analysis towards the theoretical relation between the localization errors of NMAs and networking parameters, and meanwhile derive out the closed-form solutions to the error bounds.

2.1 AP Located on the Boundary

Figure 1 shows a straight corridor with the Line-of-sight (LOS) from the AP. The N RPs (with ●'s) are uniformly calibrated with the same interval, R, in this environment (with the length of $N \times R$). The user location is described as

$$x = r_i + \sigma, 0 \leq \sigma \leq R \text{ and } 0 < i < N \tag{1}$$

We rely on the logarithmic loss model [8] to characterize the signal propagation property, as shown in (6).

$$P = P(d_0) - 10\beta\log_{10}(d/d_0) \tag{2}$$

where P and $P(d_0)$ are the RSSs collected at the locations with d and d_0 meters from the AP respectively; and β is the path loss exponent. On this basis, the distance of the RSSs collected by the user and at the n-th RP is calculated by

$$\Delta P_n = |S_n - S_u| == \begin{cases} 10\beta\log\left(n/(i + \sigma/R)\right), n \geq i + 1 \\ 10\beta\log\left((i + \sigma/R)/n\right), n \leq i \end{cases} \tag{3}$$

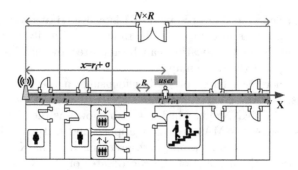

Fig. 1. AP located on the boundary.

KNN. When $K = 1$, only the i-th or $(i + 1)$-th RP can be selected as the estimated location since the location fingerprints at other RPs are farther away from the RSS collected by the user. Based on this, we derive out the results in Table 1.

Table 1. Results under $K = 1$ in KNN

Value of σ	$0 \leq \sigma \leq (-i + \sqrt{i^2 + i})R$	$(-i + \sqrt{i^2 + i})R < \sigma \leq R$
Relations of ΔP_n	$\Delta P_i \leq \Delta P_{i+1}$	$\Delta P_{i+1} < \Delta P_i$
Neighbors	The i-th RP	The $(i+1)$-th RP
Localization error	$er_1 = \|r_i - x\| = \sigma$	$er_2 = \|r_{i+1} - x\| = R - \sigma$
Error bound	$ER_1 = \sum\limits_{i=1}^{N} \int_0^{(-i+\sqrt{i^2+i})R} \sigma d\sigma + \int_{(-i+\sqrt{i^2+i})R}^{R} (R - \sigma)\, d\sigma$	

When $K = 2$, based on (3), we can easily obtain that the location fingerprints at the $(i-1)$-th, i-th, $(i+1)$-th, and $(i+2)$-th RPs are with the smallest distances from the RSS collected by the user. Thus, we derive out the results in Table 2.

Table 2. Results under $K = 2$ in KNN

Value of σ	$0 \leq \sigma \leq (-i + \sqrt{i^2 + 2i})R$	$(-i + \sqrt{i^2 + 2i})R < \sigma \leq R$
Relations of ΔP_n	$\Delta P_i \leq \Delta P_{i+1} < \Delta P_{i-1} < \Delta P_{i+2}$	$\Delta P_{i+1} \leq \Delta P_{i+2} < \Delta P_i < \Delta P_{i-1}$
Neighbors	The i-th, $(i+1)$-th RPs	The $(i+1)$-th, $(i+2)$-th RPs
Localization error	$er_1 = R/2$	$er_2 = (3R - \sigma)/2$
Error bound	$ER_2 = \sum\limits_{i=1}^{N} \int_0^{(-i+\sqrt{i^2+2i})R} \frac{R}{2} d\sigma + \int_{(-i+\sqrt{i^2+2i})R}^{R} \frac{3R-\sigma}{2} d\sigma$	

When $K = k$ (k is odd), by using the mathematical induction, we can derive out the error bound in (4). Table 3 illustrates the neighbors and the corresponding localization errors under different values of σ. In Table 3, we set $k_1 = (k - 1)/2$ and $k_2 = (k + 1)/2$ respectively.

$$ER_{\text{odd}\_(1,N)} = \sum_{i=1}^{N} \int er_1 d\sigma + \cdots + \int er_{m+1} d\sigma \cdots + \int er_{(k+3)/2} d\sigma \qquad (4)$$

where $er_{m+1} = 1/k(1 - 2m)\sigma + kmR + (k_2 - m)(k_1 - m)R$, $m \in \{0, \cdots, k_2\}$.

Table 3. Results under $K = k$ (K is odd) in KNN

Value of σ	Neighbors	errors
$0 \leq \sigma < (-i + \sqrt{(i-k_1)(i+k_2)})R$	The $(i\text{-}k_1)$-th, $\cdots$, $(i+k_2-1)$-th RPs	er_1
$\cdots$	$\cdots$	$\cdots$
$(-i + \sqrt{i(i+k)})R \leq \sigma \leq R$	The $(i+1)$-th,$\cdots$, $(i+k)$-th RPs	er_{k_2+1}

Similarly, when $K = k$ (k is even), we derive out the error bound in (9) based on the result of errors corresponding to different values of σ in Table 4. In Table 4, we set $k_3 = k/2 - 1$ and $k_4 = k/2 + 1$ respectively.

Table 4. Results under $K = k$ (K is even) in KNN

Value of σ	Neighbors	errors
$0 \leq \sigma < (-i + \sqrt{(i-k_3)(i+k_4)})R$	The $(i\text{-}k_3)$-th,$\cdots$, $(i+k_4-1)$-th RPs	er_1
$\cdots$	$\cdots$	$\cdots$
$(-i + \sqrt{(i-(k_3+1-n))(i+(k_4-1+n))})R \leq \sigma < (-i + \sqrt{(i-(k_3-n))(i+(k_4+n))})R$	The $(i\text{-}k_3+n)$-th,$\cdots$, $(i+k_4+n-1)$-th RPs	er_{n+1}
$\cdots$	$\cdots$	$\cdots$
$(-i + \sqrt{i(i+k)})R \leq \sigma \leq R$	The $i+1$-th,$\cdots$, $(i+k)$-th RPs	er_{k_4}

$$ER_{even\_(1,N)} = \sum_{i=1}^{N} \int er_1 d\sigma + \cdots + \int er_{n+1} d\sigma \cdots + \int er_{(k+2)/2} d\sigma \quad (5)$$

where $er_{n+1} = 1/k(-2n\sigma + n^2 + n + k^2/4)$, $n \in \{0, \cdots, k_4\}$.

2.2 WKNN

When $K = 1$, the WKNN becomes the KNN. When $K = 2$, based on the results in Table 2, we can easily derive out the error bound in Table 5.

Table 5. Results under $K = 2$ in WKNN

Value of σ	$0 \leq \sigma \leq (-i + \sqrt{i^2 + 2i})R$	$(-i + \sqrt{i^2 + 2i})R < \sigma \leq R$
Relations of ΔP_n	$\Delta P_i \leq \Delta P_{i+1} <$ $\Delta P_{i-1} < \Delta P_{i+2}$	$\Delta P_{i+1} \leq \Delta P_{i+2}$ $< \Delta P_i < \Delta P_{i-1}$
Neighbors	The i-th, $(i+1)$-th RPs	The $(i+1)$-th, $(i+2)$-th RPs
Localization error	$er_1 = w_1 r_i + w_2 r_{i+1} - x$	$er_2 = w_1 r_{i+1} + w_2 r_{i+2} - x$
	where $w_1 = \frac{1/(R-\sigma)}{1/(R-\sigma)+1/(2R-\sigma)}$ and $w_2 = \frac{1/(2R-\sigma)}{1/(R-\sigma)+1/(2R-\sigma)}$	
Error bound	$ER_2 = \sum_{i=1}^{N} (\int_0^{(-i+\sqrt{i^2+2i})R} er_1 d\sigma + \int_{(-i+\sqrt{i^2+2i})R}^{R} er_2 d\sigma)$	

Since the weight of each neighbor in WKNN cannot be formulated by a general expression, there is no closed-form solution to the error bound of WKNN.

However, the weights of neighbors can be easily calculated as the number of neighbors and networking parameters are determined. Hence, we only focus on the error bound of KNN in the results that follow.

2.3   AP Located at a Random Location

Figure 2 shows the layout of environment as the AP is located at a random location. In this situation, the user location is described as

$$x = r_i - mR + \sigma, m \leq N/2 \tag{6}$$

where mR is the distance between the AP and left boundary.

By assuming that the antenna of AP is omnidirectional, we assume that the RPs with the same distance from the AP have the same probability to be selected as the neighbors. On this basis, there are three cases to be discussed respectively, i.e., $0 \leq m < k_1$, $k_1 \leq m < k$, and $m \geq k$. When $K = k$ (k is odd), we derive out the error bounds with respect to different cases in Table 6.

Thus, as the number of neighbors is odd, the error bound of WKNN is calculated by

$$ER_{\text{odd}} = Er_1 + Er_2 + Er_3 \tag{7}$$

Similarly, When $K = k$ (k is even), we derive out the error bounds with respect to different cases in Table 7.

Thus, as the number of neighbors is even, the error bound of WKNN is calculated by

$$ER_{\text{even}} = Er_1 + Er_2 + Er_3 \tag{8}$$

Therefore, by assuming that the user locations obey the uniform distribution in the target environment, we can calculate the error bound in this situation as

$$ER_{\text{ave}} = \begin{cases} \frac{1}{N \cdot R} \cdot ER_{\text{odd}} & K = 2r - 1 \text{ and } r \in N^+ \\ \frac{1}{N \cdot R} \cdot ER_{\text{even}} & K = 2r \text{ and } r \in N^+ \end{cases} \tag{9}$$

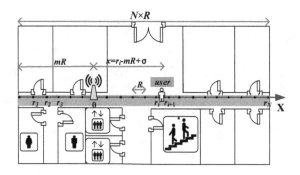

Fig. 2. AP located at a random location.

Table 6. Results under $K = K$ (K is odd) in WKNN

Case 1: $0 \leq m < k_1$
if $0 \leq i < k_1$ then $Er_1 = ER_{odd\_(1,k_1)} + ER_{odd\_(1,m)}$
if $k_1 \leq i < k_1 + m$ then $Er_2 = ER_{odd\_(k_2,k_1+m)} + \sum_{k_2}^{k_1+m} \frac{(k-m-2-2i)(m-i+1)R}{2}$
if $k_1 + m \leq i$ then $Er_3 = ER_{odd\_(k_1+m,N-m)}$
Case 2: $k_1 \leq m < k$
if $0 \leq i < k_1$ then $Er_1 = \sum_{i=1}^{k_1} \left[(k^2 - 1)/8 - i \right] R$
if $k_1 \leq i < m$ then $\begin{aligned} Er_2 = \sum_{k_1}^{m} & \left[(k^2 - 1)/8 - i + 1/2(m - k_1 - 1)(m - k_1) \right] R \\ & - \int (m - k_1)\sigma d\sigma - \int (iR + \sigma) m d\sigma \end{aligned}$
if $m \leq i$ then $Er_3 = ER_{odd\_(m,N-m)}$
Case 3: $m \geq k$
if $0 \leq i < k_1$ then $Er_1 = \sum_{i=1}^{k} \left[(k^2 - 1)/8 - i \right] R$
if $k_1 \leq i < m$ then $Er_2 = \sum_{i=1}^{k} \int (iR + \sigma) d\sigma$
if $m \leq i$ then $Er_3 = ER_{odd\_(m,N-m)}$

Table 7. Results under $K = K$ (K is even) in WKNN

Case 1: $0 \leq m < k_4$
if $0 \leq i < k_4$ then $Er_1 = ER_{even\_(1,k_4)} + ER_{even\_(1,m)}$
if $k_4 \leq i < k/2 + m$ then $Er_2 = ER_{even\_(k_4,k_3+m)} + \sum_{k_4}^{k_3+m} \frac{(k-m-2-2i)(m-i+1)R}{2}$
if $k/2 + m \leq i$ then $Er_3 = ER_{even\_(k_3+m,N-m)}$
Case 2: $k_4 \leq m < k$
if $0 \leq i < k_4$ then $Er_1 = \sum_{i=1}^{k_3} \left[(k^2 - 1)/8 - i \right] R$
if $k_4 \leq i < m$ then $\begin{aligned} Er_2 = \sum_{k_4}^{m} & \left[(k^2 - 1)/8 - i + 1/2(m - k_4 - 1)(m - k_4) \right] R \\ & - \int (m - k_4)\sigma d\sigma - \int (iR + \sigma) m d\sigma \end{aligned}$
if $m \leq i$ then $Er_3 = ER_{even\_(m,N-m)}$
Case 3: $m \geq k$
if $0 \leq i < k_4$ then $Er_1 = \sum_{i=1}^{k} \left[(k^2 - 1)/8 - i \right] R$
if $k_4 \leq i < m$ then $Er_2 = \sum_{i=1}^{k} \int (iR + \sigma) d\sigma$
if $m \leq i$ then $Er_3 = ER_{even\_(m,N-m)}$

3   Experimental Results

3.1   Localization Errors

By setting $R = 1$ m and $N = 60$, Fig. 3 compares the error bounds and simulated errors of the NMAs as the AP is located on the boundary (see Fig. 1). From this figure, we observe that the simulated errors are much close to the error bound. Furthermore, the WKNN generally exhibits higher localization accuracy compared to the KNN as expected [9].

3.2   AP Locations

By setting $R = 1$ m and $N = 60$, Fig. 4 shows the variation of error bounds with respect to the AP locations under different number of neighbors as the AP is located at a random location (see Fig. 2). From Fig. 4, we can find that the AP location has significant impact on the selection of the optimal number of neighbors corresponding to the lowest error bound. Due to the symmetry of the environment, the error bound reaches the maximum when the value m equals

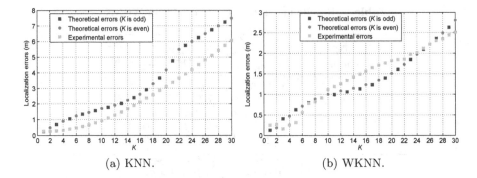

(a) KNN.   (b) WKNN.

Fig. 3. Comparison of the error bounds and simulated errors.

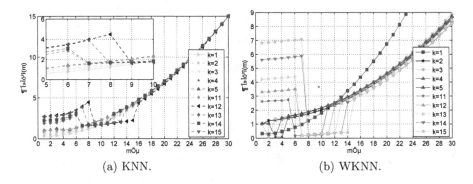

(a) KNN.   (b) WKNN.

Fig. 4. Variation of error bounds under different values of m.

to 30. The sharp variation of error bounds around $m = 7\,\mathrm{m}$ and $12\,\mathrm{m}$ is resulted from the physical constraint of the environment.

4 Conclusion

In this paper, we proposed a novel indoor WLAN deployment optimization approach based on the localization error bounds of NMAs. We present the preliminary analysis on the closed-form solutions to the error bounds of NMAs in a typical indoor environment. The purpose of this analysis is to design the effective and efficient NMAs for the indoor WLAN localization. Furthermore, we discuss the impact of networking parameters, like the environmental size, interval of RPs, number of neighbors, and AP locations, on the error bounds of NMAs. For the future work, how to optimize the WLAN deployment by using the error bound criterion in multi-floor environment forms an interesting topic.

References

1. Gezici, S.: A survey on wireless position estimation. Wirel. Pers. Commun. **44**(3), 263–282 (2008)
2. Chen, Q., Huang, G., Song, S.: WLAN user location estimation based on receiving signal strength indicator. In: The 5th International Conference on Wireless Communications, Networking and Mobile Computing, pp. 1–4. Springer, Beijing (2009)
3. Sun, G., Zhao, H., Lou, D., Yin, Y.: Research on RSSI-based location in smart space. Acta Electron. Sin. **35**(7), 1240–1245 (2007)
4. Zhou, M., Wong, A.K., Tian, Z., Luo, X., Xu, K., Shi, R.: Personal mobility map construction for crowd-sourced Wi-Fi based indoor mapping. IEEE Commun. Lett. **18**(8), 1427–1430 (2014)
5. Tran, Q., Wirawan, H., Chuan, H., Tan, A., Yow, K.: Wireless indoor positioning system with enhanced nearest neighbors in signal space algorithm. In: IEEE VTC, pp. 1–5. IEEE Xplore, Beijing (2006)
6. Xu, Y., Zhou, M., Meng, W., Ma, L.: Optimal KNN positioning algorithm via theoretical accuracy criterion in WLAN indoor environment. In: IEEE GLOBECOM, pp. 1–5. IEEE Xplore, Miami (2010)
7. Ding, G., Tan, Z., Zhang, J., Zhang, L.: Regional propagation model based fingerprinting localization in indoor environments. In: IEEE 24th Annual International Symposium on Personal, Indoor, and Mobile Radio Communications, pp. 291–295. IEEE Xplore, London (2013)
8. Herring, K.T., Holloway, J.W., Staelin, D.H.: Path-loss characteristics of urban wireless channels. IEEE Trans. Antennas Propag. **58**(1), 171–177 (2010)
9. Li, B., Wang, Y., Dempster, A.: Method for yielding a database of location fingerprints in WLAN. Communications **152**(5), 580–586 (2005)

An ASIC Fast Decoder of Rate Compatible Modulation and Its Application in Wireless Communication System

Wei Yu, Jun Wu$^{(\boxtimes)}$, Hao Cui, Zhifeng Zhang, and Haoqi Ren

College of Electronics and Information Engineering, Tongji University,
Shanghai 201804, People's Republic of China
{2014yuwei,wujun,zhangzf,renhaoqi}@tongji.edu.cn, hao.cui@live.com

Abstract. Rate Compatible Modulation (RCM) is a new rate adaptation scheme in wireless communication system, which can achieve very high spectrum efficiency both in additive white Gaussian noise (AWGN) channel and fading channel. But the high decoding complexity of RCM hinders its application in practical communication systems. This paper introduces an Application Specific Integrated Circuit (ASIC) based fast decoder of RCM to implement belief propagation (BP) algorithm in logarithm domain. Though BP algorithm has natural parallel characteristic, a partial-parallel full-pipelined architecture is designed to achieve a tradeoff between hardware resource and processing speed. In order to reduce the computing complexity and improve the throughput of the decoder, we adopt some reduced algorithms, such as piecewise function approximation, lookup tables, fixed-point computing and etc. We build a communication system to test our ASIC decoder in AWGN channel and IEEE 802.11a fading channel.

The original RCM is a rateless code modulation scheme, which can get high spectral efficiency, but is not feasible in some communication systems (such as deep space communications system) for too long transmission delay. In this paper we propose a non-rateless RCM scheme and implement it both in AWGN channel and fading channel. Through testing we confirm that the performance of our proposed scheme is very close to the original rateless scheme, but can greatly reduce transmission time.

Keywords: ASIC · Decoder · BP · Communication · Spectrum efficiency

1 Introduction

Rate adaptation is essential for wireless communication system to approach the changing capacity. Adaptive modulation and coding (AMC) and hybrid automatic repeat request (HARQ) are the most successful rate adaptation schemes currently. In AMC and HARQ, the transmitter selects the best combination of coding rate and modulation scheme to match the estimated channel condition according to feedback from the receiver. However these two schemes rely

© ICST Institute for Computer Sciences, Social Informatics and Telecommunications Engineering 2017
X.-L. Huang (Ed.): MLICOM 2016, LNICST 183, pp. 336–345, 2017.
DOI: 10.1007/978-3-319-52730-7_34

on the accurate channel state feedback and they can only achieve staircase-like spectrum efficiency.

In order to achieve seamless rate adaptation, three rateless codes were proposed recently, which are: Spinal code [5], Strider code [1] and Rate compatible modulation (RCM) [2,3]. The transmitter of rateless codes at first sends a symbol block of certain size, if the receiver can't decode the source bits, the transmitter will retransmit small blocks of symbols continuously till the receiver decodes all source bits correctly.

Spinal code uses a hash function over the message bits to produce pseudo-random bits that can be mapped directly to a dense constellation for transmission [4]. Although the performance of Spinal code is very good, but the Maximum Likelihood (ML) decoding algorithm is very complicated and time consuming, especially when the source length is large. Strider code combines a batch of conventionally encoded symbols (QPSK symbols of encoding bits that have been passed through a 1/5 rate convolutional code) linearly, and uses a decoding algorithm similar to successive interference cancellation. But unfortunately the performance of Strider code is not much better than traditional modulation and coding scheme, its spectral efficiency curve is also not smooth.

RCM is a rateless coding and modulation scheme. Symbols of RCM are incrementally generated from source bits through weighted combination without rate limit [3], and the decoding algorithm of RCM is a variant of belief propagation (BP). A large number of studies [2,3,7-9] show that RCM can achieve very high spectral efficiency both in additive white Gaussian noise (AWGN) channel and fading channel, but the complexity of traditional RCM decoder is very high. In our previous works, the fast decoding algorithm of RCM was proposed and implemented in FPGA [8]. Afterwards the partial design of the Application Specific Integrated Circuit (ASIC) decoder was described in [6], but only simulation results were given, because this chip did not tape out at that time. This paper introduces the full design of ASIC decoder chip and presents a communication system with the decoder. To the best of our knowledge, this is the first ASIC decoder of rateless coded modulation scheme. With full test we confirm that the ASIC decoder works correctly, and the spectrum efficiency is much better than conventional modulation and coding scheme in both AWGN and IEEE 802.11a fading channel.

2   System Architecture

2.1   The RCM Encoder

The encoder of RCM uses a random sparse matrix (G) to multiply a source bits vector (b) to generate coded modulation symbols (u). Each row of G only has 8 non-zero elements in random positions. These non-zero elements are random arrangement of weight set W = $\{-4, -4, -2, -1, 1, 2, 4, 4\}$. The rows number of G could be constructed as many as we need to achieve rateless effect. The encoding process can be expressed as follows.

$$u_i = \sum_{j=1}^{8} w_j \times b_{n_{ij}} \tag{1}$$

Where w_j is the non-zero element of G in the i row, n_{ij} is the index of the source bit weighed by w_j to generate symbol u_i [3]. u_i is integer and ranges from -11 to 11. Every two adjacent symbols are mapped to I and Q plane to form a complex modulation signal. So the constellation of RCM is fixed to 23×23QAM.

2.2   The RCM ASIC Decoder

The original decoding algorithm of RCM is a variation of belief propagation (BP), which includes three steps: (1) the horizontal iteration, which calculates probabilities from symbol nodes to source nodes; (2) the vertical iteration, which calculates probabilities from source nodes to symbol nodes; (3) after iterating enough times between step (1) and step (2), the hard decision of source nodes is outputted. Detailed information of the algorithm is shown in [3].

The original horizontal iteration in step (1) is a deconvolution operation (different from the horizontal iteration in BP decoding of LDPC), which includes many multiply-accumulate operations. It is not easy to be converted into logarithm field directly, so the computing complexity is very high. In order to reduce computing complexity, we proposed a fast decoding algorithm in [8], which uses lookup tables and piecewise function approximation to convert multiply-accumulate operations in arithmetic field to addition operations in logarithm field. The fast decoding algorithm can save 90% multiplication resources without noticeable performance loss.

The Architecture of the ASIC Decoding Logic. Now we design a partial-parallel and full pipelined architecture in this ASIC decoder to implement the fast decoding algorithm. The architecture of the decoding logic is shown in Fig. 1. The HUP (Horizontal Unit Processors) is used to calculate Log Likelihood Ratio (LLR) message sent from symbols node u to source bit node b. The VUP (Vertical Unit Processor) is used to calculate LLR message sent from source bit node b to symbol node u [6]. The RAG (Random Address Generators) stores columns of non-zero elements in sub-matrix of matrix G. There're 8 HUPs and 8 VUPs working parallel. Each HUP and VUP processes data of a sub-matrix (which is part of the random sparse generating matrix G) in full pipelining mode.

The Top Control Module of the ASIC Decoder. The top control module mainly includes three functions: (1) inputting RCM symbols from outside and writing them into data input memory (2) controlling iteration operation of the decoding logic; (3) reading decoding information from data output memory and outputting them to outside. These three functions are achieved by three state machines, we call them $SM1, SM2, SM3$ respectively.

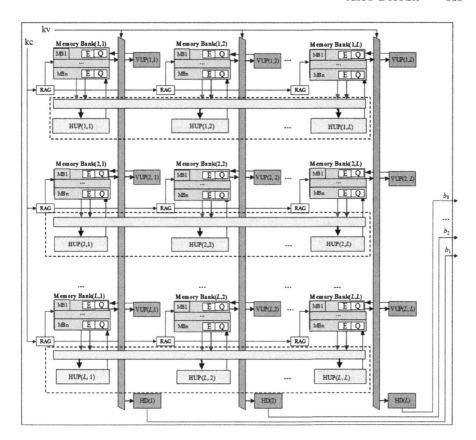

Fig. 1. Architecture of RCM ASIC decoder.

$SM1$ is shown as Fig. 2. When the decoder begins to work, the state machine is at the *Idle* state. The decoder sends ACK signal to the outside at the *Idle* state, which means the decoder can receive RCM symbols from outside. When the decoder receives $cmd1$ signal, $SM1$ jumps to state $S1$. The decoder continues to receive RCM symbols from outside at state $S1$. When one frame of RCM symbols is received completely, the $SM1$ jumps to state $S2$. The decoding logic begins to carry out decoding operation at state $S2$, the decoder also can receive RCM symbols from outside at this state. When the decoder receives dec_state signal, the $SM1$ jumps to state $S3$. The decoding logic continues to carry out decoding operation at state $S3$, but the decoder doesn't receive RCM symbols from outside any more. If the next $cmd1$ signal is received at state $S3$, the $SM1$ will jump back to state $S1$, then RCM symbols can be wrote into another data input memory (the data input memory is operated in ping pang mode).

$SM2$ is shown as Fig. 3. When the decoding process is not started, $SM2$ is at the *Idle* state. When decoding logic receives cup_enable signal from outside, $SM2$ jumps to state hup_state, the HUP module reads data from data memory

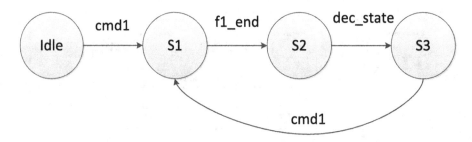

Fig. 2. State machine of SM1.

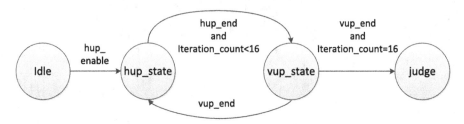

Fig. 3. State machine of SM2.

and begins to work at this state. The HUP module sets *hup_end* signal to be 1 when it ends work, and $SM2$ jumps to state *vup_state*. The VUP module begins to work at state *vup_state*. The VUP module sets *vup_end* signal to be 1 when it ends work, and $SM2$ jumps back to state *hup_state*. When the decoding logic complete iteration operation 16 times (an iteration operation includes process of state *hup_state* and state *vup_state*), $SM2$ jumps to state *judge*. The decoding information will be wrote to the data output memory at state *judge*.

$SM3$ is very similar to $SM1$, so we just describes it briefly here. When the decoder finishes decoding operation, decoding information is wrote into data output memory. If the decoder receives output data request signal (*cmd2* signal) from outside, $SM3$ reads data from data output memory and output them to the outside.

The ASIC Chip After Tape-Out. The ASIC chip uses Multi-Project Wafer (MPW) CMOS 65 nm Standard Performance technology, with 9 metal layers, occupying area of $5 \times 5 \text{ mm}^2$. When running at 300 MHz, this chip can achieve 84 Mbps throughput, and consumes power 4.21 W. Figure 4 shows the picture of this ASIC chip (we also call RCM as Random Projections Codes (RPC)). Table 1 describes main pins of RCM decoder chip and their functions.

2.3   The Non-rateless Scheme

The original RCM is a rateless code modulation scheme, which can get high spectral efficiency, but is not feasible in some communication systems for too

Table 1. Pins of ASIC decoder chip.

Name	Type	Description
data1[63:0]	Input	Data input pins, the value of high 32 bits is the channel coefficient, and the value of low 32 bits is the input RCM symbol
en1[7:0]	Input	Control signal of input data, en1[7:0] should be set to '00000001' in the 0I24 clock cycles, and be set to '00000010' in the 25I49 clock cycles
f1_end	Input	Control signal of input data, this pin should be set to 1 when one block of symbols is transmitted to decoder completely
cmd1	Input	Request signal of transmitting data to decoder
ack1	Output	If the decoder accepts data input request, then set ack1 to 1, otherwise set ack1 to 0
data2[31:0]	Output	Data Output pins, which are soft decoding value of source bits, each source bit occupies 16 pins
en2[7:0]	Input	Control signal of output data, en2[7:0] should be set to '00000001' in the 0I24 clock cycles, and be set to '00000010' in the 25I49 clock cycles
f2_end	Input	Control signal of output data, this pin should be set to 1 when the outside received one frame of soft decoding bits
cmd2	Input	Request signal of receiving data from decoder
ack2	Output	If the decoder accepts data output request, then set ack2 to 1, otherwise set ack2 to 0
wb_en	Input	When wb_en=1, the outside can configure registers inside the RCM decoder. These configuration values are transmitted from pins of data1
PDRST	Input	Reset signal of PLL
CLOCK_IO	Input	I/O clock
CLOCK_PLL	Input	PLL clock
RESET	Input	Reset signal of RCM decoder

long transmission delay such as deep space communications. In order to reduce transmission time, we propose a non-rateless RCM scheme. The transmitter sends at most 3 blocks of RCM symbols (one base block and two retransmission blocks) in the proposed non-rateless scheme.

In order to get the optimal rate adaptation of nan-rateless scheme both in AWGN channel and fading channel, we carry out massive simulations to find the relationship between the number of symbols of base/retransmission block and SNR condition (the number of source bits is fixed to be 400 in one frame in these simulations). Through simulations, we build two lookup tables shown as Tables 2 and 3, these two tables give number of RCM symbols of base block

Table 2. Number of symbols vs. SNR in AWGN channel.

SNR (dB)	5	6	7	8	9	10	11
N_{base}	1200	1100	800	650	500	400	340
$N_{retrans}$	200	200	100	100	100	80	80
SNR (dB)	12	13	14	15	16	17	18
N_{base}	300	270	240	220	210	190	180
$N_{retrans}$	80	70	70	70	70	60	60
SNR (dB)	19	20	21	22	23	24	25
N_{base}	170	160	150	140	120	110	100
$N_{retrans}$	60	60	40	40	20	20	20

Table 3. Number of symbols vs. SNR in fading channel.

SNR (dB)	5	6	7	8	9	10	11
N_{base}	1500	1500	1400	1300	1000	780	720
$N_{retrans}$	100	100	100	100	100	80	80
SNR (dB)	12	13	14	15	16	17	18
N_{base}	600	560	480	440	360	330	280
$N_{retrans}$	80	70	70	70	70	60	60
SNR (dB)	19	20	21	22	23	24	25
N_{base}	260	230	220	200	190	170	160
$N_{retrans}$	60	60	40	40	20	20	20

and retransmission block (denote as Nbase and Nretrans respectively in Tables 2 and 3) at different SNR in AWGN channel and fading channel.

2.4 The Communication System

The communication system with RCM ASIC decoder is shown as Fig. 5, which includes three parts: the personal computer (PC), the FPGA board (XILINX ML605), the PCB board with the ASIC decoder. The FPGA board uses PCIe cable to connect with the PC, and uses FMC interface to connect with the decoder board. We implement the encoder and the channel emulator in a C program runs in the PC. The PC transmits RCM symbols after AWGN/fading channel to the ASIC decoder through the FPGA board. The decoder finishes decoding and sends back decoding bits to FPGA at first, then to the PC side.

The schematic diagram of RCM PCB board is shown as Fig. 6. The PCB board not only supplies power, clock, reset and PLL configuration to the RCM decoder, but also connects all pins of RCM decoder (except power pins) with FMC interface of ML605. Considering parallel transmission of data signal, we

Fig. 4. ASIC decoder chip.

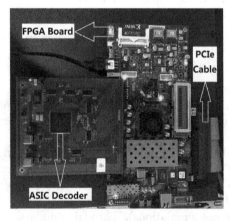

Fig. 5. Communication system.

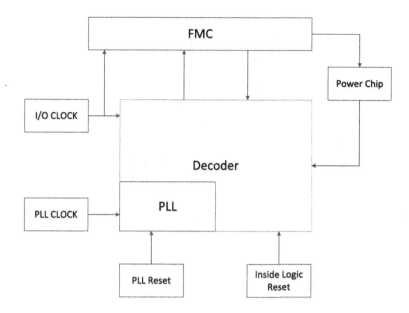

Fig. 6. The schematic diagram of RCM PCB board.

set all signal lines and control lines from the RCM decoder to FMC interface to be same length in the PCB board. The PLL CLOCK supplies outside clock to the PLL inside the RCM decoder, and the PLL configures all clocks of the whole chip. I/O CLOCK supplies outside clock to the pins of RCM decoder.

3   Results Evaluation

In order to test the average spectrum efficiency of the ASIC decoder, we generate 5000 source frames randomly, each frame include 400 bits. We test the ASIC decoder both in AWGN channel and fading channel (IEEE 802.11a fading channel mode A, which is a typical office environment for nonline-of-sight condition with 50 ns root mean square delay spread).

In the rateless scheme, the transmitter continues to send RCM symbols until the receiver can completely decode all source bits. So the average spectrum efficiency can be calculated as:

$$(N_{source} \times 5000)/N_{symbol\_total} \qquad (2)$$

Where N_{source} is the number of source bits of one frame which is 400, and N_{symbol_total} is the number of accumulation symbols when all bits of these 5000 frames can be decoded correctly.

In the non-rateless scheme, the transmitter only send at most 3 blocks of RCM symbols (1 base block and 2 retransmission blocks) according to Table 1. If the receiver can't decode all source bits correctly in the case of three transmission

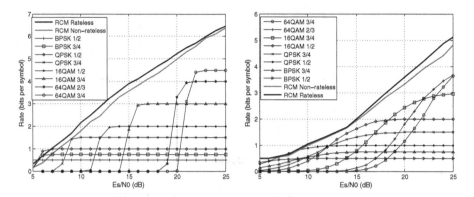

Fig. 7. Rate comparing in AWGN channel.   **Fig. 8.** Rate comparing in fading channel.

blocks, a frame error will be recorded. The average spectrum efficiency can be calculated as:

$$(1 - FER) \times (N_{source} \times 5000)/N_{symbol\_total} \tag{3}$$

Where FER is the Frame Error Ratio, N_{symbol_total} and N_{source} are the same as the rateless scheme.

We compared the spectrum efficiency (rate) of the ASIC decoder with 802.11a standard AMC, which are shown in Figs. 7 and 8. In AMC, the modulations are BPSK, QPSK, 16QAM and 64QAM and the channel code is convolutional code. Each modulation scheme has two coding rate which are 1/2 (2/3 for 64QAM) and 3/4. From these two figures we can see that the ASIC decoder works correctly; the performance of non-rateless scheme can approach the rateless scheme, and both these two schemes are not only much higher than AMC, but also varying gracefully along with channel Signal Noise Ratio (SNR) in very wide dynamic range. In the rateless scheme, the transmitter need to send small symbol blocks many times, so the whole transmission delay is very large, but the non-rateless scheme only has two retransmission blocks, so it can greatly reduce the transmission time.

4   Conclusion

In this paper we introduces an ASIC fast decoder of RCM, to the best of our knowledge which is the first ASIC decoder of rateless code. We also build a communication system to test the performance of this ASIC decoder. Through massive testing we confirm that the ASIC decoder works well, and its performance is much better than AMC both in AWGN channel and fading channel.

Acknowledgments. This work was supported in part by the National Natural Science Foundation of China under Grant 61571329 and Grant 61390513, in part by Huawei Innovation Research Plan (HIRP) Funding under Grant YB2015110117.

References

1. Aditya, G., Sachin, K.: Strider: automatic rate adaptation and collision handling. Proceedings of the ACM SIGCOMM 2011 Conference, pp. 158–169. ACM (2011)
2. Cui, H., Luo, C., Tan, K., Wu, F., Chen, C.W.: Seamless rate adaptation for wireless networking. In: MSWiM 2011, pp. 437–446 (2011)
3. Cui, H., Luo, C., Wu, J., Chen, C.W., Wu, F.: Compressive coded modulation for seamless rate adaptation. IEEE Trans. Wirel. Commun. **12**(10), 4892–4904 (2013)
4. Perry, J., Balakrishnan, H., Shah, D.: Rateless spinal codes. In: ACM Workshop on Hot Topics in Networks 2011, pp. 1–6 (2011)
5. Perry, J., Iannucci, P.A., Fleming, K.E., Balakrishnan, H., Shah, D.: Spinal codes. In: ACM SIGCOMM 2012 Conference on Applications, Technologies, Architectures, and Protocols for Computer Communication, pp. 49–60 (2012)
6. Qiu, L., Wang, M., Wu, J., Zhang, Z., Huang, X.: Design and implementation of seamless rate adaptive decoder. In: 2014 IEEE Military Communications Conference, pp. 356–361 (2014)
7. Shirvanimoghaddam, M., Li, Y., Vucetic, B.: Near-capacity adaptive analog fountain codes for wireless channels. IEEE Commun. Lett. **17**(12), 2241–2244 (2013)
8. Wang, M., Wu, J., Shi, S.F., Luo, C., Wu, F.: Fast decoding and hardware design for binary-input compressive sensing. IEEE J. Emerg. Sel. Top. Circuits Syst. **2**(3), 591–603 (2012)
9. Wang, M., Wu, J., Yu, W., Wang, H.: Efficient coding modulation and seamless rate adaptation for visible light communications. IEEE Wirel. Commun. **22**(2), 86–93 (2015)

Research on Cooperative Spectrum Sensing Algorithm

Yu Gao[1], Xin-Lin Huang[1(✉)], Si-Yue Sun[2], Xiaowei Tang[1],
and Yuan Xu[1]

[1] The Department of Information and Communication Engineering,
Tongji University, Shanghai 201804, People's Republic of China
{2459461554,962080229,2604952900}@qq.com,
xlhuang@tongji.edu.cn
[2] Shanghai Engineering Center for Micro-Satellites,
Shanghai, People's Republic of China
sunmissmoon@163.com

Abstract. The rapid development of wireless communication brings us convenience as well as scarcity of radio spectrum resources. Hence, scientists proposed cognitive radio technology to solve this problem. Spectrum sensing is a pivotal technology protecting primary users from interference of secondary users in cognitive radio, and can be achieved by different algorithms which will result in different performances. In this paper an original cooperative broadband spectrum sensing algorithm based on undersampling is proposed to reduce the hardware overhead as well as satisfying the requirement of system performance. The proposed cooperative spectrum sensing algorithm will use undersampling technology in the secondary user in order to save costs and reduce hardware overhead. On this premise, in the process of information transmission, the algorithm have adopted a method which is similar to VOFDM for signal transmission in the channel between secondary users and fusion center, so that the system can overcome the intersymbol interference caused by broadband signal and rebuild the state of primary users in the fusion center. The simulation results shows that the performance of proposed algorithm is similar to the traditional single-node spectrum sensing algorithm and "or" decision algorithm, however, worse than "and" decision algorithm. The performance loss is acceptable considering its effect of reducing hardware overhead.

Keywords: Cognitive radio · Cooperative spectrum sensing · Undersampling · Vector orthogonal frequency-division multiplexing

1 Introduction

With the development of wireless communication, wireless communication network becomes an indispensable part in our society, followed by the popularity of wireless access equipment and the increase of wireless service and applications. It is merited that such a development is limited by the lack of wireless spectrum resources.

Recent years a wireless communication technology named cognitive radio is proposed by Dr. Joseph Mitola to solve the problem mentioned above. Cognitive radio can

© ICST Institute for Computer Sciences, Social Informatics and Telecommunications Engineering 2017
X.-L. Huang (Ed.): MLICOM 2016, LNICST 183, pp. 346–355, 2017.
DOI: 10.1007/978-3-319-52730-7_35

continuously detects the channels, makes a decision of PU's existence, and finally access the idle spectrum opportunistically by using Radio Knowledge Representation Language (RKRL) [1]. In this paper the spectrum sensing part will be researched.

The current spectrum sensing technology is a PU detection technology in receivers, thus according to the number of receivers, it can be divided into single-point spectrum sensing and cooperative spectrum sensing. The computer complexity of single-point spectrum sensing is low, thus it can be easily realized. However, for its limited sensing data, the accuracy of single-point spectrum sensing is worse than that of cooperative spectrum sensing. For these reasons, single-point spectrum sensing is gradually replaced by cooperative spectrum sensing [2, 3].

In real applications, the centralized sensing model (see Fig. 1) is the most common of cooperative spectrum sensing models [4–7]. The network of centralized sensing model consists of many SUs and a fusion center (FC), which can gather SUs' sensing data, make a comprehensive decision and broadcast such spectrum decision to all SUs. Centralized sensing model has high real-time performance, although it requires a powerful computation ability.

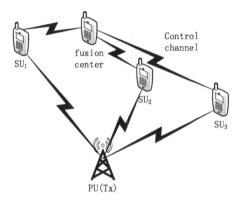

Fig. 1. Centralized sensing model

The current spectrum sensing technology is continuing to mature with the developing requirement of hardware, especially the growing high frequency at which SUs sample the target signal. To reduce the cost of receivers, a cooperative spectrum sensing algorithm based on undersampling is proposed in this paper. As the name implies, an undersampling strategy is used in SUs' receivers, on that premise, a technology similar to vector orthogonal frequency division multiplexing (VOFDM) is applied in fusion stage to restore the primeval signals.

The main contributions of this paper are as followings:

(1) The SUs of the proposed algorithm adopt an undersampling strategy to obtain the sensing data. The time of all the SUs must be synchronized so that the symbols of each time slot can be obtained by a SU.
(2) VOFDM is used in data fusion stage of the proposed algorithm to reduce the transmit distortion of broadband signals.

2 Cooperative Spectrum Sensing

2.1 Cooperative Spectrum Sensing Process

Cooperative spectrum sensing is coordinated by cognitive radio base station (CRBS), and it is assumed that all the SUs involved in cooperative sensing have the same spectrum state so that the decision made by FC is suitable for all these SUs. In real applications, however, the SUs are too decentralized to regard as sharing the same state, thus a clustering cognitive radio network is proposed (see Fig. 2), where the SUs is divided into many clusters according to geographic condition, distance and other factors and each cluster has a cluster head (CH) to control the sensing process. The SUs within a cluster can be consider sharing the same spectrum state, which can not only solve the problem that the spectrum decision of the FC is not consistent with the practical spectrum state of the SUs, but also reduce the energy consumption of multi-hop sensing information transmission to FC.

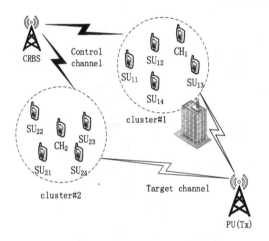

Fig. 2. The clustering cognitive radio network

After observing the target channel in a sensing period, SUs within a cluster will transmit the sensing information to CH. CH will make a locally spectrum decision about current channel state using data fusion algorithm, and broadcast the spectrum decision to SUs within the cluster through control channel. The data fusion algorithms will be discussed next.

2.2 Data Fusion Algorithms

It is mentioned above that the centralized sensing model is the most common of cooperative spectrum sensing models, and in centralized sensing model the FC would make the spectrum decision of PU's state based on SUs' sensing data. Such a decision must be made by using data fusion algorithms. Data fusion algorithms can be divided into three types according to the size of SUs' sensing data and the requirement of control channel bandwidth [8], which are as follows.

Soft combining. If the control channel bandwidth is wide, SUs will send the sensing data completely to FC, such a process is called as soft combining. Equal gain combining, Maximum likelihood ratio combining and selection combining are all belong to soft combining.

Hard combining. Hard combining is a multiple-step decision under a narrow control channel bandwidth, where SUs will make a decision of sensing data respectively and send the 1bit decision results to FC to reduce the channel overhead. The commonly used algorithms, 'and' decision, 'or' decision and 'majority' decision, are all hard combining algorithm. In 'or' decision, the FU will consider PU existing as long as there is one SU decision that the channel is occupied by PU. In 'and' decision, the FU will consider PU existing only if all the SUs decide that the channel is occupied by PU. The 'majority' decision is a compromise of the above two algorithm that the FU will consider PU existing when more than half of the SUs decide that the channel is occupied by PU. All in all, those algorithm can be reduced to 'k out of N' algorithm, where the FU will consider PU existing when more than k SUs of the all N SUs decide that the channel is occupied by PU [9]. The false alarm probability Q_f and the detection probability Q_d of 'k out of N' algorithm can be represented as:

$$Q_f = P\{H_0|H_1\} = \sum_{l=k}^{n} \binom{n}{l} P_f^l (1 - P_f)^{n-l} \qquad (1)$$

$$Q_d = P\{H_1|H_1\} = \sum_{l=k}^{n} \binom{n}{l} P_d^l (1 - P_d)^{n-l} \qquad (2)$$

where H_0 stands for the circumstance that PU is turned of and H_1 stand for the circumstance of the target channel is occupied by PU. P_f and P_d are the false alarm probability and the detection probability of a SU. Obviously, when k separately equals to 1, n and n/2, 'k out of N' algorithm would become 'and' decision, 'or' decision and 'majority' decision algorithm.

Softened hard combining. In the practical application of cooperative spectrum sensing network, the reliability of different SUs may be different under the effects of complex conditions. Thus the softened hard combining is proposed, where the SUs send a reliability parameter α (i.e., the weight of each decision result) as well as the 1bit decision results to FC to improve the performance of spectrum sensing decision.

3   Cooperative Spectrum Sensing Algorithm Based on Under-Sampling

According to the Nyquist Sampling Theorem, the sampling frequency of receivers must be greater than or equal to twice of the maximum signal frequency (i.e., $f_s \geq 2f_H$), otherwise, the spectrum aliasing will occur and the original signal will be unable to restore completely. In practical applications, the sampling frequency of receivers may be set as 4 to 10 times of the maximum signal frequency.

However, the higher the sampling frequency is, the more expensive the receivers are. Thus high sampling frequency brings us accuracy as well as huge cost. On the one hand, the low frequency samplers can be used to percept the high frequency signal in spectrum sensing system under the condition of undemanding accuracy requirement, the price of which is just lossy restore. On the other hand, the sample frequency of SUs may be limited by hardware devices. The spectrum aliasing will happen while the target channel bandwidth is wide. Thus the original signal has to be restored under a low sample frequency, which is called undersampling. In this section, a cooperative broadband spectrum sensing algorithm based on undersampling is proposed, and the algorithm is divided into two stages: SUs undersampling and sensing data fusion. The details will be discussed below.

3.1 SUs Undersampling

The first stage of cooperative broadband spectrum sensing algorithm based on undersampling is SUs undersampling. In traditional cognitive radio network, the high frequency sampling is adopted in SUs receivers. Moreover, all the SUs receivers adopt uniform and periodic sampling, i.e., the entire channel signal is gathered in each SU. It is assumed that each SU needs to gather M data in unit time to meet the requirement of perception precision, those data can be represented as:

$$Y_n = (y_{n,0}, y_{n,1}, \ldots, y_{n,M-1}), \qquad n = 0, 1, \ldots, N - 1 \tag{3}$$

where Y_n is the sensing data of the nth SU, and N is the number of SUs in a cluster. The initial state of the M data can be represented as,

$$X = (x_0, x_1, \ldots, x_{M-1}) \tag{4}$$

As shown in Eqs. (3) and (4), x_k is the PU state and $y_{n,k}$ is the PU decision state decided by the nth SU. Both of x_k and $y_{n,k}$ can be represented by $\{0,1\}$, "0" represents that the channel is idle and "1" represents that the channel is occupied by PU. The final PU decision state y_k will be made by FC according to data fusion algorithm and $y_{n,k}$.

However, in the proposed algorithm, it is assumed that there are N SUs and each of them needs to gather K data in unit time (i.e., $M = NK$) and the Eq. (4) can be rewritten as $X = (x_0, x_1, \ldots, x_{NK-1})$. The sensing data of each SU can be represented as:

$$Y_n = (y_n, y_{N+n}, y_{2N+n} \cdots, y_{(K-1)N+n}), \qquad n = 0, 1, \ldots, N - 1 \tag{5}$$

As shown in Eq. (5), the SUs are sampling the target channel alternately (see Fig. 3), and obviously the time synchronism of SUs must be accurate.

From Fig. 3, we can conclude that in the proposed cooperative spectrum sensing algorithm each time slot has only one sensing data instead of N sensing data in traditional cooperative spectrum sensing algorithm. Moreover, the SUs will gather all

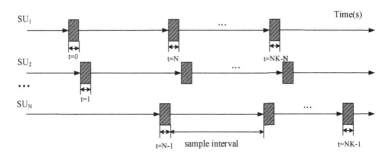

Fig. 3. Sampling process of the proposed algorithm

the K data, modulate these data, and finally send all the decision results simultaneously to the FC. Such a process will be shown in the next subsection.

3.2 Sensing Data Fusion

In the proposed algorithm, the steps of data fusion stage are similar to the multiple-input multiple-output vector orthogonal frequency division multiplexing (MIMO-VOFDM) technology [10], which is shown in Fig. 4.

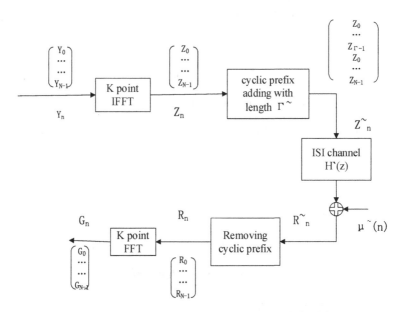

Fig. 4. The data fusion process of the proposed algorithm

In VOFDM, the modulation system of transmitter is inverse fast Fourier transform (IFFT) and the demodulation system of receiver is fast Fourier transform (FFT), such modulate and demodulate way can be used in the proposed algorithm. The transition symbols can be represented as,

$$
\begin{pmatrix}
y_0 & y_N & \cdots & y_{(K-1)N} \\
y_1 & y_{N+1} & \cdots & y_{(K-1)N+1} \\
\vdots & \vdots & \ddots & \vdots \\
y_{N-1} & y_{2N-1} & \cdots & y_{KN-1}
\end{pmatrix}
\tag{6}
$$

The K elements of the nth line in Eq. (6) is the received symbols of the nth SU, and in the transmitter of each SU, those K symbols are transformed into K new symbols by IFFT. Then the K new symbols will be transmitted in K different subchannels respectively (each subchannel has N symbols in it), just as VOFDM does. The K new symbols $Z(n)$ can be represented as:

$$
Z_k(n) = \frac{1}{\sqrt{K}} \sum_{k=0}^{K-1} Y_n(k) \exp(\frac{j2\pi nk}{K}) \qquad k = 1, 2, \ldots, K-1 \tag{7}
$$

Equation (7) is the formula of IFFT, where $Z_k(n)$ stands for the nth signal of the kth subchannel.

Due to the ISI channel, cyclic prefix must be inserted before Z_k, here we insert the first $\tilde{\Gamma}$ elements of Z_k before Z_k (or we can insert the last $\tilde{\Gamma}$ elements of Z_k before Z_k). The symbols after insertion $\hat{Z}_k$ can be represented as:

$$
\hat{Z}_k = (Z_k(0), Z_k(1), \ldots, Z_k(\tilde{\Gamma}-1), Z_k(0), Z_k(1), \ldots, Z_k(N-1)) \tag{8}
$$

where the length of cyclic prefix $\tilde{\Gamma} \geq \lceil L/N \rceil$ for the purpose of removing the ISI, and L is the order of channel transfer function $H(z)$ $(H(z) = \sum_{n=0}^{L} h(n)z^{-n})$.

For simplicity, the channel transfer function is assumed known. According to [10], the transfer function $H(z)$ can be rewritten as:

$$
\bar{H}(z) = \begin{bmatrix}
h_0(z) & z^{-1}h_{K-1}(z) & \cdots & z^{-1}h_1(z) \\
h_1(z) & h_0(z) & \cdots & z^{-1}h_2(z) \\
\vdots & \vdots & \vdots & \vdots \\
h_{K-2}(z) & h_{K-3}(z) & \cdots & z^{-1}h_{K-1}(z) \\
h_{K-1}(z) & h_{K-2}(z) & \cdots & h_0(z)
\end{bmatrix}
\tag{9}
$$

where $h_k(z)$ is the kth polynomial of $H(z)$, which can be represented as:

$$
h_k(z) = \sum_l h(Kl+k)z^{-l}, \qquad k = 0, 1, \ldots, K-1 \tag{10}
$$

And the relationship between the transfer information symbol of the kth subchannel $\hat{Z}_k$ and the received signal of the kth subchannel R_k can be formulated as:

$$R_k = \bar{H}_k \hat{Z}_k + \tilde{\xi}_k, \qquad k = 0, 1, \ldots, K-1 \qquad (11)$$

where $\bar{H}_k = \bar{H}(z)|_{z=\exp(j2\pi k/N)}$, $k = 0, 1, \ldots, K-1$, and $\tilde{\xi}_k$ is the FFT of the additive noise $\xi(n)$ and therefore has the same statistics as $\xi(n)$.

After passing through the ISI channel, the received signal needs to remove the cyclic prefix. The signal after removing cyclic prefix can be represented as:

$$\hat{R}_k = (R_k(0), R_k(1), \ldots, R_k(N-1)) \qquad k = 0, 1, \ldots, N-1 \qquad (12)$$

And finally all of the $\hat{R}_k$ need to be demodulated at the receiver by K point FFT, which can be represented as:

$$G_n(k) = \frac{1}{\sqrt{K}} \sum_{l=0}^{K-1} \hat{R}_l(k) \exp(\frac{-j2\pi nl}{K}), \qquad n = 0, 1, \ldots, K-1 \qquad (13)$$

where G_n is the reduced signal of Y_n, and after a parallel-to-serial transform process, we can get the spectrum decision of PU.

4 Simulation

In this section the feasibility and performance of proposed algorithm will be evaluated through comparing with "and" decision algorithm, "or" decision algorithm and single-point sensing algorithm. Please note that the noise power is assumed 10 mW and the SU number is assumed 5. The sensing data are observed in the durations of sensing period.

The performance of spectrum sensing algorithm can be measured by detection probability and false alarm probability. Three different spectrum sensing cases are considered, and the simulation results are plotted in Figs. 5, 6 and 7. In Fig. 5, the number of SUs is set to 5 and the SNR of received signals are set to 5 dB. In Fig. 6, the number of SUs is set to 5 and the SNR of received signals are set to 0 dB. In Fig. 7, the number of SUs is set to 5 and the SNR of received signals are set to -5 dB. We can draw a conclusion from Figs. 5, 6 and 7 that the performance of the proposed algorithm is almost the same as that of "or" decision algorithm and single-point sensing algorithm, and a little poorer that the performance of "and" decision algorithm. Considering its effect of reducing hardware overhead, such a performance loss is acceptable.

In Fig. 8, the SNR-BER performance curve of the proposed algorithm is plotted. It is shown that the BER is reducing with the increasing of SNR, and when SNR is 5 dB, the BER is $10^{-0.6}$(0.25), which means the proposed algorithm had better be used in high SNR cases.

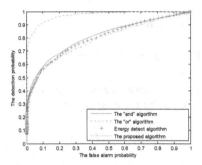

Fig. 5. Spectrum sensing performance comparisons under channel SNRs = 5 dB.

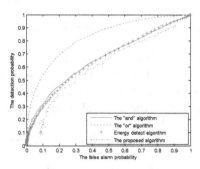

Fig. 6. Spectrum sensing performance comparisons under channel SNRs = 0 dB.

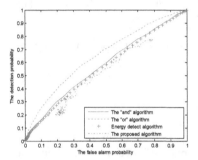

Fig. 7. Spectrum sensing performance comparisons under channel SNRs = −5 dB.

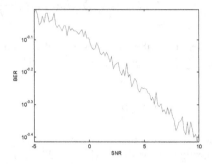

Fig. 8. Performance curve of the proposed algorithm.

5 Conclusion

In this paper, a cooperative spectrum sensing algorithm based on undersampling has been proposed. In the proposed cooperative spectrum sensing network, the under-sampling technology is used in SUs to save costs and reduce hardware overhead. Then, the algorithm have adopted a method which is similar to VOFDM for signal transmission in the channel between secondary users and fusion center, so that the system can overcome the intersymbol interference caused by broadband signal and rebuild the state of primary users in the fusion center. Under three different channel SNR cases, the simulation results show that the performance of proposed algorithm is similar to the traditional single-node spectrum sensing and the "or" decision algorithm, however, worse than "and" decision algorithm. The performance loss is acceptable considering its effect of reducing hardware overhead.

Acknowledgements. This work was supported by the "Chen Guang" project supported by Shanghai Municipal Education Commission and Shanghai Education Development Foundation under Grant No. 13CG18, the Shanghai Sailing Program under Grant No. 16YF1411000, the National Natural Science Foundation of China under Grant No. 61571329, No. 61390513 and No. 61201225, the Natural Science Foundation of Shanghai under Grant No. 12ZR1450800, and sponsored by Shanghai Pujiang Program under Grant No. 13PJD030. This paper was also supported by the Fundamental Research Funds for the Central Universities under Grant No. 20140767, the Program for Young Excellent Talents in Tongji University under Grant No. 2013KJ007.

References

1. Haykin, S.: Cognitive radio: brain-empowered wireless communications. IEEE J. Sel. Areas Commun. **23**(2), 201–220 (2005)
2. Liang, Y.C., Chen, K.C., Li, G.Y., Mahonen, P.: Cognitive radio networking and communications: an overview. IEEE Trans. Veh. Technol. **60**(7), 3386–3407 (2011)
3. Cabric, D., Brodersen, R.W.: Physical layer design issues unique to cognitive radio systems. In: Proceedings of the IEEE PIMRC, vol. 2, pp. 759–763 (2005)
4. Akyildiz, I.F., Lo, B.F., Balakrishnan, R.: Cooperative spectrum sensing in cognitive radio networks: a survey. Phys. Commun. **4**(1), 40–62 (2011)
5. Dhope, T.S., Simunic, D.: Cluster based cooperative sensing: -a survey. In: IEEE International Conference on Communication. Information & Computing Technology (ICCICT), pp. 1–6 (2012)
6. Huang, X.-L., Wang, G., Fei, H., Kumar, S., Jun, W.: Multimedia over cognitive radio networks: towards a cross-layer scheduling under bayesian traffic learning. Comput. Commun. **51**, 48–59 (2014)
7. Huang, X.-L., Wang, G., Fei, H., Kumar, S.: Stability-capacity-adaptive routing for high-mobility multihop cognitive radio networks. IEEE Trans. Veh. Technol. **60**(6), 2714–2729 (2011)
8. Hoyt, R.S.: Probability functions for the modulus and angle of the normal complex variate. Bell Syst. Tech. J. **26**(2), 318–359 (1947)
9. Akyildiz, I.F., Lo, B.F., Balakrishnan, R.: Cooperative spectrum sensing in cognitive radio networks: a survey. Phys. Commun. **4**(1), 40–62 (2011)
10. Xia, X.-G.: Precoded and vector OFDM robust to channel spectral nulls and with reduced cyclic prefix length in single transmit antenna systems. IEEE Trans. Commun. **49**(8), 44–56 (2001)

Intelligent Recognition of Traffic Video Based on Mixture LDA Model

Xiaowei Tang[1(✉)], Xin-Lin Huang[1], Si-Yue Sun[2], Hang Dong[1],
Xin Zhang[1], Yu Gao[1], and Nan Liu[1]

[1] Department of Electronics and Information Engineering, Tongji University,
Shanghai 201804, People's Republic of China
{xwtang, xlhuang, dh, mic_zhangxin,
gaoyul631643}@tongji.edu.cn
[2] Shanghai Engineering Center for Micro-satellites,
Shanghai, People's Republic of China
sunmissmoon@163.com

Abstract. In this paper, an efficient unsupervised model is proposed to recognize simple actions and complex activities in traffic scenes which is named mixture LDA model. Under this framework, we use hierarchical Bayesian models are to describe three important components in traffic video: basic visual features, simple actions, and complex activities. This model adopts an unsupervised way to learn how to recognize traffic video. Moving pixels can be divided into different simple actions and short video clips can be divided into different complex activities in a long traffic video sequence, then we can achieve the purpose of recognizing different activities in the surveillance video.

Keywords: Bayesian model · Mixture LDA model · Traffic video identification

1 Introduction

With the development of economy, video analysis and recognition technology has been applied widely in different urban public facilities and maintaining public security. There is growing demand in security control, dynamic condition records and active alarming system. Video analysis and recognition technology is playing an important role in the development of the whole society. Intelligent Transportation System (ITS) [1] is put forward by Japan and the United States at first. They called this research "intelligent vehicle system". It was used in road monitoring and intelligent vehicle research. However, along with the deepening of the research work, the system functions are extended too. The rapid development of intelligent transportation monitoring system will have a crucial impact on both transportation system and people's lifestyle.

Because of the complexity of the urban road traffic, the mature vehicle behavior recognition and detection algorithm is mainly used in expressway at present. However, most of the traffic accidents usually occur in the urban traffic route and its traffic environment are usually more complicated than expressway. Abnormal behavior recognition and detection belongs to the traffic incident detection. It refers to the

© ICST Institute for Computer Sciences, Social Informatics and Telecommunications Engineering 2017
X.-L. Huang (Ed.): MLICOM 2016, LNICST 183, pp. 356–363, 2017.
DOI: 10.1007/978-3-319-52730-7_36

unexpected events on the road including emergency brake, illegal parking, and illegal turning left or right, and illegal lane change and running through red light. Frequent abnormal vehicle events bring huge property loss and casualties to the society and usually cause traffic jams. Hence, applying vehicles abnormal behavior detection to urban traffic road can help to save lives and prevent property loss.

Present study shows that the recognition methods of video are mainly divided into two categories including supervised methods and unsupervised methods, from the perspective whether using the training data set [2].

Video recognition based on supervised methods usually first set up learning model of normal behavior, and then judge the test data whether match the model. If they two do not match, the test data will be regarded as abnormal behavior. J. Snoek used the HMM model to identify the fall event in stairs [3]. First of all, they filtered out the noise and shadows in the background by using adaptive background reduction method, then obtained the low-level features of moving objects in video with optical flow, then divided the video into several segments based on time according to the theory of condition random field, then set up a reasonable threshold referring to the normal events and determined whether the event under test belongs to abnormal event by combining the HMM model. Yin Qingbo completed the anomaly detection under supervised methods [4]. The method brought in clusters in the process of establishing the training data set. Zhu Dandong adopted the theme hidden markov model in the identification of human abnormal activities [5].

Researchers began to shift emphasis from supervised detection method to unsupervised method recently years. Unsupervised anomaly detection method can effectively make up for the disadvantages of supervised method because it doesn't need to get any prior sample sets in advance. It only needs to obtain continuous sample data sets to finish the recognition and classification of normal events, regard the events with low probability as abnormal event and finally use similarity features for abnormal judgment. Wang and others introduced HDP into the hidden dirichlet distribution to realize the behavior identification in complicated and crowned scene [6]. Hospedales used LDA model to describe the spatial correlation and used HMM model to describe time correlation [7]. He combined the two models to form a new model called Markov Clustering Topic Model (MCTM) to achieve the purpose of simplifying related sequence of time and space when describing model. D. Kuettel combined HMM model with LDA model and promoted them to the infinite dimension [8]. He proposed Dependent Dirichlet Processes HMM (DDP-HMM) relied on dirichlet process and adopted this model to detect the abnormal behaviors in video.

2 System Scheme

2.1 Recognition Model of Low-Level Visual Features

In this paper, our experiment data is a 40-second-long traffic video obtained from complicated areas. The traffic video includes many different simple actions and complex activities. There are also some hard-solving problems in the traffic video, such as changeable light, different kinds of car types and environment changes. Hence, at the first of this article, we should decide how we can divide the traffic video sequence from complicated or crowded area into simple actions and complex activities. In this section, we regard simple actions such as car going straight, car making a U-turn, people walking across the road and so on as basic element to describe more complex activities. Simple actions can always result in coherent behavior which usually couldn't quit in the middle way. We define the complex activities as the group of a variety of simple actions which take place in the meantime. For example, a car is making a U-turn while another car next to it is turning right. However, we didn't consider those complex but short activities. For example, two pedestrians are crossing the road together then turning into different directions. This article only consider those simple actions taking place in the meantime and we adopt the mixture LDA model.

We consider the moving pixels as the elementary unit to recognize the traffic video. Adopting this method can successfully refrain from tracking problems in complex area. The reason that we didn't use the overall motion characteristics is that varieties of simple actions always take place in the meantime in crowded areas, and the purpose of this article is to divide the traffic video into several kinds of motion clusters. We obtain the basic data sets by compute the location and direction of each moving pixels. When a long traffic video is given, we can divide it into several short video clips. Moving pixels which have similar simple actions always show up in one video clip. Our data sets totally have two hierarchical structures, including long traffic video sequences divided into short clips and moving pixels divided into simple actions.

Moving pixels in each frame are first obtained by using optical flow. The pixels are compared between two continuous frames. We judge the pixel as a moving pixel when the discrepancy of a pixel is larger than the threshold we set up at first. The direction of each moving pixel can also be obtained by optical flow. According to the size of the traffic video (600×800), we divide the traffic scene into cells whose size is 10×10. Then each frame totally has 60×80 cells. We use four characters to describe the direction of each cell including left, right, up, and down. Hence, we can describe the moving pixels in each frame according to the codebook whose size is $60 \times 80 \times 4$.

2.2   Mixture LDA Model

Suppose the traffic video is divided into M short video clips and these M video clips will be classified into L groups. Each group is consisted of K themes, where K is an unknown figure at first. Each theme obeys a multinomial distribution. Each group c has a Dirichlet prior which equals to α_c. For a video clip j, the probability of group label c_j is first obtained from the discrete distribution η. π_j is the probability that the theme belongs to the group j, and it can be obtained from $Dir(\pi_j|\alpha_c)$ (Fig. 1). For each simple action or complex activities i in video clip j, z_{ij} represents the probability that action i belongs to theme j, and it can be obtained from probability π_{jk}. $\beta_{z_{ji}}$ is also a discrete distribution which represents the probability of each simple action or complex activity.

Overall, π_{jk} and z_{ij} are two hidden variables in our Mixture LDA model. If α_c, β and η is given, the function relationship between these three hidden variables c_j, π_j z_j and observed variable x_j is

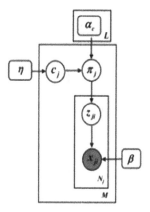

Fig. 1. Mixture LDA model [6]

$$p(x_i, z_i, \pi_j, c_j | \{\alpha_c\}, \beta, \eta)$$
$$= p(c_j|\eta)p(\pi_j|\alpha_{c_j}) \prod_{i=1}^{N} p(z_{ji}|\pi_j)p((x_{ji}|z_{ji}, \beta) \tag{1}$$

The maxi-likelihood of video clip j is

$$\log p(x_j|\{\alpha_c\}, \eta, \beta) = \log \sum_{c_{j-1}}^{L} p(c_j|\eta)p(x_j|\alpha_{c_j}, \beta) \tag{2}$$

With the help of EM algorithm, $p(x_j|\alpha_{c_j}, \beta)$ can be estimated by creating a lower bound $L_1(\phi_{jc_j}, \gamma_{jc_j}; \beta, \alpha_{c_j})$, this step is called E-step in EM algorithm.

$$\log p(x_j|\alpha_{c_j}, \beta) = \log(\int \sum_{z_j} p(x_i, z_i, \pi_j|\alpha_{c_j}, \beta) d\pi_j$$
$$= \log \int \sum_{z_j} \frac{p(x_i, z_i, \pi_j|\alpha_{c_j}, \beta)q\left(z_i, \pi_j \middle| \gamma_{jc_j}, \phi_{jc_j}\right)}{m\left(z_i, \pi_j \middle| \gamma_{jc_j}, \phi_{jc_j}\right)} d\pi_j$$
$$\geq \int \sum_{z_j} m\left(z_i, \pi_j \middle| \gamma_{jc_j}, \phi_{jc_j}\right) \log p(x_i, z_i, \pi_j|\alpha_{c_j}, \beta) d\pi_j \tag{3}$$
$$- \int \sum_{z_j} m\left(z_i, \pi_j \middle| \gamma_{jc_j}, \phi_{jc_j}\right) \log m\left(z_i, \pi_j \middle| \gamma_{jc_j}, \phi_{jc_j}\right) d\pi_j$$
$$= L_1\left(\gamma_{jc_j}, \phi_{jc_j}; \alpha_{c_j}, \beta\right)$$

As soon as the lower bound is created, we can maximize the lower bound by M-step. EM algorithm is an efficient solution to estimate the hyper parameters which is also called hidden variables. After M-step is done, we can continue to use E-step to create a lower bound and then use M-step to maximize the lower bound until the hyper parameters are estimated.

$$\log p\left(x_j | \{\alpha_c\}, \eta, \beta\right) \geq \log \sum_{c_{j-1}}^{L} p\left(c_j | \eta\right) e^{L1\left(\gamma_{jc_j}, \phi_{jc_j}; \alpha_{c_j}, \beta\right)}$$

$$= \log \sum_{c_{j-1}}^{L} m\left(c_j \middle| \gamma_{jc_j}, \phi_{jc_j}\right) \frac{p\left(c_j | \eta\right) e^{L1\left(\gamma_{jc_j}, \phi_{jc_j}; \alpha_{c_j}, \beta\right)}}{m\left(c_j \middle| \gamma_{jc_j}, \phi_{jc_j}\right)}$$

$$\geq \sum_{c_{j-1}}^{L} m\left(c_j \middle| \gamma_{jc_j}, \phi_{jc_j}\right) \left[\log p\left(c_j | \eta\right) + L1\left(\gamma_{jc_j}, \phi_{jc_j}; \alpha_{c_j}, \beta\right)\right] \qquad (4)$$

$$- \sum_{c_{j-1}}^{L} m\left(c_j \middle| \gamma_{jc_j}, \phi_{jc_j}\right) \log m\left(c_j \middle| \gamma_{jc_j}, \phi_{jc_j}\right)$$

$$= L_2\left(m\left(c_j \middle| \gamma_{jc_j}, \phi_{jc_j}\right), \{\alpha_c\}, \beta, \eta\right)$$

L_2 can reach the maximum when $m(c_j | \gamma_{jc_j}, \phi_{jc_j})$ is chosen:

$$m\left(c_j \middle| \gamma_{jc_j}, \phi_{jc_j}\right) = \frac{p(c_j | \eta) e^{L_1\left(\gamma_{jc_j}, \phi_{jc_j}; \alpha_{c_j}, \beta\right)}}{\sum_{c_j} p(c_j | \eta) e^{L_1\left(\gamma_{jc_j}, \phi_{jc_j}; \alpha_{c_j}, \beta\right)}} \qquad (5)$$

3 Simulations

3.1 Simulation Steps

In the simulation, we consider a 40-second video based on complicated and crowned scenes. Its pixel is 600×800 and its frame rate is 15 fps. The specific simulation steps and parameter setup are as follows (Table 1):

Table 1. Steps of simulation

(1) Divide the video into consecutive frames and combine these frames into video image sequences. Divide these video image sequences into 20 short video clips. Each clip is 2 s long
(2) Use optical flow method to obtain the basic features in the traffic video and put these low-level features into WS and DS, where WS is an action set and DS is a clip set
(3) Extract semantic features in low-level features obtained in step (2) by using mixture LDA model. In this step, low-level features will be clustered into different themes
(4) Process the clustering result obtained in step (3), and range the clustering result according to the size of the probability. Set up a threshold value and discard those data sets which are lower than the threshold
(5) Take data obtained in step (4) back to the video scene, so that we can get different activities and interactions

3.2 Simulation Results

In this section, low-level features obtained by optical flow will be clustered by using mixture LDA model and EM algorithm. In mixture LDA model, M video clips will be

Table 2. Simulation setup of optical flow

Document	WO	β	α	Iterations	Correlation factor	Frame rate
20	19200	0.01	50/T	20000	3	15

grouped into L clusters, and each cluster has its own Dirichlet prior α_c which decide the distribution of themes in each video clip (Table 2).

Figure 2 shows us that the 40-second video is divided into 30 themes and we can see the probability of each theme from the figure. Figure 3 shows us the probability

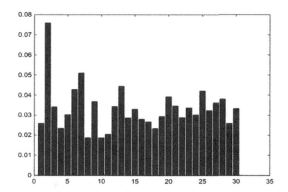

Fig. 2. Probability distribution of each topic

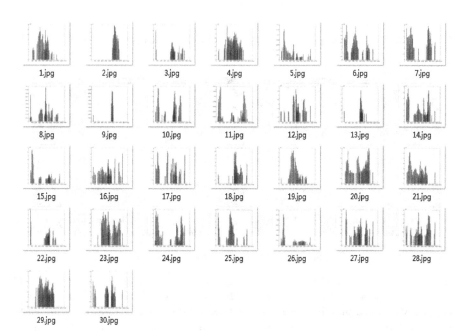

Fig. 3. Probability distribution of words in each topic

Fig. 4. Activities and interactions in each topic

Fig. 5. Topic 26 and 27 (Color figure online)

distribution of actions or activities in each theme. We can tell that Neighboring words are usually grouped into the same cluster. After the low-level features are returned to the original video scene, we can obtain Fig. 4 which shows us the different kinds of activities and interactions in each topic.

3.3 Results Analysis

In order to make it convenient to analyze simulation results, we capture a frame from the original video as the background. Figure 5 is motion pattern in topic 26 and 27 respectively. In Compared with original traffic video, we can clearly tell that the red path represents a bicycle traveling in the motor vehicle lane and turning left into the pavement directly in topic 26. In topic 27, we can see that vehicles turning left have a conflict against pedestrian passing the pavement.

4   Conclusion

In this paper, we propose the mixture LDA model to recognize and detect different motion patterns in surveillance video. From the simulation results, we can see that the model proposed in this paper can classify different kinds of activities and interactions clearly.

References

1. Lu, H., Li, R., Zhu, Y.: Intelligent transportation system standard research. Highway Traffic Sci. Technol. **7**(21), 91–94 (2004)
2. Saligrama, V., Konrad, J., Jodoin, P.M.: Video anomaly identification. IEEE Signal Process. Mag. **27**(5), 18–33 (2010)
3. Snoek, J., Hoey, J., Stewart, L., et al.: Automated detection of unusual events on stairs. Image Vis. Comput. **27**(1), 153–166 (2009)
4. Yin, Q., Zhang, R., Li, X.: Supervised clustering anomaly detection method research based on vector quantization analysis. Comput. Res. Dev. **z2**, 414–418 (2006)
5. Zhu, X., Liu, Z.: Human abnormal behavior recognition based on hidden markov models. Comput. Sci. **39**(3), 251–255 (2012)
6. Wang, X., Ma, X., Grimson, W.E.L.: Unsupervised activity perception in crowded and complicated scenes using hierarchical bayesian models. IEEE Trans. Pattern Anal. Mach. Intell. **31**(3), 539–555 (2009)
7. Hospedales, T., Gong, S., Xiang, T.: A markov clustering topic model for mining behaviour in video. In: IEEE 12th International Conference on Computer Vision, pp. 1165–1172 . IEEE (2009)
8. Kuettel, D., Breitenstein, M.D., Van Gool, L., Ferrari, V.: What's going on? Discovering spatio-temporal dependencies in dynamic scenes. In: 2010 IEEE Conference on Computer Vision and Pattern Recognition (CVPR), pp. 1951–1958. IEEE (2010)

SLNR-Oriented Power Control in Cognitive Radio Networks with Channel Uncertainty

Le Wang[1], Guoru Ding[1,2($\boxtimes$)], Guochun Ren[1], Jin Chen[1], Zhen Xue[1], Haichao Wang[1], and Yumeng Wang[1]

[1] College of Communications Engineering,
PLA University of Science and Technology, Nanjing, China
whcwl456@163.com, dingguoru@gmail.com, chenjin99@263.net,
xzalways@sina.com

[2] National Mobile Communications Research Laboratory, Southeast University,
Nanjing, China

Abstract. The majority of existing studies on power control in cognitive radio networks focus on maximization of signal-to-interference-noise ratio (SINR), while this paper firstly introduces the *signal-to-leakage-and-noise ratio* (SLNR)-oriented power control to optimize throughput in a cognitive radio network (CRN), where massive secondary connections (SCs) and a primary user (PU) coexist with each other sharing the same frequency spectrum. Considering the practical challenge that the channel gains between SCs and PU are typically uncertain, we introduce a probabilistic interference constraint to protect the PU's transmission and reformulate it according to the *Lyapunov*'s central limit theorem (CLT). Then, we apply the convex optimization theory to solve the intractable problem by excluding the probabilistic constraint. Especially, a novel algorithm based on the first-order Lagrangian is developed where the dual variables are updated simultaneously. Furthermore, we provide numerial results using different parameter, which display that the proposed method can achieve higher throughput with much lower computational complexity comparing with the existing literature.

Keywords: Cognitive radio network · Power control · Channel uncertainty · Massive secondary connections · Signal-to-leakage-and-noise ratio

1 Introduction

Spectrum resource is more and more crowded with the ever increasing demand for wireless devices and applications. Pushed by the current severe situation, cognitive radio (CR) has drawn much attention which is a promising technique to improve the efficiency of spectrum utilization [1–3]. Specifically, in the underlay CR mode, a primary user (PU) shares the same spectrum with multiple secondary connections (SCs) in a cognitive radio network (CRN) [4]. With this concept, SCs can access the licensed spectrum used by the PU provided that

© ICST Institute for Computer Sciences, Social Informatics and Telecommunications Engineering 2017
X.-L. Huang (Ed.): MLICOM 2016, LNICST 183, pp. 364–373, 2017.
DOI: 10.1007/978-3-319-52730-7_37

no harmful interference beyond tolerance is introduced. Therefore, it is widely recognized that power control becomes essential for the whole system to mitigate harmful mutual interference.

As an important issue in CR systems, power control has been studied extensively in the literature. Specifically, in [5], a heuristic algorithm under OFDMA has been proposed with an assumption that channel state information (CSI) is perfectly available. However, SCs might have not been given priority to know the signal characteristics of the PU, and thus have to rely on imperfect channel estimation. Consequently, power control for CRNs must account for channel uncertainty [6,7]. In [6], the interference constraints as a probability in a power control problem under the CR scene have been considered, where the uncertainty has been in fading channels including shadowing and Nakagami fading. Power control problems for OFDMA under channel uncertainty is also studied in [7]. However, to the authors' best knowledge, all the existing studies of throughput optimization are oriented towards the signal to interference plus noise ratio (SINR), thus there are no closed form solutions on account of the coupled nature of the corresponding optimization problem. Moreover, the majority of existing studies consider the case that the number of SCs is small (up to tens) and they do not explicitly address the mutual interference between SCs in the system.

In this paper, we propose a *signal-to-leakage-and-noise ratio* (SLNR)-oriented power control method to promote the throughput capacity of the system. Meanwhile, the interference is mitigated between massive SCs (up to hundreds of more) with channel uncertainty. Notably, the so-called SLNR is originally used to design precoders in multi-user MIMO communications [8], where leakage refers to the interference caused by the signal intended for a desired user on the remaining users in a precoding scheme. Differently, in this paper, leakage means the interference generated from one SC to all other SCs. As a result, SLNR is able to measure how much power leaks to the other SCs in the CRN. More importantly, due to the coupled interference nature of the corresponding throughput optimization problems, existing solutions based on SINR do not have closed forms. Differently, the proposed SLNR-oriented power control method in this paper can circumvent the hurdles of SINR perfectly, which leads to a decoupled optimization problem and allows an analytical closed form solution. This method has been proved to be much more effective in this paper (see Sect. 4). Specifically, there are three innovations below the part:

(i) Describe an optimal problem, where channel uncertainty and interference constraints are jointly considered. Different from other power control methods, the throughput is promoted via a novel concept which optimizes the sum SLNR instead of SINR.

(ii) Introduce Gaussian approximation based on the *Lyapunov*'s central limit theorem (CLT) to offer a conservative surrogate and propose an effective power control algorithm based on first-order Lagrangian where the dual variables are updated simultaneously.

(iii) Provide numerial results using different parameter, such as the transmit power and the interference threshold of SCs, which display that our method can outperform the state-of-the-art works in the literature.

2 System Model and Problem Formulation

Consider a CRN where a PU and massive SCs utilize the same spectrum in the underlay mode. The SCs are supposed to be randomly distributed around the PU and $\mathcal{N} = \{1, 2, \cdots, N\}$ is denoted as the set of all SCs. In addition, let p_n represent the transmit-power of the n_{th} SC. Also, let p_{max} denote the maximum transmit-power of SCs, I_{max} denote the maximum interference of SC, and I_{max}^p denote the maximum interference of PU.

The channel gain $g_{n,m}$ between the n_{th} and the m_{th} SC is known accurately [10]. Because SCs almost have no cooperation with PU through their transmissions, it is hard to precisely estimated the gain g_n^{PU} between them.

The SINR of the n_{th} SC can be obtained as follows:

$$SINR_n = \frac{p_n g_{n,n}}{\sum_{m=1, m \neq n}^{N} p_m g_{m,n} + \sigma_0^2}, \tag{1}$$

where $p_n g_{n,n}$ is the received signal power, $\sum_{m=1, m \neq n}^{N} p_m g_{m,n}$ represents the mutual interference of SCs. The whole throughput of the network is expressed as the following formulation:

$$TH_{sum} = \sum_{n=1}^{N} \log_2 (1 + SINR_n). \tag{2}$$

The SLNR of the n_{th} SC is defined as:

$$SLNR_n = \frac{p_n g_{n,n}}{\sum_{m=1, m \neq n}^{N} p_n g_{n,m} + \sigma_0^2}, \tag{3}$$

where the power of the desired signal component for SC n is given by $p_n g_{n,n}$. Meanwhile, the interference caused by SC n on SC m is given by $p_n g_{n,m}$. Therefore, $\sum_{m=1, m \neq n}^{N} p_n g_{n,m}$ represents the power leaked from SC n to all other SCs, which is the concept of *leakage* for SC n. This lies the difference from SINR.

Due to the mutual interference, the constraints of SCs are expresses as follows:

$$\sum_{m=1, m \neq n}^{N} p_m g_{m,n} = I^{(-n)} \leq I_{max}, \forall n. \tag{4}$$

In addition to this, the protection for the PU is taken into consideration. That is to say, the sum interference from all the SCs to the PU is limited under a certain threshold [10]. In order to clearly quantify this, Pr [.] is defined as the outage probability, which can be written as the following expression:

$$\Pr \left[\sum_{n=1}^{N} p_n g_n^{PU} \leq I_{max}^p \right] \geq 1 - \varepsilon, \tag{5}$$

where g_n^{PU} is a random variable which is independent and identically following a distributed exponential and the mean is θ, and the threshold of the outage probability is ε.

In this paper, in order to circumvent the hurdles of SINR, the following SLNR-oriented problem is skillfully designed to achieve the purpose of throughput maximization:

$$(\mathcal{P}) \max_{\mathbf{p}=\{p_n\}} \sum_{n=1}^{N} SLNR_n$$

$$\text{s.t.} \, C1 : p_n \leq p_{\max}, \, \forall n$$

$$C2 : \sum_{m=1, m \neq n}^{N} p_m g_{m,n} = I^{(-n)} \leq I_{\max}, \, \forall n$$

$$C3 : \Pr\left[\sum_{n=1}^{N} p_n g_n^{PU} \leq I_{\max}^p\right] \geq 1 - \varepsilon, \tag{6}$$

where $C1$ restricts the transmit-power of SC, $C2$ guarantees that all the SCs can coexist with each other and $C3$ focuses on the protection for the PU. As is mentioned in the previous definition of SLNR, not only does SLNR promote SCs benefit, but also can suppress the interference on others. In this way, the performance should be greatly improved, where the motivation lies. For problem (6), the challenge is that the channel gain is uncertain along with considering massive SCs' interference. In addition, there is not yet any satisfactory solution to the open issue so far. In the following part, we introduce a feasible way to tackle the difficulty with low complexity, which takes advantage of the convex optimization theory.

3   Algorithm Design

In this section, an algorithm based on Lagrangian techniques is developed to solve the problem $\mathcal{P}$. The objective function (3) is rewritten by $f(p_n) = \frac{p_n g_{n,n}}{\sum_{m=1, m \neq n}^{N} p_n g_{n,m} + \sigma_0^2}$ and obviously $f(p_n)$ is a concave function. According to [11], the objective function is also a concave function. The original optimization objective can be converted into the following form:

$$(\mathcal{P}*) \min_{\mathbf{p}=\{p_n\}} -\sum_n f(p_n). \tag{7}$$

Nevertheless, $C3$ does not meet the requirement of a convex function. Let $X_n = p_n g_n^{PU}$, and we assume that X_n independently follow the exponential distribution with mean $p_n \theta$. Meanwhile, assume the sum of the whole random variables is $X = \sum_n X_n$. Next, the original constraint is able to be changed into the probability form as follows:

$$\Pr\left[X = \sum_n X_n \leq I_{\max}^p\right] \geq 1 - \varepsilon. \tag{8}$$

To deal with the constraint (8), the distribution of X is very important. To study the distribution of X, we use the following Lemma 1 [12] to get Gaussian approximation.

Lemma 1. *The* Lyapunov's *central limit theorem (CLT): If $X_1, X_2, \ldots, X_n$ are independent of each other with mean $E(X_k) = \mu_k$ and variance $D(X_k) = \sigma_k^2 > 0$, we can obtain that*

$$Z_n = \frac{\sum_{k=1}^n X_k - \sum_{k=1}^n \mu_k}{B_n} \sim N(0,1) \tag{9}$$

and

$$\sum_{k=1}^n X_k \sim N\left(\sum_{k=1}^n \mu_k, B_n^2\right), \tag{10}$$

where $B_n^2 = \sum_{k=1}^n \sigma_k^2$.

Generally, X can be regard as a normally distributed random variable due to massive connections. Approximately, the mean is m and the variance is σ^2:

$$m \sim \sum_i p_i \theta$$
$$\sigma^2 \sim \sum_i (p_i \theta)^2. \tag{11}$$

As a result, the following expression is a substitute product for (8):

$$P(\mathbf{p}) = 1 - F_N(I_{\max}^p) = \frac{1}{2}\text{erfc}(\frac{I_{\max}^p - m}{\sqrt{2}\sigma}) \le \varepsilon. \tag{12}$$

where $F_N(\cdot)$ is the cumulative distribution function (CDF) of a normal distribution and its mean is m, the variance is σ^2. Moreover, $\text{erfc}(z) = \frac{2}{\sqrt{\pi}} \int_z^\infty e^{-t^2} dt$. For (12), the assumption is that $f_3(\mathbf{p}) = \frac{1}{2}\text{erfc}(\frac{I_{\max}^p - m}{\sqrt{2}\sigma}) - \varepsilon$. Inspired by the scheme proposed in [10], the problem $\mathcal{P}$ can be decomposed into a sub-problem $\mathcal{P}1$ and (12):

$$(\mathcal{P}1) \quad \min_{\mathbf{p}=\{p_n\}} \quad -\sum_n f(p_n)$$
$$\text{s.t.} \quad\quad C1, C2 \tag{13}$$

When a power allocation is given from $\mathcal{P}1$, we can check if it meets (12). If it does, the given power is optimal; or lower a step and check it again.

Next, we solve this minimization problem $(\mathcal{P}1)$ with Lagrangian techniques. Firstly, by introducing nonnegative dual variables $\boldsymbol{\lambda} = [\lambda_1, \lambda_2, ..., \lambda_N]$ and $\boldsymbol{\mu} = [\mu_1, \mu_2, ..., \mu_N]$, the Lagrange function is given by

$$L(\mathbf{p}, \boldsymbol{\lambda}, \boldsymbol{\mu}) = -\sum_n f_0(p_n) + \sum_n \lambda_n(p_n - p_{\max})$$
$$+ \sum_n \mu_n \left(\sum_{m=1, m\neq n}^N p_m g_{m,n} - I_{max}\right). \tag{14}$$

The dual function is defined as an unconstrained minimization of the Lagrangian function:

$$g(\boldsymbol{\lambda}, \boldsymbol{\mu}) = \inf_{\mathbf{p}} L(\mathbf{p}, \boldsymbol{\lambda}, \boldsymbol{\mu}). \tag{15}$$

We consider the problem in (15) for obtaining $g(\boldsymbol{\lambda}, \boldsymbol{\mu})$ with a given set of $\boldsymbol{\lambda}$ and $\boldsymbol{\mu}$. From (14), we have

$$\frac{\partial L(\mathbf{p}, \boldsymbol{\lambda}, \boldsymbol{\mu})}{\partial p_n} = \frac{g_{n,n}\sigma_0^2}{\left(\sum_{m=1, m\neq n}^{N} p_n g_{n,m} + \sigma_0^2\right)^2} + \lambda_n + \mu_n \sum_{m=1, m\neq n}^{N} g_{m,n}. \tag{16}$$

Because the dual function is always convex, a gradient-type search is guaranteed to converge to the global optimum. Problem $\mathcal{P}1$ is solved via the following first-order algorithm that utilizes the gradient of $L(\mathbf{p}, \boldsymbol{\lambda}, \boldsymbol{\mu})$ to simultaneously update the dual variables with constant Δ [11],

$$p_n^{k+1} = \left[p_n^k - \Delta \left(\frac{\partial L(\mathbf{p}, \boldsymbol{\lambda}, \boldsymbol{\mu})}{\partial p_n} \right) \right]_{\mathbf{P}} \tag{17}$$

$$\lambda_n^{k+1} = \lambda_n^k + \Delta (p_n - p_{\max}) \tag{18}$$

$$\mu_n^{k+1} = \mu_n^k + \Delta \left(\sum_{m=1, m\neq n}^{N} p_m g_{m,n} - I_{max} \right), \tag{19}$$

where k is the iteration number and Δ is the iteration step and $[\mathbf{x}]_{\mathbf{y}}$ is the projection of $\mathbf{x}$ onto the set $\mathbf{y}$. According to [11], the above gradient update can be guaranteed to converge to the optimal dual variables as long as the sequence of scalar step Δ is chosen appropriately. Finally, the entire steps to solve the optimization problem is displayed in Algorithm 1, where Step 1 to 6 solve $\mathcal{P}1$ and Step 8 to 11 check (12). The computational complexity of the algorithm is counted as follows. Solving (17) requires solving n equations by complexity $O(N)$. (17), (18) and (19) need to update $3n$ times at the same time with k iterations. Thus the complexity of the solution is measured by $O(3KN)$.

4   Simulation Results and Analysis

In Table 1, we list the key system parameters which are used in the simulations. The presented results are acquired via 1000 independent tests. In order to evaluate the performance of our proposed method, an optimal power allocation method oriented to SINR in [14] is introduced as a benchmark scheme.

Figure 1 depicts the sum throughput of different methods as a function of the density of SCs. The first observation from Fig. 1 is that the traditional SINR-oriented power control incurs a performance loss compared to the proposed SLNR-oriented power control with channel uncertainty. This phenomenon can

---

Algorithm 1. The power control algorithm with channel uncertainty

---

1: **Initialization:** Set the parameters $k = 0$, $\mathbf{p}^{(0)} = \{p_n^0\}, \Delta > 0$, ε, γ, δ.
2: **for** $k = 1, 2, \ldots$ **do**
3:     for each user, calculate $p_n^* = \left[p_n^k - \Delta \left(\frac{\partial L(\mathbf{p}, \lambda, \mu)}{\partial p_n} \right) \right]_{\mathbf{P}}$.
4:     Update λ_n^{k+1} and μ_n^{k+1} according to (18), (19).
5:     Update $p_n^{k+1} = p_n^*$.
6: **end for** Until $\left| \mathbf{p}^{k+1} - \mathbf{p}^k \right| \leq \delta$.
7: If $P(\mathbf{p}) \leq \varepsilon$, end; or, step into 8.
8: **for** $m = 1, 2, \ldots, M$ **do**
9:     Set $\mathbf{p}^{(k)} = \mathbf{p}^*(t)$.
10:     $\mathbf{p}^{(m+1)} = \mathbf{p}^{(m)} - \gamma$.
11: **end for** Until $P(\mathbf{p}) \leq \varepsilon$.

Table 1. System parameters used in simulations.

Parameter	Value	Comments
σ_0^2	$-100\,\text{dBm}$	Noise power
$p_{\max}$	$20\,\text{dBm}$	The maximum power of the SC
$I_{\max}, I_{\max}^p$	$-120\,\text{dBm}$	The interference threshold value of SC, PU
Δ	0.15	The iteration step
γ	0.02	The power step
δ	10^{-4}	The accuracy of power
ε	0.1	The threshold value of the outage probability

be attributed to the fact that not only does SLNR promote SCs benefit, but also can suppress the interference on others. In this way, the performance is greatly improved. Furthermore, we can also observe that both curves have a tendency to decline when the number of SCs becomes large, which is due to the fact that as the density increases, the mutual interference among SCs may result in several SCs out of work, considering the constraint $C2$ in the optimization.

Figure 2 discloses the sum throughput versus the SC's interference threshold level I_{max} as a function of the transmission power limit, where the number of SCs is invariable. From the figure, the sum throughput has a uptrend with the increase in transmission power. Taking $I_{max} = 10^{-10}$ as an example, it can be observed that more robust performance can be approached as the transmission power increases. Nevertheless, two conditions, the interference between SCs and the power increasing, have bind effects mutually. So the sum throughput will reach saturation state when the power increases to a certain extent. The phenomenon can be seen more noticeably when $I_{max} = 10^{-12}$ and $I_{max} = 10^{-13}$. Furthermore, the power increasing would be greatly limited if the interference threshold is too small such as $I_{max} = 10^{-15}$, and thus the sum throughput is almost invariable.

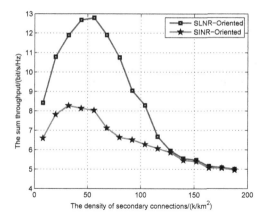

Fig. 1. The sum throughput comparison of the SLNR-oriented and SLNR-oriented power control schemes.

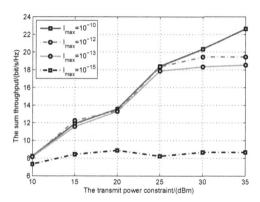

Fig. 2. The sum throughput of the CR system versus the transmission power limit for different I_{max}.

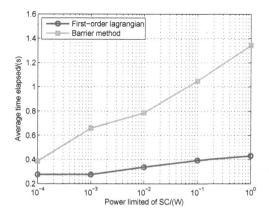

Fig. 3. Average time elapsed as a function of SC's power limit.

Figure 3 shows the average time elapsed as a function of SC's power limit. The commonly-used barrier method in the literature and the adopted first-order lagrangian method are compared. From the analysis results given in Sect. 3, we know that the adopted first-order lagrangian method has a linear complexity of $O(3KN)$. Addition to this, the results in Fig. 3 show that the time cost of our method is not only the lower one, but also it varies trivially as the SC's power limit increases.

5   Conclusion

In this paper, we have proposed a SLNR-oriented power control method to promote the throughput capacity of the system. Meanwhile, the interference is mitigated between massive SCs with channel uncertainty. Gaussian approximation based on the *Lyapunov*'s central limit theorem has been used to offer a conservative surrogate. Moreover, we have developed an effective power control algorithm based on first-order Lagrangian where the dual variables are updated simultaneously and the solution is amenable. Simulation results have validated the effectiveness of our proposed algorithm. As one future work, a subject of extension to more general channel models including correlation or feedback delay will be investigated.

Acknowledgement. This work is supported by the National Natural Science Foundation of China (No. 61501510 and No. 61301160), and Natural Science Foundation of Jiangsu Province (No. BK20150717), China Postdoctoral Science Foundation Funded Project, and Jiangsu Planned Projects for Postdoctoral Research Funds.

References

1. Huang, X., Wang, G., Hu, F.: Multitask spectrum sensing in cognitive radio networks via spatiotemporal data mining. IEEE Trans. Veh. Technol. **62**(2), 809–823 (2013)
2. Ding, G.R., Wu, Q.H., Yao, Y.D., Wang, J.L., Chen, Y.: Kernel-based learning for statistical signal processing in cognitive radio networks: theoretical foundations, example applications, and future directions. IEEE Sig. Process. Mag. **30**(4), 126–136 (2013)
3. Huang, X., Hu, F., Wu, J., et al.: Intelligent cooperative spectrum sensing via hierarchical Dirichlet process in cognitive radio networks. IEEE J. Sel. Areas Commun. **33**(5), 771–787 (2015)
4. Song, M., Xin, C., Zhao, Y., Cheng, X.: Dynamic spectrum access: from cognitive radio to network radio. IEEE Trans. Wirel. Commun. **19**(1), 23–29 (2012)
5. Ghasemi, A., Sousa, E.S.: Fundamental limits of spectrum-sharing in fading environments. IEEE Trans. Wirel. Commun. **6**(2), 649–658 (2007)
6. Dall'Anese, E., Kim, S.J., Giannakis, G.B., Pupolin, S.: Power control for cognitive radio networks under channel uncertainty. IEEE Trans. Wirel. Commun. **10**(10), 3541–3551 (2011)

7. Mokari, N., Parsaeefard, S., Saeedi, H., Azmi, P., Hossain, E.: Secure robust ergodic uplink resource allocation in relay-assisted cognitive radio networks. IEEE Trans. Sign. Process. **63**(2), 291–304 (2015)

8. Sadek, M., Tarighat, A., Sayed, A.H.: A leakage-based precoding scheme for downlink multi-user MIMO channels. IEEE Trans. Wirel. Commun. **6**(5), 1711–1721 (2007)

9. Xue, Z., Shen, L., Ding, G.R., Wu, Q.H.: Geolocation spectrum database assisted optimal power allocation: device-to-device communications in tv white space. KSII Trans. Internet Inf. Syst. (TIIS) **9**(12), 4835–4855 (2015)

10. Son, K., Bang, C.J., Song, C., Dan, K.S.: Power allocation for OFDM-based cognitive radio systems under outage constraints. In: IEEE International Conference Communications (ICC), Cape Town (2010)

11. Boyd, S.P., Vandenberghe, L.: Convex Optimization. Cambridge University Press, Cambridge (2004)

12. Milana, S., Fregolent, A., Culla, A.: Robust resource optimization for cooperative cognitive radio networks with imperfect CSI. IEEE Trans. Wirel. Commun. **14**(2), 907–920 (2015)

13. Wang, S., Shi, W., Wang, C.: Energy-efficient resource management in OFDM-based cognitive radio networks under channel uncertainty. IEEE Trans. Commun. **63**(9), 3092–3102 (2015)

14. Kha, H.H., Tuan, H.D., Nguyen, H.H.: Fast global optimal power allocation in wireless networks by local D.C. programming. IEEE Trans. Commun. **11**(2), 510–515 (2012)

15. Ben-Tal, A., Nemirovski, A.: Selected topics in robust convex optimization. Math. Prog. **112**(1), 125–158 (2008)

Research on LMMSE Channel Estimation Algorithm Using SLSM in WPM System

Weizhi Zhong[1(✉)], Sheng Su[1], Xin Liu[2], and Jianjiang Zhou[1]

[1] College of Astronautics, Nanjing University of Aeronautics and Astronautics,
210016 Nanjing, China
{zhongwz, susheng, zjjee}@nuaa.edu.cn
[2] School of Information and Communication Engineering,
Dalian University of Technology, 116024 Dalian, China
liuxinstar1984@dlut.edu.cn

Abstract. In wavelet packet modulation (WPM) system, the application of the linear minimum mean square error (LMMSE) channel estimation algorithm is limited by the nonlinearity of the fading WPM signal. To solve the problem, a simplified linear signal model (SLSM) matching with the LMMSE algorithm is established in the paper. The establishing of the SLSM is based on the fading channel and the orthogonality of WPM signals. The analysis and simulation results show that, the SLSM is matched with the LMMSE algorithm, and the SLSM based LMMSE algorithm can improve the WPM system performance effectively in frequency-selective fading environment.

Keywords: WPM system · Channel estimation · LMMSE algorithm · Simplified linear signal model

1 Introduction

Multicarrier scheme is a representative technique for high bit rate wireless communications, whose key advantages are multipath immunity, narrowband interference suppression and high band efficiency [1]. WPM system is a novel kind of multicarrier transmission method whose good time-frequency localization motivate a lot of current researches on it [2–5]. In WPM scheme, a packet structure can be divided not only in frequency domain but also in time domain and this unique division brings flexibility to wireless communication.

The wireless channel can be characterized as frequency-selective fading if the signal bandwidth is considerably wider than the coherence bandwidth; it brings a random gain for each scale of WPM signal and leads to selective fading in frequency domain. Generally, linear channel estimation method is used to compensate for this, the studies of channel compensation suitable for WPM system have been done in [6–9]. In [6, 7], least square (LS) algorithm based estimation method was proposed to improve the performance of the system in flat fading channel. In [8], parameters of multi-path Rayleigh fading channel can be obtained by using LS channel estimation and linear interpolation. In [9], Huber channel estimation algorithm was applied to achieve the best performance in non-Gaussian channel. However, LMMSE estimation algorithm,

© ICST Institute for Computer Sciences, Social Informatics and Telecommunications Engineering 2017
X.-L. Huang (Ed.): MLICOM 2016, LNICST 183, pp. 374–382, 2017.
DOI: 10.1007/978-3-319-52730-7_38

the most effective estimation method in frequency selective fading channel was not mentioned in these literatures. This is because in WPM, the inverse wavelet packet transform (IWPT) is employed for signal synthesis, which makes the relationship between the signal and the channel impulse response become more complex and this leads to the mismatch between the WPM signal model and the typical LMMSE method. As a consequence, LMMSE based channel estimation cannot be used in WPM system. Thus the SLSM which cooperates with the typical LMMSE method has been established in this paper. Theory and simulation results show that, the SLSM meets the using requirement of the LMMSE channel estimator, and the proposed estimators based on the SLSM provide significant performance gain for WPM system in frequency-selective channel.

The structure of the paper is organized as follows. After presenting the WPM system in Sect. 2, SLSM is established in Sect. 3. In Sect. 4, LMMSE estimation algorithm based on SLSM are put forward. The robustness of the LMMSE channel estimators based on the SLSM is tested in Sect. 5 and a summary and concluding remark appear in Sect. 6.

2 WPM System

2.1 WPM System Model

WPM is a kind of multicarrier transmission method [10]. The typical WPM system model is displayed in Fig. 1.

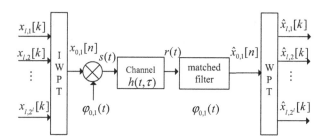

Fig. 1. Equivalent WPM base-band system model.

At the transmitter, the IWPT is applied to the input data $x_{l,m}[k]$ for synthesizing the transmitted symbols $x_{0,1}[n]$, each of which is individually amplitude modulated by scaling function $\varphi_{0,1}(t)$. Then, the synthesis WPM signal can be constructed by

$$
\begin{aligned}
s(t) &= \sum_n x_{0,1}[n]\varphi_{0,1}(t - nT_0) \\
&= \sum_{l\in\Lambda, m\in M_l} \sum_k x_{l,m}[k]\varphi_{l,m}(t - kT_l)
\end{aligned}
\tag{1}
$$

where $\varphi_{l,m}(t)$ are orthogonal subcarriers and T_l is the symbol interval. The basic principle of WPM and the performance comparison of OFDM and WPM have been described in the literature [3]. They are very different in form, therefore the effective channel estimation method LMMSE can be used in OFDM but not suitable for WPM.

3 The SLSM for WPM in Frequency Selective Fading Channel

3.1 Wireless Channel Model

Assume a multipath fading channel consisting of N resolvable paths; the channel impulse response $h(t, \tau)$ can be closely approximated as

$$h(t, \tau) = \sum_{i=0}^{N-1} \alpha_i(t)\delta(\tau - \tau_i) \tag{2}$$

In (2), N is the total number of propagation path and $\delta(\cdot)$ denotes the impulse function. For the ith path, $\alpha_i(t)$ is the complex impulse response and τ_i is the path delay. In the case, the received signal at the receiver can be expressed as

$$\begin{aligned} r(t) &= s(t) * h(t, \tau) + w(t) \\ &= \int_0^{\tau_{\max}} s(t - \tau)h(t, \tau)d\tau + w(t) \end{aligned} \tag{3}$$

where $w(t)$ is an additive white Gaussian noise with zero mean value and variance σ_w^2, and we can safely assume the noise is uncorrelated with the channel impulse response.

3.2 The SLSM

For analysis, a special case of two-path frequency selective fading channel is taken into account. In the condition, the equivalent low-pass received signal is modeled as

$$r(t) = h_1(t)s(t) + h_2(t)s(t - \tau) + v(t) \tag{4}$$

where $s(t)$ is the multiplexed WPM signal defined in (1), $h_1(t)$ and $h_2(t)$ are channel gain which are mutually independent complex Gaussian processes. At the receiver, $r(t)$ is passed through the matched filter $\varphi_{0,1}(t)$ for sampling and matching firstly. Since $h_1(t)$ and $h_2(t)$ are slow processes and can be regarded as a constant in several bit duration, the output of the match filter is

$$\begin{aligned} r_s[n] &= \int r(t)\varphi_{0,1}(t - nT_0)dt \\ &= h_1[n]x_{0,1}[n] + h_2[n]\sum_n x_{0,1}[k]R_\varphi(nT_0 - kT_0 - \tau) + v[n] \end{aligned} \tag{5}$$

where $h_1[n]$ and $h_2[n]$ are the sampled values of $h_1(t)$ and $h_2(t)$ at instant nT_0, and $R_\phi(\cdot)$ denotes the autocorrelation function of scale function $\varphi_{0,1}(t)$.

Re-arrangement of the right hand of Eq. (5) results in

$$r_s[n] = \left(h_1[n] + h_2[n]R_\varphi(-\tau)\right)x_{0,1}[n]$$
$$+ \left\{ h_2[n] \sum_{k' \neq 0} x_{0,1}[n - k']R_\varphi(k'T_0 - \tau) + v[n] \right\} \tag{6}$$
$$= h_s[n]x_{0,1}[n] + v_s[n]$$

where $h_s[n]$ and $v_s[n]$ can be expressed as

$$\begin{cases} h_s[n] = h_1[n] + h_2[n]R_\varphi(-\tau) \\ v_s[n] = h_2[n] \sum_{k' \neq 0} x_{0,1}[n - k']R_\varphi(k'T_0 - \tau) + v[n] \end{cases} \tag{7}$$

In (7), $v_s[n]$ represents the total noise including the Gaussian noise and the multi-path interference.

It is well known that LMMSE is more stable and accurate method. The applicable condition of the LMMSE is that the estimation value is the linear function of the observe value which can be modeled as

$$A = B\theta + c \tag{8}$$

where c is the noise with zero mean value and is irrelevant to θ. Obviously, the form of formula (6) does not meet the requirements, so it is not suitable for the algorithm LMMSE. Therefore, we need to simplify the formula (6), so that it can be used for LMMSE algorithm.

In frequency selective fading case, $\tau > T_0$. It is well known that $R_\varphi(\tau)$ has a good correlation ship. When $\tau > T_0$, $R_\varphi(-\tau)$ approaches to zero and $R_\varphi(0)$ is equal to one. Therefore, Eq. (5) can be transformed into the following form

$$r_s[n] = h_1[n]x_{0,1}[n] + h_2[n]x_{0,1}\left[n - \frac{\tau}{T_0}\right] + v[n] \tag{9}$$

Similarly, if the total number of propagation paths is L, and for the ith path, $\tau_i = iT_0$. At the same time assume that the channel is slow fading case, the received signal can be derived as follow

$$r_f = Xh + v \tag{10}$$

where v is the AWGN with zero mean value and variance σ_v^2, and matrix X and vector h can be described as

$$\boldsymbol{h} = [h_0, h_1, \cdots, h_{L-1}]^T \tag{11}$$

$$\boldsymbol{X} = \begin{bmatrix} x_{0,1}[1] & x_{0,1}[N] & \cdots & x_{0,1}[N-L+2] \\ x_{0,1}[2] & x_{0,1}[1] & \cdots & x_{0,1}[N-L+3] \\ \vdots & \vdots & \ddots & \vdots \\ x_{0,1}[N] & x_{0,1}[N-1] & \cdots & x_{0,1}[N-L+1] \end{bmatrix} \tag{12}$$

The received signal model described in (10) is SLSM. From (11) and (12) we can find that, the relationship between the estimation value $\hat{\boldsymbol{h}}$ and the observe value $\boldsymbol{r}_f$ satisfies the condition described in (8), which denotes that the established SLSM meets the application condition of LMMSE algorithm.

4   LMMSE Channel Estimation Adopt SLSM

LMMSE uses the statistic properties of the channel coefficients and the additive noise to reduce the mean square error, and its estimation value comes from the minimum mean square error described as follow

$$\frac{\partial J}{\partial \hat{\boldsymbol{h}}} = \frac{\partial E\left[\left(\boldsymbol{h} - \hat{\boldsymbol{h}}\right)^T \left(\boldsymbol{h} - \hat{\boldsymbol{h}}\right)\right]}{\partial \hat{\boldsymbol{h}}} = 0 \tag{13}$$

Based on (13), the estimation value $\boldsymbol{h}$ can be described as follow

$$\begin{aligned} \hat{\boldsymbol{h}}_{LMMSE} &= \boldsymbol{\mu} + \boldsymbol{P}\boldsymbol{X}^T \left[\boldsymbol{X}\boldsymbol{P}\boldsymbol{X}^T + \boldsymbol{R}\right]^{-1} \left[\boldsymbol{r}_f - \boldsymbol{X}\boldsymbol{\mu}\right] \\ &= \left[\boldsymbol{P}^{-1} + \boldsymbol{X}^T \boldsymbol{R}^{-1} \boldsymbol{X}\right]^{-1} \left[\boldsymbol{P}^{-1}\boldsymbol{\mu} + \boldsymbol{X}^T \boldsymbol{R}^{-1}\boldsymbol{r}_f\right] \end{aligned} \tag{14}$$

According to the SLSM established in (10), the $\boldsymbol{h}, \boldsymbol{r}_f, \boldsymbol{v}$ and $\boldsymbol{X}$ can be denoted as (11) and (12). In (14), $\boldsymbol{\mu}_{(L \times 1)}$ and $\boldsymbol{P}_{(L \times L)}$ are the mean value and variance of $\boldsymbol{h}$, and the superscripts $[\cdot]^{-1}$ and $[\cdot]^H$ denote matrix inversion and Hermitian transpose respectively. $\boldsymbol{R} = E\{\boldsymbol{v}\boldsymbol{v}^H\}$ is the auto-correlation matrix of $\boldsymbol{v}$. The result described in (14) has proved the matching between the proposed model and the LMMSE algorithm in theory.

5   Simulation and Results

Perfect synchronization has been done since the aim is to observe the channel compensation performance. The received signal model was shown as (10), and the simulation parameters were shown in Tables 1 and 2. In the experiment, Daubechies 12 wavelet was used to synthesize and decompose the WPM symbols, the channel was modeled as 6-ray model, and the corresponding amplitude and phase are defined in Table 2. As the channel was frequency selective fading, the block-type pilot was used [11].

Table 1. Simulation parameters

Wavelet pattern	Subcarriers number	Modulation scheme	Path number
Daubechies 12	M = 512	QPSK	L = 6

Table 2. The amplitude and phase of h

Delay time τ_i	Amplitude h_i	Phase ϕ_i
$\tau_1 = 0$	$|h_0| = 1$	1.2567
$\tau_1 = T_0$	$|h_1| = 0.6065$	0.6283
$\tau_1 = 2T_0$	$|h_2| = 0.3679$	−1.2567
$\tau_1 = 3T_0$	$|h_3| = 0.2231$	−1.2567
$\tau_1 = 4T_0$	$|h_4| = 0.1353$	0.6283
$\tau_1 = 5T_0$	$|h_5| = 0.0821$	0.6283

As shown in Figs. 2 and 3, the SLSM based estimator can improve the system performance significantly, although the performance of the LMMSE estimator is much better than that of LS, but is far less than that of the optimal estimator. This is because WPM does not use the cyclic prefix (CP), and the CP is the main method to suppress the inter-carrier interference (ICI) in OFDM system. Therefore, the ICI caused by the frequency selective fading affects the performance of the system seriously, even if the channel estimation has been done, it still can't remission the decline of the performance completely.

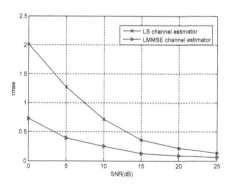

Fig. 2. The estimation error of the SLSM based typical estimators

The increase of estimation error caused by time delay and the multipath energy is shown in Figs. 4 and 5. In the simulation, two path channels were used, and the channel model was denoted as follow:

$$r_f[n] = h_0 x_{0,1}[n] + h_m x_{0,1}[n - m] + v[n] \tag{15}$$

where (15) is the particular form of (10). The time delay $\tau_m = mT$ and the ratio of the energy between the main path and the second path was defined as

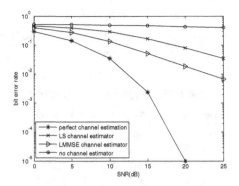

Fig. 3. BER performance of WPM system with SLSM based channel estimators in selective fading channel

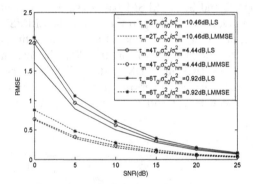

Fig. 4. The estimation error of SLSM based typical channel estimators with different time delay τ_m and $\sigma_{h_0}^2 / \sigma_{h_m}^2$

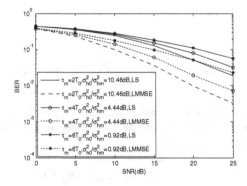

Fig. 5. BER performance of WPM system using SLSM based channel estimators in selective fading channel with different time delay τ_m and $\sigma_{h_0}^2 / \sigma_{h_m}^2$

$$\sigma_{h_0}^2 / \sigma_{h_m}^2 = 10 \lg \left(\frac{|h_0|^2}{|h_m|^2} \right) \qquad (16)$$

where h_0 was the main path.

From the simulation results we can find that, with the increase of the delay and the decrease of energy ratio, the estimation error increases and the system performance declines. The above results indicate that the SLSM based LMMSE estimation method can improve the system performance significantly, which has proved the good matching between the LMMSE and the proposed model.

6 Conclusion

The investigation shows that the nonlinear of the WPM fading signal model in frequency selective fading channel is the significant limitation to the using of the LMMSE algorithm. Therefore, the SLSM has been established and the LMMSE estimator using SLSM is employed to improve the system performance. The key features of the SLSM are: (i) It is obtained by using the orthogonality principle of the wavelet packet function and the characters of the frequency selective fading channel. (ii) The linear form of the SLSM satisfies the applicable condition of the LMMSE estimator.

The theoretical and simulation results demonstrated that the proposed SLSM has solved the mismatch problem between the nonlinear of the signal model and the linear estimator, and the SLSM based LMMSE channel estimators can significantly improve the WPM system performance in frequency selective fading environment.

Acknowledgements. This work was supported by the National Natural Science Foundations of China (61301105, 61401288 and 61601221), the Natural Science Foundations of Jiangsu Province (BK20140828), the China Postdoctoral Science Foundations (2015M581791 and 2015M580425), and the Fundamental Research Funds for the Central Universities (DUT16RC(3) 045).

References

1. Ali, S., Rahman, M., Hossain, D., et al.: Simulation and bit error rate performance analysis of 4G OFDM systems. In: Computer and Information Technology, pp. 138–143. IEEE (2008)
2. Lindsey, A.R.: Wavelet packet modulation for orthogonally multiplexed communication. IEEE Trans. SP **45**(5), 1336–1339 (1997)
3. Antony, J., Petri, M.: Wavelet packet modulation for wireless communications. IEEE Wireless Commun. Mobile Comput. **5**(2), 123–137 (2005)
4. Linfoot, S.: A novel approach to communications for DVB-T and DVB-H. In: Consumer Electronics, pp. 1–4. IEEE (2008)
5. Bajpai, A., Lakshmanan, M., Nikookar, H.: Channel equalization in wavelet packet modulation by minimization of peak distortion. In: Personal, Indoor and Mobile Radio Communications, pp. 152–156. IEEE (2011)

6. Wu, J.: Wavelet Packet Division Multiplexing. McMaster University, Canada (1998)
7. Zhou, L., Li, J., Liu, J., et al.: A novel wavelet packet division multiplexing based on maximum likelihood algorithm and optimum pilot symbol assisted modulation for rayleigh fading channels. Circ. Syst. Sig. Process. **24**(3), 287–302 (2005)
8. Yang, M., Liu, Z., Dai, J.: Frequency selective pilot arrangement for orthogonal wavelet packet modulation systems. In: Computer Science and Information Technology, pp. 499–503. IEEE (2008)
9. Yang, M., Liu, Z., Dai, J.: Robust channel estimation in WPM systems. In: Circuits and Systems for Communications, pp. 750–753. IEEE (2008)
10. Wong, K., Wu, J., Davidson, T., et al.: Wavelet packet division multiplexing and wavelet packet design under timing error effects. Sig. Process. **45**(12), 2877–2890 (1997)
11. Coleri, S., Ergen, M., Puri, A., et al.: Channel estimation techniques based on pilot arrangement in OFDM systems. IEEE Trans. Broadcast. **48**(3), 223–229 (2002)

Virtual Memory Based Radar Display and Control System

Zengshan Tian, Mingxiao Wang[✉], Mu Zhou, and Feng Qiu

Chongqing Key Lab of Mobile Communications Technology,
Chongqing University of Posts and Telecommunications,
Chongqing 400065, People's Republic of China
{tianzs,zhoumu}@cqupt.edu.cn, wangmx199111@163.com,
qiufeng245@outlook.com

Abstract. Since the graphics processor of common X86 platform does not support the display function of multi-layer graphics, this paper proposes to combine the Qt Graphics-view framework with OpenGL pixels operation function and utilizes the graphics memory under the Linux system to achieve the layered and efficient display based on the radar information. The proposed approach can effectively realize the main functions of radar display and control system, including the radar Plan Position Indicator (PPI) display, target glint in warning area, and Automatic Identification System (AIS) target management and display. The experimental results prove that the proposed approach has the advantages of high efficiency, smooth image update, and low hardware requirements. In addition, the proposed approach has been successfully applied to a typical shipborne radar navigation system.

Keywords: Radar · PPI display · Qt · OpenGL · AIS

1 Introduction

Since the radar system has the advantages of low environmental requirement and target detection with real-time capacity, high positioning accuracy, and long ranging property, it is indispensable for ship navigation. The main features of modern shipborne navigation radar display systems are the digital information processing, high-performance information display, and friendly interactive experience [1]. Most of the implementation platform is based on the existing System on Chip (SOC) platform or embedded platform by selecting the Advanced RISC Machines (ARM) as the core. As one of the most important parts for human-computer interaction in the radar system, the radar display and control terminal undertakes the fundamental tasks of original radar image display, radar control, and tracking target display. In addition, the modern radar can integrate the radar images with the Global Position System (GPS), Automatic Identification System (AIS), compass, and log to guarantee the well navigation performance.

Since the radar Plan Position Indicator (PPI) display involves a large amount of image data with high refresh rate, the radar display terminal requires high-grade processor and graphics card. Many research institutes mainly focus on the advanced embedded processors combined with specialized embedded graphics processor, which generally has multiple images display layers and 2D/3D graphics acceleration. The

© ICST Institute for Computer Sciences, Social Informatics and Telecommunications Engineering 2017
X.-L. Huang (Ed.): MLICOM 2016, LNICST 183, pp. 383–392, 2017.
DOI: 10.1007/978-3-319-52730-7_39

three typical types of software development approaches used for radar display terminal are summarized as follows. (1) Based on the embedded Linux system, the advanced GUI development environment, like the Qt and GTK +, is adopted to develop the display and control interface and render the secondary radar information, and meanwhile the Framebuffer technology is used to draw the radar original image through the direct operation of video memory [2]. (2) The third-party plug-in which supports hardware acceleration, like the OpenGL and DirectFB, is combined with the GUI development environment, like the echo display method based on OpenGL texture mapping [3] and DirectFB library, to write driver for the sake of achieving the image layering and hardware acceleration [4]. And (3) On the X86 platform, the DirectDraw technology based on the Windows system is adopted to achieve the multi-layer image display and acceleration, by using the Qt or VC++ to develop the radar interface. For example, the direct video technology with Visual C++ and DirectX programming method is used to realize the radar image display [5].

Another common approach to develop the radar display and control system is to select the Field Programmable Gate Array (FPGA) as the main controller, with an external Video Graphics Array (VGA) conversion chip and Digital Signal Processor (DSP) chip or ARM processor to process the radar data, and then rely on the FPGA processing of image data fusion to display the radar images [6].

Aiming at the deficiencies of the approaches mentioned before, this paper proposes to combine the Qt Graphics-view framework with OpenGL hardware acceleration to achieve the layered and efficient display for radar information. This approach is based on the X86 platform with low configuration, and does not require the graphics processor to support the multi-layer image display function on hardware. Since the graphics memory, Qt multi-thread, and OpenGL pixel operation interface are used to develop the cross-platform radar display and control software, this approach can effectively compensate the radar display and control terminal development vacancies on the low configure X86 platform based on the Linux system.

2   Rendering of Radar Images

2.1   Qt Graphics-View Framework

The Qt Graphics-view framework in Fig. 1 can manage and display a large number of custom 2D graphic items and interact with each other by using the view component to visualize the graph and support the scaling and rotation. It includes event propagation architecture, with the ability of accurately doubling the precision interaction with the items in scene. The UI controls can be easily embedded in the QGraphicsScene and use the Qt's QSS to separate the interface beautification and software code which can be used to customize many GUI elements.

2.2   OpenGL Pixels Operation Interface

The OpenGL is the interface of developing high-quality image in SGI's graphics workstation and has become a representative 3D graphics interface [7]. The OpenGL

Fig. 1. Qt graphics-view framework.

provides three basic functions of processing image data, namely the frame buffer pixel read function glReadPixels, frame buffer pixel write function glDrawPixels and frame buffer pixel duplicate function glCopyPixels. The glDrawPixels reads a rectangular array of pixels from the processor memory and writing the data to frame buffer at the raster position specified by glRastePos* [8]. During the process of pixel transfer from the memory to frame buffer, the glPiexlZoom is used to set the image scaling and rotation and glPixelStore is used to set the display range of image.

2.3 Radar Image Display

Since most of the Intel integrated graphics on the X86 platform do not support the image layered display, the directly operating frame buffer to realize the radar image display will not benefit the drawing of radar secondary information, like the AIS targets, ARPA targets, and navigation routes. The proposed approach is based on the characteristics of Qt Graphics-view framework which can coexist with the OpenGL by using the image layering method and creating the virtual graphics memory, to realize the raw radar data drawing. The implementation steps are described as follows.

Step 1: Create the QGLWidget in the main window and use the member function makeCurrent to set its RenderContext to be the current value.

Step 2: Create the custom view and scene object which are inherited from the QGraphicsView and QGraphicsScene respectively, and then set the QGLWidget as the window of the custom view object. The background RenderContext is set to be the OpenGLContext, with the purpose of avoiding the CPU rendering and improving the efficiency of drawing.

Step 3: Create the virtual memory in the custom view object by a two-dimensional array, namely radar video buffer in processor memory, which size is $1024 \times 1024 \times 4$ bytes. We need to initialize it with the background color and modify the array to implement the real-time change of radar image. The subsequent drawing of radar image only requires changing the pixel values in virtual memory area, which is similar to the process of the direct operation of memory.

Step 4: In the function drawBackgroud of custom view object, use the QPainter member function beginNativePainting and endNativePainting to make the pure OpenGL rendering, and then use the function InitGL, ResizeGL and PaintGL to execute the OpenGL operations.

Step 5: In the function PaintGL, use the OpenGL function glDrawPixels to draw the virtual memory data into the screen as a background layer.

Step 6: Select the azimuth circle, heading line, AIS targets and tracking targets as QGraphicsItem added to the custom scene object, and then use the function setZValue to set the items' display order.

Our system uses the GL_RGBA format when calling the function glDrawPixels and the data format are GL_UNSIGNED_BYTE because using the same image format and data format with frame buffer can reduce the workload of OpenGL implementation. When zooming in the radar image, we use the function glPixelStore* to set the corresponding parameters in order to capture the image display area, and meanwhile set the scale factor and display position on the screen by the function glPixelZoom and glRasterPos*.

3 Design of Radar Display and Control System

Our system relies on the Linux system to receive the raw radar data which are processed by the FPGA, and then parse and display the data. In addition, it is also responsible for the deployment and control of the entire radar system, secondary information display, recording and managing the AIS targets, and providing a graphical interactive interface.

3.1 Interface Design

Our system combines the GPS, AIS, compass, and log to enhance the function and enrich the display content. The display and control system interface includes the radar display area, information monitoring area, display control menu, function control menu, and install menu. Among them, the information monitoring area overlaps the menu area and is switched by the control button for the sake of ensuring that the interface is concise. The interface is designed in Fig. 2.

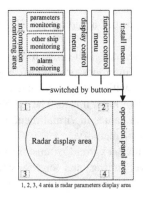

Fig. 2. Radar display and control interface.

The radar display area shows raw radar data, tracking targets, and AIS targets. The parameters monitoring area shows the ship information, like the latitude, longitude, course, and speed, while other ship monitoring area shows the tracking and AIS targets information, like the name or label, Maritime Mobile Communication Service Identifier (MMSI), speed, heading, Time to Closest Point of Approach (TCPA), and Distance to Closest Point of Approach (DCPA). The alarm monitoring area shows all the types of alarm information in a real-time manner. The display control menu contains the color choices, trail display, vector line length, brightness adjustment and adjustment buttons. The function control menu contains the warning area selection, route points, TCPA, DCPA and function setting buttons. The install menu is used for the radar installation and requires the password for entering.

3.2 Radar Video Display

The conventional radar PPI display depends on the afterglow effect of fluorescent material to display the echo signal [9], whereas the modern radar is based on the raster scanner. Since the echo signal is described by polar coordinates, it is required to perform the coordinates-transformation before drawing. Our system combines the coordinates index look-up table method and uniform motion model [1] to exhibit the relations between the raw radar data and drawing positions in the virtual memory [10]. The polar coordinates represent the target position with the distance and azimuth (ρ, θ). The conversion from the polar coordinates into rectangular coordinates (x, y) is shown in (1).

$$\begin{cases} x = \rho \sin \theta \\ y = \rho \cos \theta \end{cases} \tag{1}$$

Where θ is the angle between the target and ship's heading.

The purpose of coordinate index look-up table method is to calculate each point's corresponding quantization angle and display radius based on its two-dimensional index in the virtual memory, and then create the two-dimensional coordinate conversion table. Based on (2), we first create a two-dimensional data list $Index[r][\theta]$ ($0 \leq r < MaxRng$, $0 \leq \theta < MaxAzi$), where $MaxRng$ is the maximum display radius and $MaxAzi$ is the maximum quantization angle, and then use (3) to calculate each point's index idx and store it into the list Index. When drawing the radar data, we need to look up the table according to its polar coordinates to obtain the corresponding index in the virtual memory, and assign a color value.

$$\begin{cases} r = \sqrt{\left((x_1 - R)^2 + (y_1 - R)^2 \right)} \\ \theta = \mathrm{atan}(x_1 - R, y_1 - R) * MaxAzi / 2\pi \end{cases} \tag{2}$$

where R is the half of the height of screen.

$$idx = y_1 2R + x_1 \tag{3}$$

Since the uniform motion model selects the true motion of radar operation as the off-center relative motion, it is not necessary to calculate the two-dimensional coordinate transformation table when the observing mode is changed. The size of the two-dimensional coordinate index table should be doubled from the size of actual screen by the reason that the radar display range is doubled under the off-center condition. Thus, when the radar is off-center in anywhere, we can obtain the actual screen location of each echo point from the look up table with simple coordinates-transformation.

The radar display area of our system is a square area with the size of 1024 × 1024 pixels, and the actual PPI display area is a circle with the radius is equaling to 500. After the raw radar data are parsed, we store the range, angle, index, and video data into the structure temEchoData as defined below.

```
Typedef struct temEchoData{
  quint16 range;
  quint16 angle;
  quint16 packetNum;
  quint8 echo[1000];
}EchoData;
```

Since the angle of two adjacent frames echo data are not fixed in FPGA sampling, we use the angle difference of two frames echo data to draw a sector. Our system selects the previous frame echo data as the actual drawing data, and selects the current frame data's angle as the end angle of the sector drawing. In addition, the refresh rate of the image is determined by the division angle means the radar image span of each drawing, and our system choose 6°. When the division angle is large, both the refresh rate and hardware consumption will be low, but the image will not be smooth. When the division angle is little, the image will be smooth, but the hardware consumption will be high.

3.3   Radar Warning Area Alarm

When using the shipborne navigation radar, the warning area alarm is a convenient and practical function, which achieves the automatic detection and alarm in the designated area and reduces the workload of radar operator. We aim to detect whether there are targets in warning area, and make the warning area targets glint when the radar image is in real-time refresh.

Since the radar image display system uses the virtual memory combined with angle dividing, the blinking effect of targets can be achieved by changing target content in the virtual memory between two adjacent refresh of radar images. However, the significant attention should be paid when the target content is changed since the logical relationship between the target content and actual echo data is used to modify the virtual memory.

3.4 AIS Targets Management

The modern common radar is limited by the hardware constrain, and meanwhile the number of automatic tracking target based on the raw radar data is generally less than 100 and error tracking occurs due to the existence of noise. As a new type of digital navigation system, the AIS broadcasts to the nearby ships and coast stations via Very High Frequency(VHF) and enables the nearby ships and coast stations to master the dynamic and static information of ships which greatly increased the accuracy of the target identification and tracking.

The AIS target management mainly includes the AIS target information extraction, target dynamic refresh, alarm detection, display position locating, and quantity management. Considering the characteristics, like the large amount of calculation about the AIS target data, frequent calculation of display position, and stability of refresh time, our system defines two different coordinate system when calculate the AIS target screen coordinates. The first one is the rectangular coordinate system between the AIS target and the corresponding ship, and while the second one is the screen coordinate system between the screen and the corresponding ship.

The AIS target information is processed with fixed time, to create a QHash table to store the parsed data of AIS information with MMSI number as index. During the processing of AIS target information, we refresh the AIS target state, calculate the alarm information relevant to the corresponding ship, update the historical points, and analyze whether the AIS target is lost.

Since the number of AIS targets in the harbor or busy waterway is large, handling all the AIS targets and displaying them on the screen will cause difficulties and waste hardware resources. To solve this problem, we limit the number of AIS targets by removing the long-range AIS targets based on the distance between the AIS targets and the corresponding ships. The steps of AIS target screening is as follows.

Step 1: By assuming that the screened number of AIS targets is k, we create two arrays, namely Rng[k] and Mmsi[k], as the sorting containers.

Step 2: We select k targets from AIS targets storage list and store them into array Rng in ascending order of distance, and then store the MMSI number of the corresponding objects into Mmsi.

Step 3: Based on the AIS targets storage list, if the distance is less than the first element in Rng, we ignore it, and otherwise insert it into Rng with the corresponding position, remove the minimum, and insert the target's MMSI number into Mmsi with the corresponding position. Finally, the MMSI number saved in Mmsi is the AIS targets index to be removed.

4 Test Results

In our testing, the hardware platform is the Celeron 2 GHz CPU, 1 GHz memory, intel integrated graphics and flash 8G hard disk.

The control and display system is developed in Ubuntu12.04 and the test system is TimeSys Linux. The Ethernet cable is selected as the transmission channel between the baseband process board and display and control terminal. The terminal and baseband process board are separated for the sake of facilitating the miniaturization of radar display terminal.

4.1   Real-Time Performance Testing

The system operation interface is shown in Fig. 3. The yellow and gray images represent the actual target and clutter in the circular display area respectively. The antenna rotation period is 2.5 s and image refresh angle is 6°. The image refreshes 24 times per second with smooth running effect. The number of echo data to be processed reaches the millions level. The system is able to perform coordinate conversion, drawing display, and target detection and trail control, as well as respond quickly to a variety of radar operations.

Fig. 3. Operation interface of radar display and control system.

In the Qt environment, the time consumption of rendering the 1024 × 1024 pixels is shown in Table 1. From this table, we can find that the drawing efficiency of OpenGL is higher than other methods and it's fully meet the real-time requirements.

Table 1. Time consumption by different drawing approaches

Approaches	QPainter	QGraphicsItem	OpenGL
Time consumption(ms)	134	625	15

4.2   Alarm Function Testing

The radar warning area function is to detect the targets and glint alarm automatically based on the original echo image. Our system detects the values of virtual memory, and when the warning area appeared targets than glint alarm immediately, and there is no false alarm or leak detection. Figure 4 shows the normal display image after delimiting the warning area surrounded by the white line. Figure 5 shows the image under the condition that the warning area targets are hided. Our system achieves the glint effect of targets by rapidly alternating the image refresh.

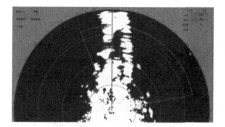

Fig. 4. Normal display images of warning area.

4.3 AIS Targets Screening Testing

In this section, we simulate 10 AIS targets to test the AIS targets management and screening function. The display interfaces before and after screening are shown in Fig. 6. In the figure, the green triangle is the AIS target, while the triangle surrounded by a box is the selected AIS target. When setting the number of AIS display targets as 4, we can find our system reserves four AIS targets which are closest to the corresponding ships. The distances from AIS targets to the corresponding ship before and after scanning are shown in Table 2.

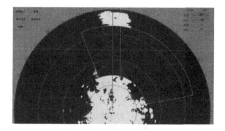

Fig. 5. Hidden images of warning area.

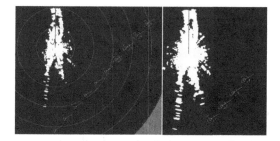

Fig. 6. AIS targets display before and after scanning.

Table 2. Distance from AIS targets and ship before and after scanning

Target IDs	1	2	3	4
Before(Nm)	3.226	3.698	4.267	4.902
After(Nm)	3.176	2.873	2.774	2.901

5 Conclusion

Since the hardware graphics processor on the common X86 platform does not support the multi-layer image display, this paper proposes a new radar image drawing method that selects the graphic memory combined with the Qt Graphics-view framework and OpenGL hardware acceleration function to achieve the real-time radar image display, glint alarm in warning area, and AIS targets management and integrated display. This method does not require the graphics processor to support the multi-layer image display, and meanwhile it has the advantages of cross-platform, low hardware requirements and well display performance. Based on the extensive testing result, this method is proved to be able to work well under the condition that the refresh rate of radar image equals to 36 per second.

Acknowledgments. This work was supported by the Program for Changjiang Scholars and Innovative Research Team in University (IRT1299), National Natural Science Foundation of China (61301126), Special Fund of Chongqing Key Laboratory (CSTC), and Fundamental and Frontier Research Project of Chongqing (cstc2013jcyjA40041, cstc2015jcyjBX0065).

References

1. Wei, B., Guo, Y., Mo, H.: The realization of echo displaying for marine navigation radar under high-rotating speed and multiple operating modes. Sci. Technol. Eng. **13**(17), 4962–4967 (2013)
2. Ren, Q., Yang, J.: Implementation and application of FrameBuffer in radar display. Electron. Sci. Technol. **22**(6), 61–63 (2009)
3. Zhang, P.: A method of radar echo display based on OpenGL. Shipboard Electron. Countermeasure **34**(3), 39–42 (2011)
4. Liang, W.: Research on the DirectFB graphics engine transplant based on BCM7241 platform. Comput. Telecommun. **4**, 53–55 (2015)
5. Wu, W.: Design and implementation of radar control system software on Direct3D under windows. Sci. Technol. Inf. **1**, 88–89 (2014)
6. Ying, S., Zhang, X.: A radar display terminal based on a partial-screen-updating method. In: IEEE International Conference on Embedded Software and Systems Symposia, pp. 28–31 (2008)
7. He, Y., Zhang, G., Zhang, Q.: Design and simulation of radar terminal display based on OpenGL. Informatization Res. **38**(2), 15–18 (2012)
8. OpenGL Architecture Review Board, Dave, S., Mason, W., Jackie, N., Tom, D.: OpenGL Programming Guide(Version 4). Posts & Telecom Press, Bei Jing (2005)
9. Fan, W.: An efficient algorithm for radar PPI display. Mod. Radar **37**(2), 41–45 (2015)
10. Li, B., Liu., D.: Research and realization of coordinate conversion in radar video display. In: Ninth International Conference on Computational Intelligence and Security, pp. 277–279 (2013)

Robust Spectral-Temporal Two-Dimensional Spectrum Prediction

Guoru Ding[1,2(✉)], Siyu Zhai[2], Xiaoming Chen[2], Yuming Zhang[2], and Chao Liu[2]

[1] National Mobile Communications Research Laboratory,
Southeast University, Nanjing, China
dingguoru@gmail.com
[2] PLA University of Science and Technology, Nanjing, China
13222759316@163.com, chenxm23732@126.com, zhangym_2000@163.com,
liuchao20121601@163.com

Abstract. With the development of mobile network, the limited spectrum resources are being running out of. Therefore, there is a harsh need for us to be able to know the current spectrum state as well as predict the future spectrum state. Though a number of studies are about spectrum prediction, some fundamental issues still remain unresolved: (i) the existing studies do not account for anomaly data, which causes serious performance degradation, (ii) they do not account for missing data, which may not hold in reality. To address these issues, in this paper, we develop a robust spectral-temporal spectrum prediction (R-STSP) framework from corrupted and incomplete observations. Firstly, we present data analytic of real-world spectrum measurements to analyze the impact of anomalies on the rank distribution of spectrum matrices. Then, from a spectral-temporal spectrum perspective, we formulate the R-STSP as a matrix recovery problem and develop an optimization method to efficiently solve it. We apply the formulated R-STSP to real-world VHF spectrum data and the results show that R-STSP outperforms state-of-the-art schemes.

Keywords: Spectrum prediction · Anomaly data · Missing data · Matrix completion and recovery

1 Introduction

The rapid development of mobile network is running out of the limited spectrum resource, which is a signal that we need more probable spectrum usage to adapt to this trend [1,2]. To achieve this goal, we need to know the current spectrum state as well as predict the future spectrum state. Spectrum sensing helps us determines the current spectrum state using various signal detection methods [3–5], while spectrum prediction gives us the future spectrum data. Spectrum prediction's applications in wireless networks has many merits, for example,

© ICST Institute for Computer Sciences, Social Informatics and Telecommunications Engineering 2017
X.-L. Huang (Ed.): MLICOM 2016, LNICST 183, pp. 393–401, 2017.
DOI: 10.1007/978-3-319-52730-7_40

it increases system's throughput of spectrum access and reduces the time delay in spectrum sensing and so on.

As is mentioned above, spectrum sensing is usually used to obtain spectrum state. However, due to the limitations of hardware processing speed, the cost of equipment and network deployment cost in real world, we can just only get sparse spectrum data on time, frequency or space through spectrum sensing. On the other hand, the measured spectrum data analysis both at home and abroad [6,7] indicates that any spectrum data does not exist in isolation. They have a close correlation in time, frequency and space dimensions. The sparsity of spectrum sensing data sample caused by the limitations of hardware processing speed, the cost of equipment and network deployment cost can be overcome by fully modeling, analyzing, mining and using the correlations in all dimensions and then predicting spectrum state. The current domestic and international researches on spectrum prediction have made staggered results. The early researches on spectrum prediction mainly focused on the time domain. With the deepening of the data analysis based on real-world spectrum measurements, spectrum correlation phenomenon (the relationship between different channels of spectrum) gradually attracts researchers' attention [8,9] and spectrum prediction algorithm based on spectrum's frequency domain correlation also constantly emerges. The related researches can be found in [6,10,11].

There are mainly three challenges in spectrum prediction, namely anomalies, measurement errors and missing data. Anomalies and the wrong data are common, for the error in the process of electromagnetic wave transmission is unavoidable [12]. Data missing is also inevitable for three reasons. Firstly, the data missing in the transmission process is normal [13]. Secondly, the limitation of measuring equipment brings the fact that it is unrealistic to measure all the spectrum bands [14]. Thirdly, the existing measurement algorithm is not perfect [15].

In this paper, we consider the spectrum state matrix. The columns correspond to time slots and the rows correspond to spectrum bands. Next, we analyze real-word spectrum data to excavate the correlation structure between time slots and spectrum bands. Then from a two-dimensional perspective, we regard spectrum prediction as a matrix recovery optimization problem from incomplete and corrupted historical data. We develop an alternating direction optimization method to solve it. Finally, we apply the algorithm to real-world VHF spectrum data and the results show that it outperforms state-of-the-art schemes.

2   System Model

As stated before, we consider a spectrum matrix $\mathbf{X} \in \mathcal{R}^{F \times T}$. The rows correspond to frequency bands and the columns correspond to time slots. Each element $x_{f,t}, f \in \{1, ..., F\}, t \in \{1, ..., T\}$ represents the spectrum state in the t-th time slot on the f-th frequency band. Each row $\mathbf{x}_{f,.} := [x_{f,1}, x_{f,2}, ..., x_{f,T}], f \in \{1, ..., F\}$ represents the state evolution of T successive time slots over the f-th frequency band. Each column $\mathbf{x}_{.,t} := [x_{1,t}, x_{2,t}, ..., x_{F,t}]', t \in \{1, ..., T\}$ represents

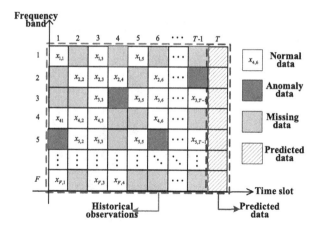

Fig. 1. System model.

the state distribution of F frequency bands in the t-th time slot. As is shown in Fig. 1, the abscissa axis represents time slot, the vertical axis represents frequency band. The data from column 1 to column $T-1$ is historical data, among which there exist anomaly data and missing data. What we do is to predict the data in column T from a spectral-temporal 2D perspective by exploiting the relationships among historical data and predicted data. To achieve this objective, there are two critical issues:

- There are many factors contributing to practical spectrum data matrices, including signals, anomalies and noise.
- Unlike the conventional matrix completion or interpolation that elements are missing uniformly and randomly, an entire column of the matrix is known in the case of spectrum prediction.

As for the first issue, we consider the original dataset as a mixture of all these effects and then decompose the original spectrum matrix into a low-rank component, a sparse component and a dense noise component, which capture the major effects of signals, anomalies and noise, respectively. As for the second issue, we utilize some essential properties of spectrum matrices and add the time series forecasting into the matrix interpolation.

3   Analysis of Datasets

3.1   Real-World Spectrum Measurement Dataset

As shown in Fig. 2, in this paper we use a software defined radio NI USRP N2920 to perform real-world spectrum measurement in the basement of a 10-floor building. The frequency band spans from 50 MHz to 75 MHz with a frequency resolution 25 kHz. In total 1000 bands are measured and each band is measured 100 times. Therefore, the data size is 1000×100. The spectrum measurement in terms of power spectral density (dbm/25 kHz) is shown in Fig. 3.

Fig. 2. A real-world spectrum measurement platform.

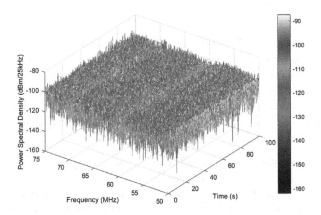

Fig. 3. The spectrum measurement data across various frequency bands and time slots.

3.2   Rank Analysis

From the knowledge of linear algebra we know that a higher correlation of a matrix generally means low rank. For each data matrix, we first make a processing by subtracting from each row its mean value. Then we apply singular value decomposition (SVD) to analyze the rank distribution of all mean-centered spectrum data matrices. In Fig. 4, we plot the normalized singular values in a descending order for VHF bands and for both the cases with and without anomalies. For comparison, we also analyze an i.i.d Gaussian random signal dataset.

From Fig. 4, it is suggested in the case of spectrum data matrices without anomaly, the energy is always contributed by the top several singular values in measured practical data matrices, which reflects the fact that practical spectrum data matrices show approximate low-rank structure, and this is quite different from the Gaussian random signal dataset. In the case with anomaly, we use the

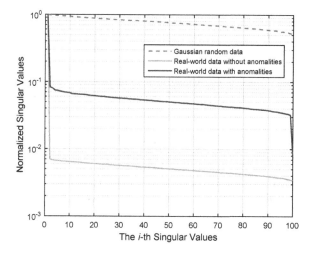

Fig. 4. Normalized singular values of spectrum datasets.

standard anomaly injection method [16] and inject anomalies to a portion of the entries in the original matrices. As a result, the presence of anomalies has a destructive effect on the approximate low-rank structure.

4 Problem Formulation and Algorithm Design

4.1 Problem Formulation

We use $x_{f,t}$ to express the measured spectrum data in the f-th frequency band over the t-th time slot, then we have

$$x_{f,t} = z_{f,t} + a_{f,t} + v_{f,t}, f = 1, ..., F, t = 1, ..., T, \tag{1}$$

where $z_{f,t}$ denotes the signal of interest, $a_{f,t}$ denotes the anomaly component and $v_{f,t}$ denotes the additive noise component. As for $z_{f,t}$, because the signal of interest is not always present, so we have

$$z_{f,t} = h_{f,t} \cdot p_{f,t} \tag{2}$$

where $h_{f,t}$ is a function indicating the presence or absence of the signal and $p_{f,t}$ is the signal strength in the t-th time slot and the f-th frequency band. If the signal is present, then $h_{f,t} = 1$. If the signal is absent, then $h_{f,t} = 0$.

Introduce the matrix $\mathbf{X}_T := [x_{f,t}]$, $\mathbf{Z}_T := [z_{f,t}]$, $\mathbf{A}_T := [a_{f,t}]$, $\mathbf{V}_T := [v_{f,t}]$ $\in \mathrm{R}^{F \times T}$, then Eq. (1) can be further rewritten as follows:

$$\mathbf{X}_T = \mathbf{Z}_T + \mathbf{A}_T + \mathbf{V}_T \tag{3}$$

where $\mathbf{Z}_T$ represents a low-rank signal component, $\mathbf{A}_T$ a sparse anomaly component and $\mathbf{V}_T$ a dense noise component. The low-rank property of the signal has been observed from Fig. 4. The introduction of dense noise component makes the low-rank structure of $\mathbf{Z}_T + \mathbf{V}_T$ approximate and the injection of sparse anomaly component $\mathbf{A}_T$ makes the approximate low-rank structure does not hold at more.

To further model missing data, introduce the operator $P_{\Omega_T}(\cdot)$, which sets the entries of its matrix argument not in Ω_T to zero, and keeps the rest unchanged, then the spectrum state data can be further given as

$$P_{\Omega_T}(\mathbf{X}_T) = P_{\Omega_T}(\mathbf{Z}_T + \mathbf{A}_T + \mathbf{V}_T) \tag{4}$$

As is stated before, the objective in this paper is to predict the data in column (*i.e.*, $\mathbf{x}_{.,T}$) from a two-dimensional perspective, based on the incomplete and historical data $P_{\Omega_{T-1}}(\mathbf{X}_{T-1})$. Now this objective falls into the field of joint (low-rank) Matrix Completion and (sparse) Matrix Recovery (MCMR).

4.2   Algorithm Design

Consider the fact that the spectrum data in the T-th time slots are completely unknown, conventional MCMR methods cannot function well, so we first forecast a few frequency bands of large evolution regularity $f \in S_{LER}$. Specifically, for any band $f \in S_{LER}$, the forecast spectrum state is given as follows:

$$\bar{x}_{f,T} = \begin{cases} TSF(P_{\Omega_{T-1}}(\mathbf{X}_{T-1})) & f \in S_{LER} \\ 0 & \text{otherwise} \end{cases} \tag{5}$$

where TSF stands for various time series forecasting functions. After studying the evolution trajectories of TV and ISM spectrum, we find that there are always several bands in each service that their spectrum evolution trajectories are highly predictable.

Based on $\bar{x}_{f,t}$, the spectrum matrix for further processing is as follows:

$$P_{\Omega_T}(\bar{\mathbf{X}}_T) = \left[P_{\Omega_{T-1}}(\mathbf{X}_{T-1}), \bar{\mathbf{x}}_{.,T} \right] \tag{6}$$

In addition, a natural estimator leveraging the low-rank property of $\mathbf{Z}_T$ and the sparsity property of $\mathbf{A}_T$ attempts to fit the incomplete data $P_{\Omega_T}(\bar{\mathbf{X}}_T)$ to $\mathbf{Z}_T + \mathbf{A}_T$ in the least-squares error sense. Meanwhile, the estimator minimize the rank of $\mathbf{Z}_T$ measured by its nuclear norm $\|\mathbf{Z}_T\|_*$ and the number of nonzero entries of $\mathbf{A}_T$ measured by its l_1 norm $\|\mathbf{A}_T\|_1$. Therefore, we have

$$\min_{\mathbf{Z},\mathbf{A}} \frac{1}{2} \left\| \Gamma_{\Omega_T}(\bar{\mathbf{X}}_{\mathbf{T}} - \mathbf{Z} - \mathbf{A}) \right\|_F^2 + \lambda_T^* \|\mathbf{Z}\|_* + \lambda_T^1 \|\mathbf{A}\|_1, \tag{7}$$

where rank-controlling parameter $\lambda_T^* \geq 0$ and sparsity-controlling parameter $\lambda_T^1 \geq 0$. In order to provide a effective resolution to the above problem, we face the following challenges: (i) This is a non-smooth optimization problem due to the fact that the nuclear and l_1 norms are not differentiable from the very beginning; (ii) The scale of the problem can easily become very large since the quantity of optimization variables $2 * F * T$ grows with time.

To address the above challenges, first we introduce a constraint that rank($\hat{\mathbf{Z}}$) $\leq r$, where is the estimate obtained in Eq. (6) and r is the upper bound rank of the signal part in Eq. (2). Next we factorize the matrix as $\mathbf{Z} = \mathbf{PQ}'$ through a bilinear decomposition. $\mathbf{P}$ and $\mathbf{Q}$ are $F \times r$ and $T \times r$ matrices, respectively. Furthermore, consider the following alternative property of the nuclear norm [17,18].

$$\|\mathbf{Z}\|_* := \min_{\mathbf{P},\mathbf{Q}} \frac{1}{2}\{\|\mathbf{P}\|_F^2 + \|\mathbf{Q}\|_F^2\}, s.t. \mathbf{Z} = \mathbf{PQ}' \tag{8}$$

Apply Eq. (7) to Eq. (6) and we have

$$\arg\min_{\mathbf{P},\mathbf{Q},\mathbf{A}} \frac{1}{2}\left\|\Gamma_{\Omega_T}(\bar{\mathbf{X}}_{\mathbf{T}} - \mathbf{PQ}' - \mathbf{A})\right\|_F^2 + \frac{\lambda_T^*}{2}\{\|\mathbf{P}\|_F^2 + \|\mathbf{Q}\|_F^2\} + \lambda_T^1\|\mathbf{A}\|_1, \tag{9}$$

Obviously, on condition that $rank(\hat{\mathbf{Z}}) \leq r$, the separable Frobenius-norm regularization in Eq. (8) does not damage the optimality relative to Eq. (6). So far, the optimization in Eq. (8) can be solved by the standard method introduced in [19].

5   Experiment Results

In this section, spectrum measurements are used to validate the effectiveness of the proposed robust spectral-temporal two-dimensional spectrum prediction (R-STSP) scheme over the joint (two-dimension) spectral-temporal spectrum prediction (J-STSP) scheme [20].

We quantify the spectrum prediction performance in terms of prediction error. Root mean square error (RMSE) in dB is used to quantify the prediction error, which is defined as:

$$\text{RMSE}(T) = 10\log_{10}\left(\frac{\|\mathcal{P}_{\bar{\omega}_T}(\hat{\mathbf{z}}_T - \mathbf{z}_T)\|_2^2}{\|\mathcal{P}_{\bar{\omega}_T}(\mathbf{z}_T)\|_2^2}\right), \tag{10}$$

where $\mathbf{z}_T$ and $\hat{\mathbf{z}}_T$ are the ground-truth and predicted spectrum data in the T-th time slot, respectively. $\bar{\omega}_T$, the complementary set of ω_T, contains the indices of missing/incomplete observations, while the corresponding sampling operator $\mathcal{P}_{\bar{\omega}_t}(\cdot)$ sets the entries not in $\bar{\omega}_t$ to zero, and keep the rest unchanged.

Figure 5 shows the cumulative distribution functions (CDFs) of RMSE in dB for the two schemes. It shows that: (i) the prediction performance of both J-STSP and R-STSP decrease with an increasing percentage of anomaly data; (ii) R-STSP always outperforms the J-STSP under different configurations; (iii) the prediction performance of the proposed R-STSP is improved with a decreasing percentage of anomaly data.

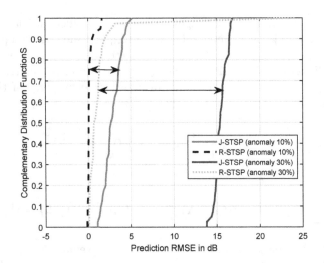

Fig. 5. Illustration of RSTSP with anomaly data.

6   Conclusion

This paper considered spectrum prediction as a matrix recovery optimization problem from incomplete and false historical data. We developed an optimization method to solve it. Finally, we apply the algorithm to real-word VHF spectrum data and the results show that R-STSP outperforms state-of-the-art schemes. One future work is to further develop online algorithms to perform real-time prediction and reduce the delay.

Acknowledgement. This work is supported by the National Natural Science Foundation of China (Grant No. 61501510 and No. 61301160), Natural Science Foundation of Jiangsu Province (Grant No. BK20150717), China Postdoctoral Science Foundation Funded Project, and Jiangsu Planned Projects for Postdoctoral Research Funds.

References

1. Bangerter, B., Talwar, S., Arefi, R., Stewart, K.: Networks and devices for the 5G era. IEEE Commun. Mag. **52**(2), 90–96 (2014)
2. Wang, T., Li, G., Ding, J., Miao, Q., Li, J., Wang, Y.: 5G spectrum: is China ready? IEEE Commun. Mag. **53**(7), 58–65 (2015)
3. Axell, E., Leus, G., Larsson, E.G., Poor, H.V.: Spectrum sensing for cognitive radio: state-of-the-art and recent advances. IEEE Signal Process. Mag. **29**(3), 101–116 (2012)
4. Huang, X., Hu, F., Wu, J., et al.: Intelligent cooperative spectrum sensing via hierarchical dirichlet process in cognitive radio networks. IEEE J. Sel. Areas Commun. **33**(5), 771–787 (2015)

5. Huang, X., Wang, G., Hu, F.: Multitask spectrum sensing in cognitive radio networks via spatiotemporal data mining. IEEE Trans. Vehic. Technol. **62**(2), 809–823 (2013)
6. Yin, S., Chen, D., Zhang, Q., Liu, M., Li, S.: Mining spectrum usage data: a large-scale spectrum measurement study. IEEE Trans. Mob. Comput. **11**(6), 1033–1046 (2012)
7. Wellens, M.: Empirical modelling of spectrum use and evaluation of adaptive spectrum sensing in dynamic spectrum access networks. Ph.D. dissertation, Department of Wireless Networks, RWTH Aachen University (2010)
8. Hossain, K., Champagne, B.: Wideband spectrum sensing for cognitive radios with correlated subband occupancy. IEEE Signal Process. Lett. **18**(1), 35–38 (2011)
9. Hossain, K., Champagne, B., Assra, A.: Cooperative multiband joint detection with correlated spectral occupancy in cognitive radio networks. IEEE Trans. Signal Process. **60**(5), 2682–2687 (2012)
10. Kumar Acharya, P.A., Singh, S., Zheng, H.: Reliable open spectrum communications through proactive spectrum access. In: Proceedings of the First International Workshop on Technology and Policy for Accessing Spectrum (2006)
11. Li, H., Qiu, R.C.: A graphical framework for spectrum modeling and decision making in cognitive radio networks. In: Proceedings of the 2010 IEEE Global Telecommunications Conference (GLOBECOM) (2010)
12. Zhang, L., et al.: Byzantine attack and defense in cognitive radio networks: a survey. IEEE Commun. Surv. Tutor. (2015) doi:10.1109/COMST.2422735
13. Meng, J., et al.: Collaborative spectrum sensing from sparse observations in cognitive radio networks. IEEE J. Sel. Areas Commun. **29**(2), 327–337 (2011)
14. Nguyen, T.V., Shin, H., Quek, T.Q.S., Win, M.Z.: Sensing and probing cardinalities for active cognitive radios. IEEE Trans. Signal Process. **60**(4), 1833–1848 (2012)
15. Cheng, H.T., Zhuang, W.: Simple channel sensing order in cognitive radio networks. IEEE J. Sel. Areas Commun. **29**(4), 676–688 (2011)
16. Lakhina, A., Crovella, M., Diot, C.: Diagnosing network-wide traffic anomalies. In: Proceedings of ACM SIGCOMM (2004)
17. Srebro, N., Rennie, J., Jaakkola, T.S.: Maximum-margin matrix factorization. In: Proceedings of the Advances in Neural Information Processing Systems, Vancouver, Canada (2004)
18. Srebro, N., Shraibman, A.: Rank, trace-norm and max-norm. In: Auer, P., Meir, R. (eds.) COLT 2005. LNCS (LNAI), vol. 3559, pp. 545–560. Springer, Heidelberg (2005). doi:10.1007/11503415_37
19. Mardani, M., Mateos, G., Giannakis, G.B.: Decentralized sparsity-regularized rank minimization: algorithms and applications. IEEE Trans. Signal Process. **61**(11), 5374–5388 (2013)
20. Ding, G., Wang, J., Wu, Q., Yu, L., Jiao, Y., Gao, X.: Joint spectral-temporal spectrum prediction from incomplete historical observations. In: Proceedings of the GlobalSIP (2014)

Spectrum Sensing and Spectrum Allocation Algorithms in Wireless Monitoring Video Transmission

Xin-Lin Huang[1]([✉]), Yu-Bo Zhai[1], Si-Yue Sun[2], Qing-Quan Sun[3], and Shu-Qi Hu[1]

[1] Tongji University, Shanghai 201804, People's Republic of China
{xlhuang,102677}@tongji.edu.cn, hsqhjrhrz@163.com
[2] Shanghai Engineering Center for Micro-satellites, Shanghai, China
sunmissmoon@163.com
[3] California State University San Bernardino, San Bernardino, CA, USA
quanqian12345@gmail.com

Abstract. Video monitoring system is an important measure to guarantee people's safety. Wireless video monitoring system has been widely used with advantages of simpler construction and higher flexibility compared with traditional wired video monitoring system. Cognitive radio network is introduced into the wireless video monitoring system in this paper to improve the spectrum utilization. Cognitive radios have enabled users to utilize licensed bands opportunistically without harmful interference to licensed users. The basic concept of wireless video monitoring system is recalled in this paper first. Then we analyze the system model of the cognitive radio network based video monitoring system. We provide a centralized cooperation spectrum sensing algorithm and a priority and channel ranking based spectrum allocation algorithm. Simulation results show that our algorithms have better performance than algorithms without the consideration of priority or channel ranking.

Keywords: Cognitive radio · Video monitoring system · Cooperation spectrum sensing · Spectrum allocation

1 Introduction

With the rapid development of national economy and the improvement of our living standard, there are increasingly stringent requirements for the improvement of security facilities. As a major component of security system, video monitoring system can provide real-time surveillance video for us to find and deal with emergencies in time as well as record and store the event process, which makes great contribution to the safety, property and social stability. Therefore, video monitoring system has been gradually applied to all walks of life.

Traditional video monitoring system is a wired video system, where transmission is realized by prepositioned cables. Despite the advantage of wide bandwidth and small interference, wired video surveillance system has lots of flaws which cannot be ignored. Monitory points have to be set within the scope of the cable network, which reduces the

© ICST Institute for Computer Sciences, Social Informatics and Telecommunications Engineering 2017
X.-L. Huang (Ed.): MLICOM 2016, LNICST 183, pp. 402–411, 2017.
DOI: 10.1007/978-3-319-52730-7_41

flexibility of the system. Construction and installation have to take the cable pipes into consideration, which increases the cost and operating time of the system as well as the difficulty to rapidly maintain the system when failure occurs [1, 2]. Due to the existence of these defects, the access mode of video surveillance system is gradually turned into wireless. With the development of wireless communication technology, wireless video monitoring system with high flexibility and strong extendibility has increasingly broad application prospects.

However, spectrum resources are quite limited in wireless communication. The wireless spectrum is mainly allocated by a stationary policy so far. The government unifies the spectrum resources and allocates specific frequency bands to particular communication services. Spectrum resource allocation is imbalanced due to the fixed strategy. Lots of unauthorized spectrum is overloaded with some licensed spectrum being idle at the same time. A research from Federal Communications Commission (FCC) shows that the spectrum utilization of authorized frequency bands is from 15% to 85% with temporal and geographical changes [3]. Another measurement report from National Radio Network Research Test-bed (NRNRT) indicates that the average spectrum utilization below 3 GHz spectrum band is just 5.2% [4].

Wireless video monitoring system usually works in the ISM band without permission to pre-allocated frequency bands. The realization of the system has been limited because of the finite available spectrum resources and the need of ensuring not to cause interference to other users. Hence, we propose the cognitive radio network based wireless video monitoring system in this paper to improve spectrum utilization to get more efficient and extensive applications [5].

The rest of this paper is arranged as follows. A brief introduction of the cognitive radio network and the framework of cognitive radio based video surveillance system are provided in Sect. 2. A centralized cooperation spectrum sensing algorithm and a priority and channel ranking based spectrum allocation algorithm are proposed in Sects. 3 and 4, respectively. The performance of the algorithm is verified in Sect. 5. Section 6 concludes this paper.

2 System Model

2.1 Cognitive Radio Network

The traditional fixed spectrum allocation scheme has been unable to satisfy the increasing needs for high quality communication with the rapid growth of wireless communication services. Therefore, dynamic spectrum access technology arises to provide more wireless spectrum for cognitive radio users, resulting in cognitive radio network.

The cognitive radio technology is the most important part in cognitive radio network, which allows the secondary users (SUs) to detect idle spectrum bands and share them with other users without causing harmful interference to primary users (PUs), thus improving the average utilization of spectrum resources. Cognitive radio can help the secondary users with the following functions [3, 6]: Determine available spectrum bands and discover the occupation of licensed bands by original authorized users timely (Spectrum Sensing), select the optimal idle channel (Spectrum Management), share the available spectrum bands with other cognitive users fairly (Spectrum Sharing), move out of current channel immediately when a primary user is detected or the channel quality becomes unacceptable (Spectrum Mobility).

The relationship between the above four functions can be concluded as a basic cognitive cycle [7, 8], as shown in Fig. 1. Secondary users detect the radio environment by spectrum sensing to estimate the channel occupancy of primary users and other secondary users. The spectrum quality report will be given in the spectrum analysis step with the help of spectrum sensing results. Spectrum decision helps to assign best available channel to cognitive users on the basis of channel quality and different user requirements [9, 10].

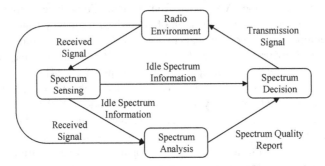

Fig. 1. The basic cognitive cycle [7, 8].

2.2 Framework of Cognitive Radio Based Wireless Video Monitoring System

The cognitive radio network based video monitoring system proposed in this paper mainly consists of front end video capture system with cognitive radio devices, wireless transmission system, sensing data fusion center and monitoring center. The system structure diagram is given in Fig. 2.

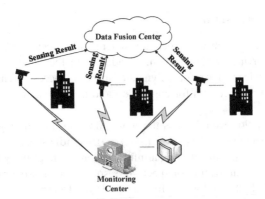

Fig. 2. The structure diagram of the cognitive radio based wired video monitoring system.

The video capture system is composed of a plurality of capturing front, which capture and process real time video signals by preset cameras. Cognitive radio equipment detect the radio environment and send spectrum sensing results to the fusion center. Available channels are allocated to different capturing front according to fusion results as well as the priority of users. Compressed encoded video stream is transmitted through the assigned channels to corresponding monitoring center to be decoded and displayed on the monitoring interface.

3 Centralized Cooperative Spectrum Sensing Algorithm

Spectrum sensing enables SUs to identify radio environment and detect spectrum holes, which is a crucial technique in cognitive radio network. Spectrum sensing algorithms can be classified into non-cooperative detection and cooperative detection. Each SU selects their own sensing method such as energy detection, match filter detection and cyclostationary feature detection and makes their own spectrum decision in non-cooperative spectrum sensing, which is easy to implement. Cooperative spectrum sensing can reduce the uncertainty of single SU caused by interference and noise, thus improving the accuracy of spectrum decision. We use centralized cooperative spectrum sensing algorithm in this paper where all users transmit their sensing data to the fusion center, who makes global spectrum decision according to all sensing data and assigns channels to users.

The users need to process the sensing signal to build a Bayesian model before sending to the fusion center. We will use the same sensing signal acquisition model as in our previous work [11, 12], which is summarized as follows.

We choose $Y(k)$ to be our observations of channel k, which obeys exponential distribution, as expressed in Eq. (1):

$$Y(k) \sim Exponential(\lambda_k) = Gamma(l = 1, \lambda_k) \tag{1}$$

$$\lambda_k \triangleq 1/\sigma_k^2 + \sigma_0^2 \tag{2}$$

λ_k is the parameter of the exponential distribution, where σ_k^2 and σ_0^2 are the variance of Gaussian distribution of signal variable and noise variable, respectively. Detailed derivation process is provided in [11]. We use Gamma distribution as the conjugate prior of λ_k to build a Bayesian model, that is:

$$\lambda_k \triangleq Gamma(a_k, b_k), p(\lambda_k) = \frac{b_k^{a_k}}{\Gamma(a_k)} \lambda_k^{a_k-1} e^{-\lambda_k b_k} \tag{3}$$

The posterior distribution of λ_k can be represented as in Eq. (4) giving n observations according to the Bayesian theory:

$$p(\lambda_k | \{Y(k)\}_n) = Gamma(a_k + nl, b_k + n\overline{Y(k)}) \triangleq Gamma(a_k', b_k') \tag{4}$$

$$\widehat{(1/\lambda_k)}|\{Y(k)\}_n = E[(1/\lambda_k)|\{Y(k)\}_n] = \frac{b'_k}{a'_k - 1} \tag{5}$$

where $\overline{Y(k)}$ is the main value of n observations and $\widehat{(1/\lambda_k)}$ represents the Bayesian estimation of channel state parameter.

Our previous work [11] focus on large-scale cognitive radio network where we need to group the users with the same channel state distribution parameters and fuse the sensing data within each group in a distributed manner. In this paper, we discuss a small-scale cognitive radio network where all video monitoring users share the same channel conditions. The fusion center obtains the global channel state parameter based on Eq. (4) according to the observations from all users by updating parameters (a'_k, b'_k).

4   Priority Based Spectrum Allocation Algorithm

High-definition real-time video monitoring system can restore event scenarios accurately after an emergency to ensure the safety of people's life and property. However, large amount of transmission data has brought great challenges to the limited spectrum resources. Limited bandwidth may not be able to meet the HD video transmission requirements for all video surveillance users in practical applications.

Prioritizing the users according to the urgent degree of emergency events is a good solution to above problem. Add intelligent analysis module to video monitoring system to discover unexpected events, determine the emergency degree of the incident and set higher priority for users with more urgent events. Under normal circumstances with no emergency, users with lower priority can sacrifice video clarity to reduce data transmission rate and save resources for other users. In case of an emergency, alarm and improve the priority of corresponding user immediately to provide greater bandwidth to achieve real-time transmission and storage of high-definition video, thus ensuring timely tracking and accurately recording emergencies.

We combine wireless video monitoring system and cognitive radio network with centralized cooperative spectrum sensing algorithm in Sect. 3, where the data fusion center processes the sensing data of all users to obtain globally consistent spectrum decision results and allocates channels according to the priority of user events. The specific steps of the algorithm are shown in Table 1.

We will realize the algorithm and analyze simulation results in the next section.

5   Simulation Result

In our simulation scenarios, we considered a cognitive radio network in a 15 km × 15 km square area which is further divided into 9 grids with the size of 5 km × 5 km. Each PU works on one of 32 channels located in the middle of one grid randomly. It can be considered as a PU works on multiple channels if multiple PUs happen to be in the same location. The probability of the channel occupancy of each PU is 1/3 and transmission signals go through a Rayleigh channel and attenuate according to the free

Table 1. The channel allocation algorithm

---

1) All video monitoring users with cognitive radio equipment in the same network detect the radio environment at the same time and send their sensing results, the observations $Y(k)$ of channel k mentioned in Section 3, to data fusion center.

2) Data fusion center uses observations of all users to update the distribution parameters (a'_k, b'_k) of the state parameter λ_k of each channel according to Equation (4) and utilizes the updated distribution parameters to estimate the channel state parameter according to Equation (5).

3) For each channel k, compare λ_k with a preset threshold. Channel k is considered unavailable when λ_k is larger than the threshold, and vice versa.

4) All available channels from step 3 will be ranked according to the large of λ_k. Channels with larger λ_k will be considered to have better conditions than those with smaller state parameters.

5) Data fusion center assigns the available channels to users based on their event priority by providing more channels with better conditions for users with higher priority.

---

space propagation model. Video monitoring users are located randomly in the middle grid. We considered a channel unavailable if the received signal from corresponding PU is larger than the noise power according to the interference temperature model proposed by FCC [13]. The noise threshold is set to be −90 dBm.

We compare three algorithms in this paper. The first one is the channel ranking and priority based spectrum assignment algorithm proposed in the previous section where all channels are ranked based on their channel state parameter and users with higher priority will get more and better channels. Instead of sorting the channels first, we will compare the channel state parameter with a preset threshold to determine channel availability in second and third algorithms. We will take user priority into consideration in second algorithm where high priority users will get more available channels randomly selected from spectrum pool. In the third algorithm, each user will get the same number of available channels.

There are 32 channels which can be allocated to 10 cognitive video monitoring users if available. The probability of a user to have the highest priority, lower priority and lowest priority is 0.3, 0.2 and 0.5, respectively. Each of them will get three, two and one channel. We use the total "benefits" that all users can get to measure the performance of each algorithm. If a channel assigned to a user is actually available, the user will obtain certain benefits. The benefits each user can obtain from each idle channel are set to be 1.5, 1.25 and 1 for the highest priority, lower priority and lowest priority users, respectively.

The comparison of the performance of three algorithms is shown in Figs. 3 and 4, with receiver SNR being set to be 10 dB and 20 dB, respectively. The abscissa represents decision threshold. As we can see, the performance of the two priority based algorithms is much better than the third algorithm without considering the user priority. Furthermore, the performance of the algorithm without channel ranking changes with the decision threshold and is poorer than the first channel ranking based algorithm under most thresholds. All channels are sorted and allocated to users based on their channel state parameter in the first algorithm without comparing to a threshold, so its performance doesn't vary with threshold.

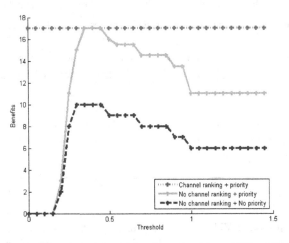

Fig. 3. The comparison of the three algorithms (SNR = 10).

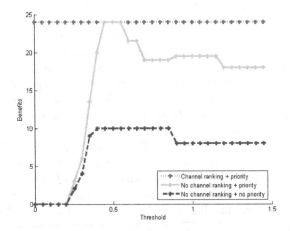

Fig. 4. The comparison of the three algorithms (SNR = 20).

We can also see from Figs. 3 and 4 that receiver SNR will influence algorithm performance. Figure 5 shows the effect of receiver SNR on the performance of our algorithm. The receiver SNR will influence the spectrum sensing accuracy, thus affecting algorithm performance. We can see that algorithm performance gets better when SNR increases.

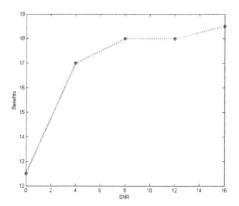

Fig. 5. The effect of receiver SNR on the performance of our algorithm.

Figure 6 compares the accuracy of cooperative and non-cooperative spectrum sensing algorithms. The correct detection probability of cooperative spectrum sensing algorithm is apparently higher than non-cooperative spectrum sensing algorithm under the same false alarm probability. Figure 7 shows the effect of user number on the performance of our algorithm. The number of users who can get the channels is fixed on 10 while the number of users who participate in cooperative spectrum sensing changes. We can see from the two figures that the uncertainty of single-user spectrum sensing reduces and the algorithm performance gradually reaches the optimal when the number of cooperative spectrum sensing users increases.

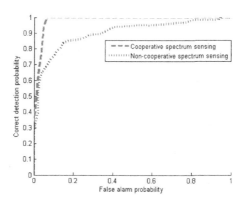

Fig. 6. The comparison between cooperative and non-cooperative spectrum sensing.

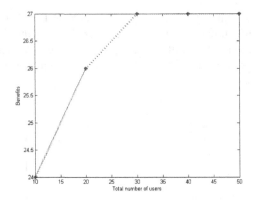

Fig. 7. The effect of user number on the performance of the algorithm.

6   Conclusion

In this paper, we first introduced cognitive radio network into wireless video monitoring system to increase spectrum efficiency. Then we proposed a centralized cooperation spectrum sensing algorithm and a priority and channel ranking based spectrum
allocation algorithm where we sort the channels to provide more and better channels to
higher priority users. Our simulation results proved that compared with algorithms
without consideration of priority or channel quality, the performance of our channel
assignment algorithm is much better.

Acknowledgments. This work was supported by the National Natural Science Foundation of
China (No. 61390513 and No. 61201225), Shanghai Sailing Program (No. 16YF1411000),
"Chen Guang" project supported by Shanghai Municipal Education Commission and Shanghai
Education Development Foundation (No. 13CG18), the Program for Young Excellent Talents in
Tongji University (No. 2013KJ007).

References

1. Junji, M.: Wireless Network Based Video Monitor System Design and Realization.
   University of Electronic Science and Technology of China, Xi'an (2010). (in Chinese)
2. Xun, L.: Key Technology Research of Wireless Video Monitoring System. Jilin University,
   Jilin (2012). (in Chinese)
3. Federal Communications Commission: Spectrum policy task force report, FCC 02-155
   (2002)
4. McHenry, M.: Report on spectrum occupancy measurements. Shared Spectrum Company
   (2005)
5. Huang, X.-L., Wang, G., Hu, F., Kumar, S., Wu, J.: Multimedia over cognitive radio
   networks: towards a cross-layer scheduling under Bayesian traffic learning. Comput.
   Commun. **51**, 48–59 (2014)

6. Huang, X.-L., Wang, G., Hu, F., Kumar, S.: Stability-capacity-adaptive routing for high-mobility multihop cognitive radio networks. IEEE Trans. Veh. Technol. **60**(6), 2714–2729 (2011)

7. Akyildiz, I.F., Lee, W.Y., Vuran, M.C., et al.: NeXt generation/dynamic spectrum access/cognitive radio wireless networks: a survey. Comput. Netw. **50**(13), 2127–2159 (2006)

8. Kushwaha, H., Xing, Y., Chandramouli, R., et al.: Reliable multimedia transmission over cognitive radio networks using fountain codes. Proc. IEEE **96**(1), 155–165 (2008)

9. Huang, X.-L., Wang, G., Hu, F.: Multitask spectrum sensing in cognitive radio networks via spatiotemporal data mining. IEEE Trans. Veh. Technol. **62**(2), 809–823 (2013)

10. Huang, X.-L., Wang, G., Hu, F., Kumar, S.: The impact of spectrum sensing frequency and packet-loading scheme on multimedia transmission over cognitive radio networks. IEEE Trans. Multimedia **13**(4), 748–761 (2011)

11. Huang, X.-L., Fei, H., Jun, W., Chen, H.-H., Wang, G., Jiang, T.: intelligent cooperative spectrum sensing via hierarchical dirichlet process in cognitive radio networks. IEEE J. Sel. Areas Commun. **33**(5), 771–787 (2015)

12. Huang, X.-L., Wu, J., Li, W., Zhang, Z., Zhu, F., Wu, M.: Historical spectrum sensing data mining for cognitive radio enabled vehicular ad-hoc networks. IEEE Trans. Dependable Secure Comput. **13**(1), 59–70 (2016)

13. Federal Communications Commission: Notice of inquiry and notice of proposed rulemaking. Public Notice FCC, 03-289 (2003)

Author Index

Printed in the United States
By Bookmasters